Geotechnical Engineering

Established as a standard textbook for students, this new edition of *Geotechnical Engineering* provides a solid grounding in the mechanics of soils and soil-structure interaction.

Renato Lancellotta gives a clear presentation of the fundamental principles of soil mechanics and demonstrates how these principles are applied in practice to engineering problems and geotechnical design. This is supported by numerous examples with worked solutions, clear summaries and extensive further reading lists throughout the book. Thorough coverage is given to all classic soil mechanics topics such as boundary value problems and serviceability of structures and to topics which are often missed out of other books or covered more briefly including: the principles of continuum mechanics, Critical State Theory and innovative techniques such as seismic methods.

Geotechnical Engineering is suitable for soil mechanics modules on undergraduate civil engineering courses and for use as a core text for specialist graduate geotechnical engineering students. It explores not only the basics but also several advanced aspects of soil behaviour, and outlines principles which underpin more advanced professional work therefore providing a useful reference work for practising engineers. Readers will gain a good grasp of applied mechanics, testing and experimentation, and methods for observing real structures.

Renato Lancellotta is Professor of Geotechnical Engineering, Director of the Geotechnical Laboratory and since 1990 has been Head of the Ph.D. Programme on Geotechnical Engineering at the Technical University of Torino, Italy.

Geotechnical Engineering

Second Edition

Renato Lancellotta

Taylor & Francis
Taylor & Francis Group

LONDON AND NEW YORK

First English language edition published 1995 by A. A. Balkema

Second English language edition published 2009
by Taylor & Francis
2 Park Square, Milton Park, Abingdon, Oxon, OX14 4RN

Simultaneously published in the USA and Canada
by Taylor & Francis
270 Madison Ave, New York, NY 10016

*Taylor & Francis is an imprint of the Taylor & Francis Group,
an informa business*

© 2009 Zanichelli editore S.p.A., via Irneroo 34, 40126 Bolonga [7653]

Typeset in Sabon by Keyword Group Ltd
Printed and bound in Great Britain by
CPI Antony Rowe, Chippenham, Wiltshire

British Library Cataloguing-in-Publication Data
A catalogue record for this book is available
from the British Library

Library of Congress Cataloging in Publication Data
Lancellotta, Renato.
    Geotechnical engineering / Renato Lancellotta. – 2nd ed.
        p. cm.
    'First English language edition published 1995 by A.A. Balkema'–T.p. verso.
    Includes bibliographical references and index.
    I. Engineering geology. 2. Soil mechanics. 3. Rock mechanics. I. Title.

    TA705.L265 2008
    624.1'51–dc22   2007045353

ISBN10: 0-415-42003-2 (hbk)
IBSN10: 0-415-42004-0 (pbk)
IBSN10: 0-203-97783-4 (ebk)

ISBN13: 978-0-415-42003-7 (hbk)
ISBN13: 978-0-415-42004-4 (pbk)
ISBN13: 978-0-203-92783-0 (ebk)

To Donata, Paolo and Giulia

Contents

Figures and tables

Figures

Tables

Preface

This book is about the mechanics of soils and structures interacting with soils. It has been primarily designed as a classroom teaching book, but it also covers more advanced aspects, usually included in upper level courses on civil and geotechnical engineering. To provide a clear separation of basic and more advanced arguments, the material is presented in a series of focused sections, and having included much more material has the advantage of giving the teacher the possibility to adapt the book to his personal view, when it is being used in a one or two-semester course.

There is no necessity for a motivation to study soil mechanics. Since the beginning of time man has been a builder, he had to deal with earth pressures, groundwater flow, stability of earth constructions, settlement and stability of structures interacting with soils. Man learned to solve complex problems by trial and error, and, paradoxically, when he was successful, this ability remained invisible.

The behaviour of soil is today approached on sound and rational bases, using concepts of continuum mechanics, and experimental evidences prove that it is a significant and powerful procedure.

To give this book a self-contained character, all basic concepts of continuum mechanics and mathematical prerequisites have been collected in Chapters 2 and 3, but readers who have a good grasp of these prerequisites can obviously omit these chapters and move from Chapter 1 to Chapter 4.

Soils are natural materials and the solution of engineering problems requires the knowledge of the properties of real soils. These properties have to be investigated, rather then being prescribed, so that it is necessary to be familiar with testing techniques and to be aware of the relevance of field investigation.

However, both phases of programming and interpreting experimental tests require a conceptual framework, and for this reason reference has been made throughout this book to the Critical State Theory. The central idea of this theory is that volume changes play a role so important as the effective stresses and it is the relation between the initial state and the critical state that has a major influence on soil behaviour. Moreover, this theory also provides a link between soil mechanics and previously grasped concepts as elasticity and plasticity, that in turn are extensively used in two final chapters, aimed at applications.

The material contained in this book has been organized and developed following this line of thought, and, having the students in mind, many worked examples have been disseminated throughout the book in order to show how basic principles are applied to solve problems of engineering interest.

The author is indebted to several colleagues at Technical University of Torino: Professor Michele Jamiolkowski, for encouragement and guidance; Professor Daniele Costanzo, Dr Sebastiano Foti, Dr Guido Musso and Dr Lodovica Tordella, for helpful criticism provided when writing the Italian version; Roberto Maniscalco, who drew the figures with remarkable skill and care.

The author is also indebted to Zanichelli and Taylor & Francis for their cooperation and assistance in preparing the final version of the book. Finally, the author thanks authors and publishers and especially the Council of the Institution of Civil Engineers of London for permission to reproduce many figures from *Géotechnique*.

Renato Lancellotta
Torino, November 2007

Chapter 1

Origin, description and classification of soils

The beginning reader will quickly realize that soils are unusual engineering materials, mainly if compared with materials described in structural mechanics. The reason is that soils are natural materials, formed by the weathering of rocks, and the behaviour of soils is a legacy of natural processes, from their origin to the actual state. This gives soils a character of inhomogeneity and anisotropy, and basic parameters, such as strength, stiffness and hydraulic conductivity, need to be measured instead of being specified and may vary over a wide range.

The discrete particles that make up soils are not strongly bounded together, they are free to move relatively among themselves and, when a soil element deforms, the overall deformation is essentially the result of relative sliding between particles and rotation of particles. Therefore, it is not surprising that soil behaviour is highly non-linear and irreversible.

Furthermore, it must be realized that the voids or pores between particles are filled with water, or there may be more than one fluid, typically water and air at near-surface depths, but there could be water and liquid and gaseous hydrocarbon in certain circumstances. It follows that soils are multi-phase materials, their behaviour being influenced by the interaction between solids and fluids.

From this concise description, and mainly from the consideration that we have to deal with natural materials, it appears that a first step to learn about the behaviour of soils is certainly that of having a knowledge about the origin of soils. For this reason, in this chapter we will study the weathering processes of rocks, as well as the main features of natural soil deposits. Then, we say that basic parameters, such as strength, stiffness and hydraulic conductivity, need to be measured by specific tests rather then being specified. However, an accurate description of soils can help the engineer by providing him with a general guidance about the expected soil behaviour. Therefore, a second step is to learn how to describe soils, by using a common language, in order to benefit from any description. Finally, we need to learn how to classify soils, presuming that soil samples within the same class have a similar behaviour.

1.1 Soil formation

Composition, structure and properties of a natural soil element are the result of its geological history. This history includes weathering, transportation, deposition and post-depositional changes. The actual state of homogeneity and anisotropy of any soil deposit is related to this formational history and to subsequent changes, summarized in Figure 1.1 and discussed in detail in the sequel.

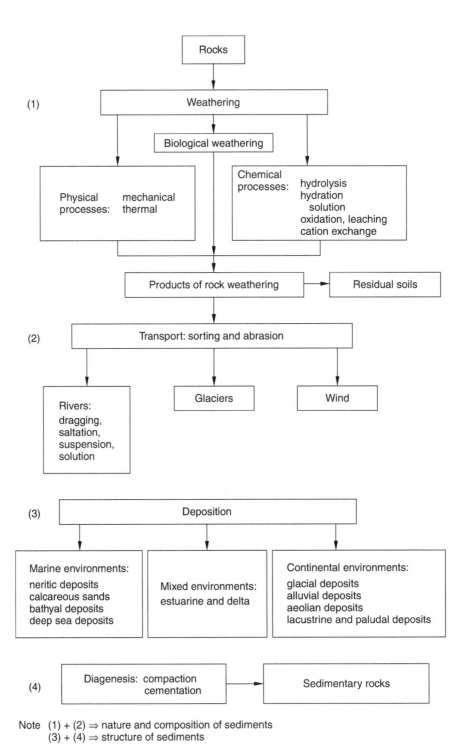

Figure 1.1 Soil formation.

1.1.1 Weathering

Weathering is the physical, chemical and biological process by which debris is formed. Physical processes include the breakdown of rocks into particles. This process of disintegration is promoted by the presence of discontinuities (cracks, joints and faults), deriving from tectonic events as well as from stress relief due to uplift, erosion or changes of fluid pressure. When water enters rocks through cracks and fissures, the ice formed by freezing exerts high pressures and is able to disintegrate the outer layer of rocks. In addition, expecially in hot climates, the rock surface is subjected to a wide range of temperatures (up to 60° C), giving rise to cycles of expansion and contraction. The result is that outer layers can be pulled away and this process is called *exfoliation*.

Chemical weathering, or decomposition, is the breakdown of minerals into new compounds, by the action of chemical agents such as acids in the rain and flowing water and in the air. Weathering can also be attributed to plants and animals.

Understanding chemical weathering requires to digress on some basic aspects of soil mineralogy, so that we start by recalling that minerals are defined as naturally occurring inorganic substances, which have a definite chemical composition and a regular atomic structure to which its crystalline form is related.

Mineral composing rocks are mainly a combination of elements listed, in decreasing order of abundance, in Table 1.1, and, since oxygen and silicon preponderate, **silicates** are the most common minerals and make up about 95% of the earth's crust. The remaining minerals, with the exception of **carbonates**, do not appear in such quantities to affect both physical and mechanical properties of common rocks and soils.

The fundamental unit of silicate minerals is the **tetrahedron** SiO_4, formed by 4 oxygen ions grouped around a single tetravalent silicon ion (Figure 1.2). The bonding energy Si-O is half the bonding energy of the oxygen ion, so that this latter can be linked to another available silicon ion, or the excess of four negative valencies can be balanced when the SiO_4 group is linked to metal ions (Mg, Fe, Ca).

This originates arrangements, or *structures*, of tetrahedra, and silicate minerals are then grouped according to these arrangements.

As an example, when the tedrahedra are linked by sharing three of four oxygen ions, and their bases lie in a common plane, the **sheet** structure of *phyllosilicates* is formed.

Table 1.1 Composition of continental crust

Element	Symbol	Percentage of weight (%)
Oxygen	O	46.6
Silicon	Si	27.7
Aluminium	Al	8.1
Iron	Fe	5.0
Calcium	Ca	3.6
Sodium	Na	2.8
Potassium	K	2.6
Magnesium	Mg	2.1
Other elements	—	1.5

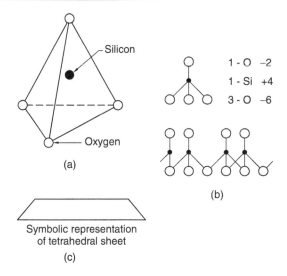

(a)

1 - O   −2
1 - Si +4
3 - O   −6

(b)

Symbolic representation
of tetrahedral sheet

(c)

Figure 1.2 Tetrahedral unit.

Sheet structures are found in **micas** and flaky minerals such as chlorite, talc and clay minerals.

If the tetrahedron is sharing all its oxygen with adjacent tetrahedra, this originates a three-dimensional structure, called three-dimensional **framework**. The mineral **quartz** (SiO_2) is a representative of framework structure.

Another example of framework is provided by **feldspars**, the most abundant group in igneous rocks. In this case, some of the tetravalent Si^{4+} is replaced by the trivalent Al^{3+} and this substitution releases one negative charge, which is satisfied by a cation (sodium or potassium).

Important non-silicate minerals are **calcite** ($CaCO_3$) and **dolomite** ($CaMg(CO_3)_2$), essential components of detrital calcareous sediments (limestone, dolomite and dolomitic limestone) .

The processes (most commonly) involved in chemical weathering are represented by the reactions of minerals with water, oxygen and organic acids.

Quartz (an essential constituent of granites, of most sands and sandstones, also found abundantly in gneisses, quartzites, in some schists and other metamorphic rocks) and muscovite (which occurs in granites and other acid rocks as well as in sedimentary rocks as micaceous sandstones) are very stable minerals. On the contrary, feldspar minerals are sensitive to weathering and their alteration forms clay minerals.

Examples are provided by *orthoclase* (potassium feldspar), which alters (through hydrolysis, i.e. hydrogen ions in percolating water replace mineral cations) to **kaolinite** ($Al_2Si_2O_5(OH)_4$), with potassium K^+ and silica SiO_2 being removed in solution.

Other examples of chemical weathering processes consist of *hydration* (absorption of water molecules into the mineral structure), *solution* (dissociation of minerals into ions), *oxidation* (the combination of oxygen with a mineral to form oxides and hydroxides), *leaching* (the migration of ions produced by all the previously mentioned processes), *cation exchange* (absorption of positively charged cations in solution onto the negatively charged surface of clays).

The rate of chemical weathering depends upon the presence of water and is greater in wet climates than in dry climates. It is also correlated to the activity of vegetation, to the production of carbon dioxide and to the frequency of rain.

When the products of rock weathering (disintegration and alteration) are not transported as sediments but accumulate in place, a **residual soil** originates. As can be inferred from their geographical distribution, residual soils are predominant in warm, humid regions, where they can develop significant thicknesses, up to about 30 m.

1.1.2 Sediment transport

Water and wind currents, as well as glaciers and gravity, are responsible for transporting soils from their place of weathering. If the current turbulence is greater than the settling velocity, the smaller sediments are transported in *suspension*. The largest particles are moved by *dragging* along the bottom, while intermediate particles are moved by *saltation*. Finally, soluble materials can be transported in *solution* and may precipitate successively. Therefore transportation has the effect of **sorting** the particles (in general the particles are coarser towards the place of origin) and of **abrasion**, thus modifying the shape of particles.

1.1.3 Sedimentation

When the current velocity reduces or the ice melts, deposition of sediments occurs. The properties of the aggregate are linked to the environment of deposition, which represents the condition of accumulation and consolidation. For this reason it is important to distinguish between *continental* (glacial, alluvial, aeolian, paludal or lacustrine), *mixed* (estuarine, deltaic) and *marine environments*. Main features of these deposits will be discussed in Section 1.3.

1.2 Clay particles

From the previous description it can be inferred that natural soils are a mixture of particles of different sizes and engineers use terms such as *cobble, gravel, sand, silt* and *clay* to specify particular ranges of particle size. In this respect, the term *clay* is used to indicate particles smaller than $2 \mu m$. But the same term is also used more generally to describe natural soils in which the properties of the clay size range predominate.

In this sense a *clay soil* may contain little more than 10% of clay size particles, but it is so described because the presence of clay size particles significantly affects its behaviour. Finally the term *clay* can be used to refer to specific minerals, as is the case in the sequel.

Clay minerals are *hydrous aluminium silicates* plus other metallic ions, and can form as either primary or secondary minerals. They can only be seen with an electron microscope, since they are very small crystals, of platy-like shape and of colloidal size. On the basis of X-ray diffraction, it has been found that they are formed of two-dimensional sheets, which are stacked one upon another. These sheets are of two kinds: a *tetrahedral sheet*, in which adjoining tetrahedra are linked together by sharing oxygen (Figure 1.2) and an *octahedral sheet*, in which adjoining octahedra are linked by sharing hydroxyls (Figure 1.3).

There are dozens of clay minerals, depending on the way the basic sheets are stacked together and depending on the cations present in the tetrahedral and octahedral sheets. However, since the objective here is to show the essential features of their microstructure in order to elucidate qualitatively its influence on soil behaviour, it is sufficient to describe only a few examples of clay minerals.

Kaolinite is made up of layers of one tetrahedral sheet and one octahedral sheet (for this reason kaolinite is also called a 1:1 clay mineral), as represented in Figure 1.4. The basic layer has a thickness of 0.72 nm, and the bonding (hydrogen bonds) between successive layers is of sufficient strength to prevent hydration, so that there is no interlayer swelling and layers can be stacked up to make large crystals. A kaolinite crystal typically contains 70 to 100 layers.

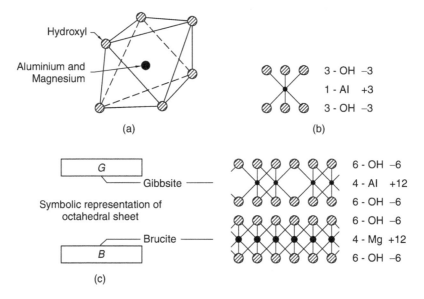

Figure 1.3 Octahedral unit.

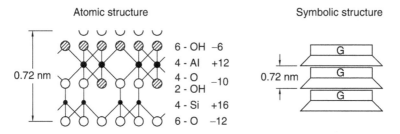

Figure 1.4 Kaolinite.

Montmorillonite presents a more complicated structure (Figure 1.5), the single layer (of thickness of 0.96 nm) being composed of an octahedral sheet between two tetrahedral sheets (i.e. it is a 2:1 clay mineral). The bonding between the silica sheets is represented by van der Waals' forces, so it is a weak bond. In addition, there is a negative charge deficiency in the octahedral sheet, due to *isomorphous substitution* of Al^{3+} with Mg^{2+}, so that exchangeable ions (sodium or calcium) lie between the layers, or are attached at the edges of the crystal. Molecular water may also occur between layers. As a consequence, the crystals can be rather small (the thickness can be of the order of 1 nm) and clay soils containing montmorillonite are susceptible to swelling, with important engineering implications.

Illite (Figure 1.6) is another example of 2:1 clay mineral, similar to montmorillonite, but in this case the layers are strongly bonded by a potassium atom, which fills the hexagonal hole in the tetrahedral sheet, and crystals have a thickness of about 10 to 30 nm.

From the above description of microstructural features we can deduce that the seat of positive charges and the seat of negative charges do not coincide, so that a clay particle exhibits a surface negatively charged. In addition, because Al^{3+} can partly substitute Si^{4+} in the tetrahedral and Mg^{2+} can substitute Al^{3+} in the octahedral

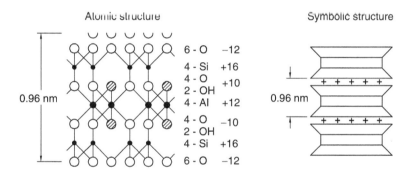

Figure 1.5 Montmorillonite.

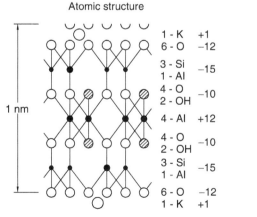
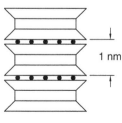

Figure 1.6 Illite.

sheet (this substitution is called *isomorphous substitution* because of the same form of the atom substituting the other), a net unit charge deficiency results.

It can then be expected that the interaction of adjacent soil particles, as well as the interaction of a particle with its surrounding ambient, will be influenced by this net charge deficiency and by this charge distribution.

Since the magnitude of the electrical charge is directly related to the surface of the particles, a measure of the surface area per unit mass of the particle is an indicator of the influence of surface forces relative to mass forces on the behaviour of a soil particle. This measure is termed **specific surface.**

In colloidal chemistry, a specific surface of 25 m^2/g is considered to be the lower limit of colloidal behaviour, because particles characterized by lower values are mainly influenced by mass forces. Comparing particles of our interest, it can be observed that silt particles have a specific surface lower than 1 m^2/g, whereas the values of a kaolinite particle range from 10 to 20 m^2/g and that of montmorillonite up to 800 m^2/g.

To neutralize the net charge deficiency, soil particles attract cations to its surface, as well as dipolar molecules of water. Furthermore, since cations are hydrated, they also contribute to the attraction of water. The first few molecular layers of water are so strongly bonded to the clay crystal by hydrogen bonding (hydrogen of water is attracted by oxygen or hydroxyls on the surface) that water cannot be removed under the action of hydraulic gradients. This water is termed **adsorbed water** and is part of the clay structure. It is important to recognize that, when in the sequel the attention will be devoted to water as a liquid phase in the pores, reference is made to water 'free' to move if driven by hydraulic gradients.

The cations ($Ca^{++}, Mg^{++}, Na^{+}, K^{+}$) attracted to the particle surface are weakly held on the particle and can be replaced by other cations. For this reason they are called **exchangeable cations.**

In addition to this tendency to being attracted, the cations also tend to move away because of their thermal energy, so that their concentration reduces with distance from the particle surface, with a tendency towards the concentration in the pore water, as shown in Figure 1.7.

The charged particle surface and the cations located in the adsorbed water layer together form the so-called **diffuse double layer.** The association of the clay particle with its double layer controls the interaction with other particles. In fact, the particles

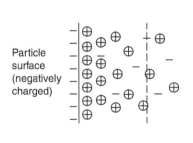
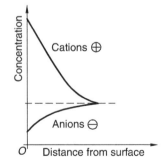

Figure 1.7 Diffuse double layer.

start to repel each other when the double layers come into contact. But, on the contrary, van der Waals' forces also act as attractive forces between the particles.

Depending on the net attractive or repulsive forces, clay particles during deposition tend to be attracted to each other, giving rise to a *flocculate* structure, or tend to move away, giving rise to a *dispersed* structure.

1.3 Soil deposits

Back to the fundamental processes listed in Figure 1.1, the weathering and the transport processes are responsible for the *nature of sediments* (i.e. dimension, shape, grain-size distribution and mineralogical composition); the environment of sedimentation and the post-depositional events are responsible for the *structure* of a soil element and of all macroscopic features of the deposit (bedding, discontinuities, inclusions).

Bedding is the term used to indicate the original layering in sediments (*stratification* is also used as synonymous). Each layer is indicated as *stratum* or *bed* and the interfaces between beds are indicated as *bedding planes*.

The **environment of deposition** is the complex of physical, chemical and biological conditions under which the sediments accumulate and consolidate.

Environments above the maximum tidal excursion are described as *continental environments*; *marine environments* are those below the minimum tidal level; *mixed environments* are those falling within the tidal excursions.

The *sediments*, which with time accumulate to form a natural deposit, are classified on the basis of their origin as *detrital* or *clastic sediments* (composed of minerals and rock fragments), *chemical* (formed from the precipitation of salts dissolved in water) and *biochemical (organic) sediments* (formed from the skeletal remains of plants and animals).

The deposition of the sediments is episodic in its character, since it depends on the rhythms of transport as well as on their interaction with tidal and volcanic cycles, so that the thickness of sediments deposited during each cycle can range from a few millimetres to many metres.

In this context, the term **formation** is used to indicate a bed or an assemblage of beds, with well-defined upper and lower boundaries, related to a specific environment of deposition.

1.3.1 Glacial deposits

A glacier, during its expansion, erodes the bottom and the sides of a valley, and when it retreats, it leaves deposits of sediments called *glacial drifts*. Today glaciers cover about 10% of the Earth's surface (90% of which is in Antarctica), but they were more extensive during the Pleistocene Epoch (which started two million years ago). Their extension fluctuated several times and the last extensive one occurred 20,000 to 15,000 years ago (about 30% of the Earth's surface was covered). Melting of glaciers results in deposition of debris and creates a variety of landforms: *bottom moraines* result from materials dropped directly from ice melting; *lateral moraines* originate from debris eroded from valley sides, debris avalanches and rockfalls; medial *moraines* originate when a tributary glacier merges with the main glacier; the arc-like ridge at the end of the glacier is called *terminal moraine*.

A common feature of moraine deposits is the spreaded grain size distribution curve. Since the included materials range from cobbles to clays, these deposits are difficult to characterize in terms of geotechnical properties.

Glacial lake deposits are usually composed of fine-grained materials and a relevant example is represented by **varved clays**, alternating layers of grey inorganic silts and darker silty clays (of a thickness less than 10 mm), transported in freshwater lakes by melt water. Usually the *varve* consists of a lower part of coarsest particles deposited in summer and a finer-grained upper part that sedimented in winter.

1.3.2 Alluvial deposits

The Latin term *alluvium* was being used in antiquity to indicate the material left in place by a river after a flood event. For this reason alluvial deposits include detrital materials (gravel, sand, silt, clay and mixtures of these) deposited permanently or in transit by flowing water. This definition applies to deposits of streams in their channel as well as in the floodplain. The alternating regime and pattern of flow dictate the main features of these deposits: lateral discontinuity and heterogeneity, presence of lenticular beds, of cross-bedded or evenly bedded sands, silts and gravels and the spread of fine silts and clays across the floodplain.

1.3.3 Lacustrine and paludal deposits

The deposition in fresh and quiet water gives rise to deposits of fine-grained materials, characterized by lateral uniformity and regular stratigrafical changes with depth.

In particular, paludal deposits include plastic silts, clays and organic matter, and, because these materials exhibit low strength and high compressibility, it can be expected that these deposits pose difficult engineering problems.

1.3.4 Estuarine and delta deposits

Mixed continental and marine deposits include estuarine and delta deposits. In both cases, common aspects are the large spatial variability and the unique character of each case under consideration. The depositional features of estuarine deposits depend on the tidal range and the river flow changes, and usually facies change from marine to fluvial progressing upstream. Delta deposits are formed by coastal and nearshore sediments of fluvial origin and, since the shape of the delta depends on many factors, a pronounced spatial variability results.

1.3.5 Aeolian deposits

Wind can transport only small particles, i.e. mostly sand and silt particles, so it is a very selective transport agent. Usually, sand starts to roll and slide when the wind speed reaches values of about 18 km per hour and, as the speed picks up, particles can lift and jump. Dunes, composed of uniform sand of rounded particles, are the prevalent landform associated with wind transport. Silt fractions are transported in suspension and deposited as *loess*. By taking into account the conditions of deposition, aeolian deposits exhibit a rather low degree of compaction (the relative density, as defined in the sequel, ranges from 50 to 60%), and a collapsing structure, i.e. they are susceptible to large decreases in bulk volume when saturated.

1.3.6 Marine deposits

Marine deposits include sediments laid down in different environments: continental shelf, continental slope, continental rise and deep sea. The *continental shelf* (or neritic environment) is characterized by a gentle slope of 1 to 1000 and water depth up to 200 m. Sediments are transported from continents and include sand, silt and clay, their distribution depending on local geology.

The *bathyal environment* includes the *continental slope*, characterized by a much steeper gradient (on average 1 to 40) if compared with the continental shelf, and the *continental rise*, with gradients ranging from 1 to 1000 to 1 to 700. The water depth ranges from 200 to about 2000 m. Subacqueous transport of sediments (fine sand, silt and mud) can occur down slope through debris flow and turbidity currents. *Turbidity currents* (triggered by earthquakes, hurricanes or large discharge of river sediments) are also known as density flow, since they represent a transition from mass to fluid flow. The deposition of the transported material occurs according to the settling velocity of the particles, so that a graded bedding results, with the coarsest particles at the bottom and the finer particles towards the top. The later consolidated forms created by turbidity currents are known as *turbidites*.

In tropical and subtropical areas, *calcareous sands* are widely distributed. They are formed from the remains of corals, shells and algae. These calcareous sediments deserve special attention, because they are composed of weak angular particles and their aggregate is characterized by uneven cementation, high void ratio and high compressibility. In particular, the particle crushing means that the knowledge gained with silica sands cannot be applied to these deposits.

Deep sea deposits include brown clays (also referred as red clays because of their colour), derived from wind and volcanic dust (*terrigenous sediments*), and calcareous and siliceous oozes, a term that indicates any soft, mud deposit, derived from shell and skeletal remains of marine organisms and plants (also described as *pelagic sediments*).

An extreme example of how the environment may influence the behaviour of clays is represented by the so-called **quick clays**. These post-glacial marine clays (widespread in Canada and in the Scandinavian countries) were deposited in salt water, but the deposit was successively uplifted above the sea level.

As a consequence, these clays have been leached by the flow of fresh water that removed the salt in the pore water, and this reduction of the electrolyte concentration caused a reduction of the shear strength (Skempton and Northey (1953) reproduced this process in the laboratory). In its actual condition, if subjected to any source of disturbance (even small), the clay may lose its strength and its behaviour will be that of a soil-water slurry.

1.4 Phase relations

As discussed in the introductory section, soils are multi-phase materials consisting of solid particles with voids filled with water and air. Because of this discrete nature, it can be expected that mechanical properties of engineering relevance, such as shear strength

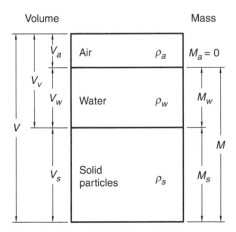

Figure 1.8 Idealized representation of phases.

and compressibility, depend on the degree of packing of the particles, so that it is of interest to define the amount of packing. This can be conveniently done by introducing the idealization represented in Figure 1.8 and by making use of some scalar quantities such as the porosity, the void ratio or the specific volume.

The porosity 'n' is defined by the ratio of the volume of voids V_v to the total volume V

$$n = \frac{V_v}{V} \tag{1.1}$$

and is usually expressed as a percentage.

In soil mechanics, it is customary to refer to the **void ratio** 'e', the ratio of the volume of voids V_v to the volume of solid particles V_s

$$e = \frac{V_v}{V_s}. \tag{1.2}$$

Alternatively, reference is made to the **specific volume**, this latter being the volume of a soil sample containing the unit volume of soil particles

$$v = 1 + e. \tag{1.3}$$

When referring to the case of uniform spheres, the minimum void ratio e_{min}, corresponding to the densest packing (rhombic packing) is equal to 0.35, and the maximum void ratio e_{max}, corresponding to the loosest state (cubic packing), is equal to 0.92.

The limiting values for coarse-grained soils are remarkably close to these theoretical values, the typical range being 0.43 to 0.67 when considering well graded sands and 0.51 to 0.85 for uniform sands (see also Table 1.2).

The actual packing of an assembly of sand and gravel particles is conveniently expressed in terms of the loosest and the densest states through the **relative density** D_R

$$D_R = \frac{e_{max} - e_o}{e_{max} - e_{min}} 100, \tag{1.4}$$

Table 1.2 Typical properties of some natural soils

Description	Porosity n (%)	Void ratio e	Water content (at S = 1) w (%)	Density (Mg/m³)	
				ρ_d	ρ
Loose uniform sand	46	0.85	32	1.4	1.8
Dense uniform sand	34	0.51	19	1.7	2.1
Loose well-graded sand	40	0.67	25	1.6	2.0
Dense well-graded sand	30	0.43	16	1.8	2.2
Soft clay	55	1.2	45	1.2	1.8
Stiff clay	37	0.6	22	1.7	2.1
Soft very organic clay	75	3.00	110	0.7	1.4

where e_o is the actual value of the void ratio and the values of e_{max} and e_{min} depend primarily on the particle roundness and the grain size distribution.

According to their definition, it is apparent that porosity and void ratio are not independent quantities and are related to each other by the following relations

$$n = \frac{e}{1+e} \qquad (1.5)$$

$$e = \frac{n}{1-n}. \qquad (1.6)$$

The void ratio of a soil element, as well as any other equivalent parameter, is not the object of any direct measurement, being deduced from other quantities, object of relatively simple measurements.

One of the most important quantities is the **water content** 'w', which defines the amount of water present in the voids, with respect to the amount of solids (Figure 1.8):

$$w = \frac{M_w}{M_s}. \qquad (1.7)$$

Note that water content is zero for a dry soil and is usually lower than 100%. Typical values for clay soils range from 20% to 70%, but in some cases it may be as high as 300%, as is the case of the Mexico City clay.

The procedure used for the determination of **water content** (see ASTM Standards, D2216) is to determine the weight of water removed by drying a sample of moist soil in a drying oven at a controlled temperature of $100 \pm 5°$ C. The weight of soil after oven-drying is used as the weight of the solid particles (see Example 1.1). Note that in principle every mixture sample should be checked that it is dried to a constant weight; in practice, drying a sample overnight (or 15 to 16 hours) is sufficient.

Care is required when selecting the sample used for the determination of water content, avoiding outer portions as well as the bottom and the top of a core sample, because it is subjected to evaporation and disturbance. It is also important to remember that the deposition of the sediment is episodic in its character, so that changes of water content can also occur over small spatial intervals, as is shown in Figure 1.9.

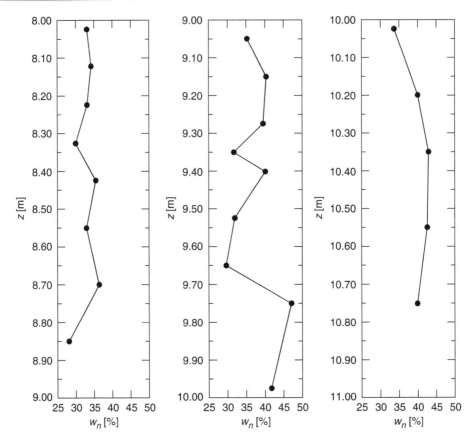

Figure 1.9 Water content distribution within a continuous sample of Pisa clay (Lancellotta *et al.*, 1994).

The water content plays an important role in soil mechanics, since mechanical parameters such as the undrained shear strength of clay soils can be related to it (this is a topic discussed in Chapters 5, 7 and 8).

The **density** of water, particles and the bulk density of a soil sample are defined respectively by the following relations:

$$\rho_w = \frac{M_w}{V_w} \tag{1.8}$$

$$\rho_s = \frac{M_s}{V_s} \tag{1.9}$$

$$\rho = \frac{M_s + M_w}{V}. \tag{1.10}$$

Siliceous sands have ρ_s equal to 2.65 Mg/m^3 and clay from 2.65 to 2.8 Mg/m^3 (see Table 1.3), whereas organic soils can have a lower value of 2.5 Mg/m^3. In the absence

Table 1.3 Specific gravity of some minerals

Minerals	Specific gravity $G_s = \rho_s / \rho_w$
Quartz	2.65
Calcite	2.72
Illite	2.84
Kaolinite	2.61
Montmorillonite	2.74

of direct measurements, it is customary to assume $\rho_s = 2.65\,\mathrm{Mg/m^3}$ for sand and $\rho_s = 2.70\,\mathrm{Mg/m^3}$ for clay.

Parameters of practical use are the dry **density**:

$$\rho_d = \frac{M_s}{V} = \rho_s(1-n) \tag{1.11}$$

and the **unit bulk weight** (weight per unit volume):

$$\gamma = \rho g, \tag{1.12}$$

where g is the acceleration of gravity (9.81 m/s²). Common values of γ range from 18 to 22 kN/m³.

Finally, the **buoyant unit weight** is defined by:

$$\gamma' = \gamma - \gamma_w. \tag{1.13}$$

The above are basic definitions and several relations can be derived from them. In the sequel we give a derivation of the most used ones and Examples 1.1 and 1.2 provide suggestions about the most convenient procedure to follow in order to solve practical problems.

(a) First, by referring to definitions:

$$\rho = \frac{M_s + M_w}{V}, \quad w = \frac{M_w}{M_s} \quad \text{and} \quad \rho_d = \frac{M_s}{V},$$

a substitution process gives the relation:

$$\rho = \rho_d(1+w). \tag{1.14}$$

(b) Let us now introduce the **degree of saturation** S as the percentage of the void volume which contains water.

$$S = \frac{V_w}{V_v}, \tag{1.15}$$

so that $S = 0\%$ indicates a dry soil and $S = 100\%$ a saturated soil.

The bulk density can then be expressed as:

$$\rho = \rho_s(1 - n) + \rho_w Sn,\tag{1.16}$$

as well as

$$\rho = \rho_d(1 + w) = \rho_s(1 - n)(1 + w).$$

Therefore, it follows

$$\rho_s(1 - n)w = \rho_w Sn$$

and then

$$S \cdot e = G_s \cdot w.\tag{1.17}$$

where

$$G_s = \rho_s/\rho_w\tag{1.18}$$

is the specific gravity of solids.

Example 1.1
Given the mass of a *saturated* soil sample (150 g) and its mass when oven dried (90 g), *find* the water content. Suppose that the sample, used for a triaxial test, has a diameter of 38 mm and the height of 76 mm, *find* the void ratio.

Solution. By referring to its definition, the water content is given by:

$$w = \frac{150 - 90}{90} \times 100 = 66.67\%.$$

The volume of the sample is equal to $86.19\,\text{cm}^3$, so that the bulk density and the dry density are given by:

$$\rho = \frac{150}{86.19} = 1.74\,g/cm^3 (= Mg/m^3) \quad \text{and} \quad \rho_d = \frac{90}{86.19} = 1.04\,g/cm^3 (= Mg/m^3).$$

Since $\rho = \rho_d + n\rho_w$, it follows $n = 0.70$ and $e = \dfrac{n}{1 - n} = 2.33$.

Example 1.2
Given $\rho = 1.76\,\text{Mg/m}^3, w = 10\%$ and $\rho_s = 2.70\,\text{Mg/m}^3$, *find* the dry density, the void ratio and the degree of saturation.

Solution. The relation $\rho = \rho_d(1 + w)$ gives $\rho_d = 1.6\,\text{Mg/m}^3$. Further, since $\rho_d = \rho_s(1 - n)$, it follows $n = 1 - \dfrac{1.60}{2.70} = 0.407$ and $e = \dfrac{n}{1 - n} = 0.687$.
Finally, by using the relation $\rho = \rho_d + Sn\rho_w$, we obtain $S \cdot n = 0.16$ and $S = 0.39$.

1.5 Description and classification of soils

Civil engineers need a knowledge of soil behaviour in order to predict the performance of structures interacting with soils and under what conditions a structure may reach failure. Therefore, engineers are primarily interested in basic parameters, such as strength, compressibility and hydraulic conductivity.

The assessment of these parameters requires specific tests, but an accurate *description* of the soil can help the engineer by providing a general guidance about the expected soil behaviour.

Essentially, **description** is what you see and how soil reacts to simple tests, and any comprehensive description should include the characteristics (size, shape, grain-size distribution, plasticity) of soil material, as well as the structural features (bedding, fissures, joints) of the soil mass, which can be inferred in the field.

In order to benefit from any description a standard language is essential, and for this reason a detailed scheme is provided by BS 5930-(1981), reproduced in Table 1.4. More specifically, Table 1.4 illustrates (a) how soils can be distinguished on the basis of appearance of particles and visibility to naked eye; (b) how mixed soils can be classified; (c) how the soil strength can be preliminarily assessed through qualitative terms: loose or dense for coarse grained soils, soft or stiff for fine grained soils; (d) what are the relevant structural features.

The term **classification** refers to the procedure of sorting soil samples in different classes, with the assumption that elements within the same class are expected to exhibit a similar behaviour. There are two key aspects when selecting classifications properties: (i) these properties must have a precise meaning and their determination must be suitable for simple tests, to be also used in the field; (ii) the selected properties must be independent of the state of the sample, that means independent of the environment or the stress state.

These requirements are fulfilled by the *size*, the *shape* and the *mineral composition* of particles.

On the basis of particle sizes, engineers consider it useful to classify soils into **coarse-grained soils** (gravel and sand) and **fine-grained soils** (silt and clay). In fact, although the fundamental principles governing soil behaviour are the same, when considering parameters of engineering relevance, such as the *hydraulic conductivity*, this may vary over 10 orders of magnitude and is related to particle sizes. As an additional example, fine-grained materials usually compress more than coarse-grained materials.

Since natural soils are an assortment of particles of many different sizes, the description requires to define the range and the frequency distribution of particle sizes. Both these objectives are reached by performing a **sieve analysis** (see ASTM D422).

A sample of dry soil (of about 500 g) is mechanically shaken through a series of sieves (see Table 1.5) and the percentage retained or passing through each sieve is weighted. The results are then plotted as a cumulative curve against the sieve size, as shown in Figure 1.10, and because the range of possible particles is of the order of 10^6 (from over 100 mm to less than 0.001 mm), the grain size distribution is usually represented versus the logarithm of the average grain diameter.

Note that Figure 1.10 also gives guidance for the name given to any particular range of particles (*clay, silt, sand* and *gravel*). As an example, *fine sand* ranges from 0.06 to 0.2 mm and *coarse sand* from 0.6 to 2 mm. Furthermore, in addition to terms specified

Table 1.4 Description of soils

	Basic soil type	Particle size, mm	Visual identification	Particle nature and plasticity	Composite soil types (mixtures of basic soil types)	
Very coarse soils	BOULDERS	— 200	Only seen in pits or exposures.	Particle shape:	Scale of secondary constituents with coarse soils	
	COBBLES	— 60	Often difficult to recover from boreholes	Angular subangular	Term	% of clay or silt
Coarse soils (over 65% sand and gravel sizes)	GRAVELS — coarse	— 20	Easily visible to naked eye; particle shape can be described; grading can be described.	Rounded Flat Elongate	slightly clayey } GRAVEL or } SAND slightly silty	under 5
	GRAVELS — medium	— 6	Well graded: wide range of grain sizes, well distributed. Poorly graded: not well graded. (May be uniform; size of most particles lies between narrow limits; or gap graded; an intermediate size of particle is markedly under-represented.)	Texture:	– clayey } GRAVEL or – silty } SAND	5 to 15
	GRAVELS — fine	— 2		Rough Smooth Polished	very clayey } GRAVEL or very silty } SAND	15 to 35
	SANDS — coarse	— 0.6	Visible to naked eye; very little or no cohesion when dry; grading can be described.		Sandy GRAVEL } Sand or gravel and Gravelly SAND } important second constituent of the coarse fraction	
	SANDS — medium	— 0.2	Well graded: wide range of grain sizes, well distributed. Poorly graded: not well graded. (May be uniform: size of most particles lies between narrow limits; or gap graded: an intermediate size of particle is markedly under-represented.)		(See 41.3.2.2) For composite types described as: clayey: fines are plastic, cohesive: silty: fines non-plastic or of low plasticity	
	SANDS — fine	— 0.06				
Fine soils (over 35% silt and clay sizes)	SILTS — coarse	— 0.02	Only coarse silt barely visible to naked eye; exhibits little plasticity and marked dilatancy; slightly granular or silky to the touch, Disintegrates in water; lumps dry quickly; possess cohesion but can be powdered easily between fingers.	Non-plastic or low plasticity	Scale of secondary constituents with fine soils	
	SILTS — medium	— 0.006			Term	% of sand or gravel
	SILTS — fine	— 0.002			sandy } CLAY or gravelly } SILT	35 to 65
	CLAYS		Dry lumps can be broken but not powdered between the fingers; they also disintegrate under water but more slowly than silt; smooth to the touch; exhibits plasticity but no dilatancy; sticks to the fingers and dries slowly; shrinks appreciable on drying usually showing cracks. Intermediate and high plasticity clays show these properties to a moderate and high degree, respectively.	Intermediate plasticity (Lean clay) High plasticity (Fat clay)	– CLAY: SILT	under 35
					Examples of composite types (Indicating preferred order for description) Loose, brown, subangular very sandy, fine to coarse GRAVEL with small pockets of soft grey clay Medium dense, light brown, clayey, fine and medium SAND	
Organic soils	ORGANIC CLAYS, SILT or SAND	Varies	Contains substantial amounts of organic vegetable matter.		stiff, orange brown, fissured sandy CLAY	
	PEATS	Varies	Predominantly plant remains usually dark brown or black in colour, often with distinctive smell; low bulk density.		Firm, brown, thinly laminated SILT and CLAY Plastic, brown, amorphous PEAT	

Table 1.4 (Cont'd)

Compactness/strength		Structure			Colour
Term	Field test	Term	Field identification	Interval scales	
Loose	By inspection of voids and particle packing.	Homogeneous	Deposit consists essentially of one type.	Scale of bedding spacing	Red Pink
Dense		Inter-stratified	Alternating layers of varying types or with bands or lenses of other materials. Interval scale for bedding spacing may be used.	**Term** / **Mean spacing mm** Very thickly bedded / over 2000 Thickly bedded / 2000 to 600	Yellow Brown Olive Green Blue White
Loose	Can be excavated with a spade; 50 mm wooden peg can be easily driven.	Heterogeneous	A mixture of types.	Medium bedded / 600 to 200 Thinly bedded / 200 to 60	Grey Black etc.
Dense	Requires pick for excavation; 50 mm wooden peg hard to drive.	Weathered	Particles may be weakened and may show concentric layering.	Very thinly bedded / 60 to 20 Thickly laminated / 20 to 6	Supplemented as necessary with:
Slightly cemented	Visual examination; pick removes soil in humps which can be abraded.			Thinly laminated / under 6	Light Dark Mottled etc. and
Soft or loose	Easily moulded or crushed in the fingers.	Fissured	Break into polyhedral fragments along fissures. Interval scale for spacing of discontinuities may be used.		Pinkish Reddish Yellowish Brownish etc.
Firm or dense	Can be moulded or crushed by strong pressure in the fingers.				
Very soft	Exudes between fingers when squeezed in hand.	Intact	No fissures.		
Soft	Moulded by light finger pressure.	Homogeneous	Deposit consists essentially of one type.	Scale of spacing of other discontinuities	
Firm	Can be moulded by strong finger pressure.	Inter-stratified	Alternating layers of varying types. Interval scale for thickness of layers may be used.	**Term** / **Mean spacing mm**	
Stiff	Cannot moulded by fingers. Can be indented by thumb.			Very widely spaced / over 2000	
Very stiff	Can be indented by thumb nail.	Weathered	Usually has crumb or columnar structure.	Widely spaced / 2000 to 600	
Firm	Fibres already compressed together.			Medium / 600 to 200 Closely spaced / 200 to 60	
Spongy	Very compressible and open structure.	Fibrous	Plant remains recognizable and retains some strength.	Very closely spaced / 60 to 20	
Plastic	Can be moulded in hand, and smears fingers.	Amorphous	Recognizable plant remains absent.	Extremely closely spaced / under 20	

Source: Reproduced with permission from BS5930: 1981.

Table 1.5 US standard sieve sizes and their corresponding opening dimension

US standard sieve number	Sieve opening (mm)
4	4.75
10	2.00
20	0.85
40	0.425
60	0.25
100	0.15
140	0.106
200	0.075

in Figure 1.10, we use the term *cobbles* for sizes from 60 to 200 mm and *boulders* for sizes from 200 to 600 mm.

The sieve analysis is impracticable for sieve openings of less than about 0.075 mm, thus fine-grained soil particles must be separated by sedimentation. The sedimentation analysis is based on the **Stoke's law**: the terminal velocity v of falling spheres in a viscous fluid depends on the sphere diameter D and on the densities of the sphere in suspension and of the fluid:

$$v = \frac{\rho_s - \rho_w}{18\eta} gD^2 ,$$

where η is the dynamic viscosity of the fluid (equal to $1 \cdot 10^{-3}$ N·s/m² if the fluid is water at a temperature of 20°C). As smaller particles settle more slowly than coarser ones, a hydrometer determines the density of the suspension at different times, and this allows one to compute the percentage of particles of a given equivalent diameter. Note that because of the supposed spherical shape of particles, the reliability of this analysis is restricted to particle size in the range from 2 to 50 μm.

According to the grain-size distribution curve a given soil is called *well-graded* if it has a good representation of particle size over a wide range, so that the gradation curve is generally smooth and upward concave (see the glacial till in Figure 1.10); a *uniform soil* is characterized by particles of about the same diameter (see the almost vertical curve of Ticino sand); the absence of particles of intermediate size (a curve with a horizontal step) defines a *gap-graded soil.*

In sedimentary petrology it is common to use statistical parameters to describe the grain size distribution. In soil mechanics it is usual to refer to D_{10}, the grain size corresponding to 10% of passing, to locate the curve and an approximate way of describing the shape of the curve is through the uniformity coefficient:

$$C_U = \frac{D_{60}}{D_{10}}, \tag{1.19}$$

where D_{60} is the grain size corresponding to 60% of passing.

Note that when applied to the previous illustrated example, the uniformity coefficient is revealed to be a misnamed term, because the smaller the number the

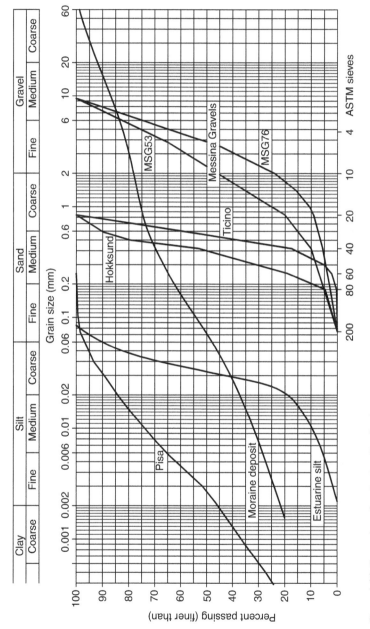

Figure 1.10 Examples of grain size distribution curves.

more uniform the gradation. As suggested by Holtz and Kovaks (1981), it is actually a coefficient of disuniformity.

Finally, we observe that samples for sieve analysis need not be undisturbed, but they need to be representative, i.e. they should be complete with no fraction of the deposit being missed.

1.6 Atterberg limits and plasticity chart

The expected behaviour of coarse-grained soils can be inferred from the grain-distribution curve, the shape of particles and the degree of packing. On the contrary, the behaviour of fine-grained soils depends on the amount of clay sized particles and on the mineral composition. On the basis of this latter observation, to establish a criterion to classify fine-grained soils, we can formulate the following working hypothesis: if the mineral composition affects the behaviour of the particles and their interaction with water, in given circumstances the water content must reflect the mineral composition. According to such a hypothesis, a reliable procedure requires to single out a clearly defined and standardized physical condition and to make measurements of water content, in correspondence of which a given behaviour is observed.

Moving from the observation that a clay sample can be in a liquid, plastic, semi-solid or solid state (this physical state is called **consistency**), depending on its water content, the Swedish agricultural scientist Albert Atterberg defined, in the early 1900s, the water contents corresponding to the transitional stages indicated in Figure 1.11.

The upper and lower limits of water content within which a clay element exhibits a plastic behaviour are defined as **liquid limit** and **plastic limit** and the term **plasticity** must be intended as the ability of a soil sample to be worked and remoulded, i.e. the ability of undergoing plastic deformations without cracking or crumbling.

Casagrande (1932) developed the following standard procedure for the determination of the liquid limit. A clayey soil paste (a mass of about 100 g from passing the sieve 40) is mixed with water to form a creamy paste and is placed in a metal cup (see Figure 1.12). It is levelled off and a V-groove is then cut through the sample and the cup is tapped, by counting the numbers of blows required just to close the groove. The sample is then removed and its water content is measured. The remaining sample is remixed to a different water content and the test is repeated. Water content is finally plotted against the number of blows and the liquid limit is defined as the water content at which the groove is closed after being tapped 25 times (Figure 1.12).

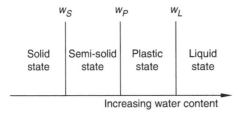

Figure 1.11 Consistency of a clay soil.

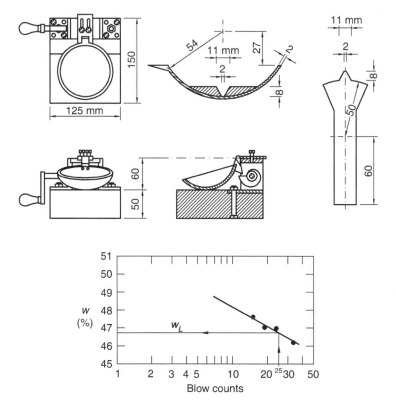

Figure 1.12 Casagrande apparatus for the determination of the liquid limit.

An alternative procedure is represented by the so-called fall-cone test (BS 1377, 1991). This test consists of a miniature penetrometer with a 30° cone apex, 35 mm long. The cone and the sliding shaft, which is attached, have a mass of 80 g. Performing the test requires mixing up with distilled water a number of samples, each of different water content. The paste is then placed in a cylindrical metal cup, 55 mm internal diameter and 40 mm deep, and levelled to give a smooth surface. The cone is then released in order to penetrate into the paste for a period of five seconds, and the depth of penetration is measured. The cone penetration is plotted against the water content and the best linear fitting is drawn (note that the Critical State Theory predicts that the water content is linearly related to the logarithm of the cone penetration). From this plot, the liquid limit is obtained as the water content corresponding to a cone penetration of 20 mm.

The plastic limit is defined as the water content below which the remoulded soil sample ceases to behave as a plastic material. The related test consists of finding the water content at which the sample starts to crumble when rolled into a pencil shape of 3 mm diameter (see Figure 1.13). The range over which the soil remains plastic is defined by the **plasticity index** PI:

$$PI = w_L - w_P. \tag{1.20}$$

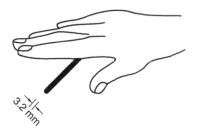

Figure 1.13 Rolling-out test for the determination of plastic limit.

The current natural water content is related to Atterberg limits through the **liquidity index** LI:

$$LI = \frac{w - w_P}{PI}. \tag{1.21}$$

Note that the liquidity index plays for fine-grained soils the same role the relative density plays for coarse-grained soils. Soft clays have a liquidity index near to unit, whereas stiff clays may have values near to zero and quick clays have a liquidity index greater than 1.0.

The arguments presented in Section 1.2 suggest that the properties of a clay depend on physical and chemical characteristics of the constituent minerals. The determination of these characteristics would require the use of an X-ray spectrometer, thermal analysis and more sophisticated techniques, which are not part of conventional laboratory procedures. Atterberg limits fulfil the requirement of being simple tests to account for these characteristics. However, the colloidal properties depend on both the amount of clay (i.e. the *clay fraction, CF*), and the mineralogy. Therefore, it can be expected that, at a given clay fraction, clays with higher *PI* will be more colloidally active than clays with a lower *PI*. Moving from this observation, Skempton (1953) suggested introducing a parameter called **activity**:

$$A = \frac{PI}{CF}, \tag{1.22}$$

and, based on this parameter, clay samples are distinguished as follows:

inactive clays, if A < 0.75

normal clays, if 0.75 < A < 1.25

active clays, if A > 1.25.

Kaolinite exhibits very low activity, in the range from 0.33 to 0.46. Illite shows values near to unit, but when testing illite it must be observed that it usually occurs in conjunction with other minerals. Montmorillonite exhibits activity up to 7 when sodium cations are present, but the activity can be lowered to 1.5 through a base exchange from sodium to calcium.

By using Atterberg limits, fine-grained soils can be classified according to **plasticity chart** (Figure 1.14), developed by Casagrande (1948). The chart is divided into

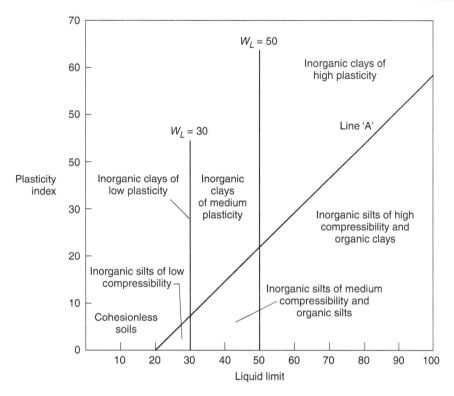

Figure 1.14 Plasticity chart.

six regions by the inclined line 'A', of equation:

$$PI = 0.73 \cdot (w_L - 20) \tag{1.23}$$

and by two vertical lines $w_L = 30$ *and* $w_L = 50$.

Soils represented by points above line 'A' are **inorganic clays** of low$(w_L < 30)$, medium $(30 < W_L < 50)$ and high plasticity $(W_L < 50)$. Points below line 'A' represent inorganic silts, organic silts and organic clays. **Inorganic silts** are defined by low, medium or high compressibility, if the liquid limit is lower than 30, between 30 and 50, or higher than 50%.

If a pat of soil of about 6 mm thick is allowed to dry, it can be broken or crumbled between fingers. Inorganic clays have relatively high dry strength; inorganic silts have little or no dry strength and can be easily crumbled.

Organic silts are represented in the region with a liquid limit between 30 and 50 and **organic clays** correspond to a liquid limit higher than 50%. Organic soils have a dark brown, dark grey or bluish grey colour and the presence of H_2S and CO_2 (deriving from decomposition of organic matter) gives them a distinctive odour.

The plasticity chart suggested by Casagrande can also be used to have a preliminary qualitative identification of the predominant clay minerals, by comparing the location of the tested sample with those of known minerals (Holtz and Kovaks, 1981). If the

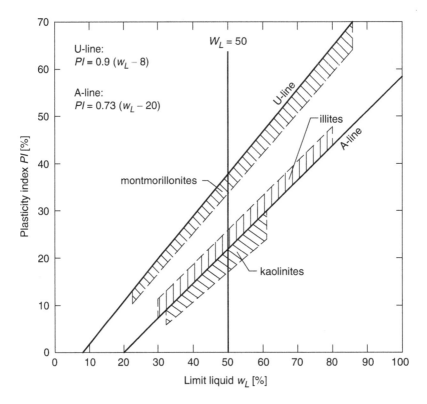

Figure 1.15 Clay minerals on plasticity chart (adapted from Mitchell, 1993, and Holtz and Kovaks, 1981).

tested sample plots near the U-line (see Figure 1.15), the clay fraction is predominantly montmorillonite. Illites are located just above the A-line. Inactive kaolonites plot just below the A-line and their expected behaviour is that of silts of medium to high compressibility.

Finally, if the location plots to the left of the U-line, data should be further checked.

1.7 Summary

Soils are formed by weathering processes of rocks and are therefore particulate systems. Due to this particulate nature, when a soil element deforms, the overall deformation is essentially the result of relative sliding between particles and rotation of particles, and, as these are non-linear and irreversible processes, soil behaviour is highly non-linear and irreversible.

Voids or pores between particles are filled with water, or there may be more than one fluid, so that soils are multi-phase materials, their behaviour being influenced by the interaction between solids and fluids.

Composition, structure and properties of a natural soil element are the result of the history of a soil deposit, which includes weathering, transportation, deposition and post-depositional changes.

The *weathering* and the *transport processes* are responsible for the *nature of sediments* (i.e. dimension, shape, grain-size distribution and mineralogical composition); the *environment of sedimentation* and the *post-depositional events* are responsible for the *structure* of a soil element and for all macroscopic features of the deposit (bedding, discontinuities, inclusions).

The term *classification* refers to the procedure of collecting soil samples in different classes, with the assumption that elements within the same class are expected to exhibit a similar behaviour.

On the basis of particle size, soils are grouped into *coarse-grained soils* (gravel and sand) and *fine-grained soils* (silt and clay).

Mechanical properties of engineering relevance, such as shear strength and compressibility, depend on the degree of packing of the particles, so that it is of interest to define the amount of packing through the introduction of scalar quantities: the porosity ($n = V_v/V$), the void ratio ($e = V_v/V_s$) or the specific volume ($v = 1 + e$). The actual packing of an assembly of sand and gravel particles is conveniently expressed in terms of the loosest and the densest states through the *relative density*.

The behaviour of fine-grained soils depends on the amount of clay and on the mineral composition. *Atterberg limits* reflect both these aspects and are used to classify fine-grained soils, through the plasticity chart. The upper and lower limits of water content within which a clay element exhibits plastic behaviour are defined as *liquid limit* and *plastic limit* and the ability of undergoing plastic deformations at constant volume without cracking or crumbling is intended in soil mechanics by *plasticity*.

The current state, in terms of Atterberg limits, is defined by the *liquidity index*: $LI = (w - w_p)/PI$, which plays for fine-grained soils the same role the relative density plays for coarse-grained soils.

1.8 Further reading

Further reading on index properties of soils:

T.W. Lambe and R.V. Whitman (1969) *Soil Mechanics*. Wiley.
R.D. Holtz and W.D. Kovaks (1981) *An Introduction to Geotechnical Engineering*. Prentice-Hall.

Reading on clay mineralogy and related engineering aspects includes the following books:

J.K. Mitchell and K. Soga (2005) *Fundamentals of Soil Behaviour* (2nd edn). Wiley.
R.E. Grimm (1962) *Applied Clay Mineralogy*. McGraw-Hill.

Geotechnical students should be conscious that understanding geology is essential to properly interpret site investigations and to recognize geological hazards. A list of textbooks on Geology includes:

F.G.H. Blyth and M.H. De Freitas (1974) *A Geology for Engineers*. Edward Arnold.
C. Pomerol, Y. Lagabrielle and M. Renard (2005) *Eléments de Géologie* (13th edn). Dunod.
J. Dercourt and J. Paquet (2002) *Géologie. Objects and methods* (11th edn). Dunod.
F. Press and R. Siever (1986) *Earth* (4th edn). Freeman.
R.F. Flint (1971) *Glacial and Quaternary Geology*. Wiley.
C.D. Ollier (1969) *Weathering*. Longman.
B.C.M. Butler and J.D. Bell (1988) *Interpretation of Geological Maps*. Longman Group UK Limited.

To generate interest on Geology, an introductory delightful reading is the book by M. Marthaler (2001) *Le Cervin est-il Africain? Une histoire géologique entre les Alpes et notre planète*, Editions L.E.P. Loisirs et Pédagogie S.A., Losanne.

Further readings on more specific aspects:

The book by P. Doyle, M.R. Bennet and A.N. Baxter (2001) *The Key to Earth History: An introduction to stratigraphy*, Wiley, provides an introduction to stratigraphy and stratigraphical methods used to explain the process and pattern of Earth history.

The book by M.E. Tucker (2001) *Sedimentary Petrology* (3rd edn), Blackwell, is a reference textbook which deals with sediments and sedimentary rocks, examining composition, sedimentary structures, diagenesis and depositional environments.

Some relevant ASTM Standards:

ASTM D422. Particle Size Analysis of Soils.
ASTM D821/ASTM 2217. Grain Size Analysis.
ASTM D854. Specific Gravity of Soils.
ASTM D2216. Laboratory Determination of Water Content of Soil and Rock.
ASTM D2487. Classification of Soils for Engineering Purposes.
ASTM D4253. Max Index Density of Soils Using a Vibratory Table.
ASTM D4254. Min Index Density of Soils and Calculation of Relative Density.
ASTM D4318. Liquid Limit, Plastic Limit and Plasticity Index of Soils.

Chapter 2

Continuum mechanics

In Chapter 1 we commented on the particle-continuum duality, which characterizes the approach in geotechnical engineering: fundamental aspects of soil behaviour cannot be properly understood without a knowledge of the discrete nature of soils, but mathematical models used in practice to predict the behaviour of geotechnical structures are based on the continuum medium idealization of soils.

For this reason learning continuum mechanics is an essential step and familiarity with the language of continuum mechanics is recommended.

In order to pursue this goal, this chapter presents a graduated approach to the subject and we suggest to the beginning reader the following path:

1 Sections 2.1, 2.2, 2.3 and 2.4 should be carefully mastered, because they focus on the language and on mathematical prerequisites.
2 Sections 2.5, 2.6, 2.7, as well as 2.11, 2.12, 2.13 and 2.14 may be postponed, since they offer a treatment of finite deformation and basic laws of continuum mechanics, as usually thought in upper level courses.
3 Sections 2.8, 2.9 and 2.10 give a presentation of stress and infinitesimal strain concepts that should be clearly understood before moving to other chapters.

2.1 The language of continuum mechanics: indicial notation

The aim of this section is to introduce the language used in continuum mechanics. To do this in a graduate approach, let's start by considering entities we are already familiar with, such as *scalars* (specified by just one number) and *vectors* (specified by three numbers, representing the components with respect to a given basis).

We learned from linear algebra that in a *cartesian coordinate system* (Figure 2.1) a vector $\mathbf{v}$ can be expanded in the form:

$$\mathbf{v} = v_1 \mathbf{e}_1 + v_2 \mathbf{e}_2 + v_3 \mathbf{e}_3, \tag{2.1}$$

where v_i are the cartesian components of $\mathbf{v}$ and $\mathbf{e}_1, \mathbf{e}_2, \mathbf{e}_3$ are the unit base vectors.

We now observe that the same vector $\mathbf{v}$ can be represented in a more convenient and more compact form, if we introduce the so-called index notation. This requires

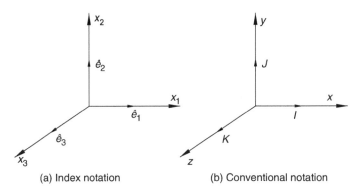

(a) Index notation   (b) Conventional notation

Figure 2.1 A cartesian coordinate system.

to grasp the **summation convention**: *if an index is repeated twice in a term, then that index is to be given all possible values (i.e. 1, 2, 3) and the results added together.*

According to this rule, the previous expansion (2.1) can be written in the compact form:

$$\mathbf{v} = v_k \mathbf{e}_k. \tag{2.2}$$

The repeated index k is called a **dummy index**, because it can be replaced by any other index, so that the expansion (2.2) can equally be written as $\mathbf{v} = v_j \mathbf{e}_j$.

However, scalars and vectors do not exhaust the quantities of interest in continuum mechanics, and to explain concepts such as stress or strain we need to introduce more general quantities, called *tensors*.

In rather abstract terms, tensor quantities are specified in Euclidean space by 3^n numbers, if n is the *order* or *rank* of the tensor. Accordingly, a symbol with a single index, either subscript or superscript:

$$a_i, a^i,$$

is a first-order quantity, specified in the Euclidean space E^3 by 3^1 components, and vectors are examples of such quantities.

A **second-order symbol** is characterized by two indices, according to one of the following forms:

$$a_{ij}, a^{ij}, a^j_i, a^i_j.$$

In this book reference will be made to an **orthonormal basis,** namely a basis composed of mutually orthogonal unit vectors, so that there will be no need to distinguish among *covariant components* (subscript indices), *contravariant components* (superscripts indices) or *mixed components*, and we shall always use subscript indices (see Malvern, 1969).

In the Euclidean space E^3, to every second-order quantity correspond 3^2 components, which can be rearranged as follows:

$$a_{ij} \Rightarrow \begin{matrix} a_{11} & a_{12} & a_{13} \\ a_{21} & a_{22} & a_{23} \\ a_{31} & a_{32} & a_{33} \end{matrix} \; .$$

When dealing with second-order symbols, examples related to the summation convention are the following:

$$T_{ii} = T_{11} + T_{22} + T_{33}$$
$$b_i = T_{ij}a_j = T_{i1}a_1 + T_{i2}a_2 + T_{i3}a_3 \tag{2.3}$$

and we can observe that no index can appear more than twice, but more than one pair of dummy indices can appear in a term. Moreover, if an index appears once in each term of an equation the index is called **free index**. In this case the equation is assumed to hold for all values the index can take (i.e. 1, 2, 3). This rule provides a big advantage in term of compactness, when expressing equations in three dimensions. In fact, writing:

$$a_i = b_i + c_i$$

corresponds to a system of three equations, and writing:

$$T_{ij} = \lambda E_{kk}\delta_{ij} + 2\mu E_{ij}$$

corresponds to nine equations, because both indices i, j can assume values 1, 2, 3.

The previous examples also allow to state that *if an equation is correctly written, every term must have the same letter for the free indices.*

2.1.1 The Kronecker delta and the Levi-Civita symbol

In order to simplify equations in index notation, it is necessary to introduce two symbols, corresponding to special operators.

The **Kronecker delta**, defined by:

$$\delta_{pq} = \begin{cases} 1 & if \quad p = q \\ 0 & if \quad p \neq q \end{cases}, \tag{2.4}$$

is usually thought as the index representation of the unit (or identity) matrix. But an important property of the Kronecker delta to be highlighted is the *substitution property*, because its definition implies:

$$a_i\delta_{ik} = a_k, \tag{2.5}$$

since in the l.h.s. the only surviving term is that for which $i = k$.

According to this property, we can write the **dot product** of two vectors **a** and **b** as:

$$\mathbf{a} \cdot \mathbf{b} = a_i \mathbf{e}_i \cdot b_j \mathbf{e}_j = a_i b_j \mathbf{e}_i \cdot \mathbf{e}_j = a_i b_j \delta_{ij} = a_i b_i, \tag{2.6}$$

since for the base vectors we have:

$$\mathbf{e}_i \cdot \mathbf{e}_j = \delta_{ij}. \tag{2.7}$$

We leave as an exercise to prove that:

$$\begin{aligned} b_{jk}\, \delta_{km} &= b_{jm} \\ b_{ij}\, c_{km} \delta_{ir}\, \delta_{ks} &= b_{rj}\, c_{sm} \end{aligned}.$$

The **Levi-Civita symbol** (also called the permutation symbol) is defined as follows:

$$\varepsilon_{ijk} = \begin{cases} 0 & \text{when any two indices are equal} \\ 1 & \text{when } ijk = 123 \text{ or } 231 \text{ or } 312 \\ -1 & \text{otherwise} \end{cases}. \tag{2.8}$$

By using this symbol, the **cross product** of two vectors, defined as:

$$\mathbf{a} \times \mathbf{b} = (a_2 b_3 - a_3 b_2)\, \mathbf{e}_1 + (a_3 b_1 - a_1 b_3)\, \mathbf{e}_2 + (a_1 b_2 - a_2 b_1)\, \mathbf{e}_3, \tag{2.9}$$

can be expressed by the more compact notation:

$$\mathbf{a} \times \mathbf{b} = a_i \mathbf{e}_i \times b_j \mathbf{e}_j = a_i b_j \mathbf{e}_i \times \mathbf{e}_j = \varepsilon_{ijk} a_i b_j \mathbf{e}_k, \tag{2.10}$$

since for the unit base vectors we have:

$$\mathbf{e}_i \times \mathbf{e}_j = \varepsilon_{ijk} \mathbf{e}_k. \tag{2.11}$$

Finally, it can be proved by direct expansion that the product of permutation symbols can be expressed in terms of Kronecker delta:

$$\varepsilon_{miq} \varepsilon_{jkq} = \delta_{mj} \delta_{ik} - \delta_{mk} \delta_{ij}. \tag{2.12}$$

2.2 Tensors

In the following we consider the three-dimensional Euclidean space E^3 and we define *vectors* the element of the associated vector space V. Then, a **second-order tensor T** is defined as the linear mapping of V into V:

$$\mathbf{T} := V \rightarrow V,$$

which relates any arbitrary vector **a** in V to another vector **b** in V:

$$\begin{aligned} \mathbf{b} &= \mathbf{T}\mathbf{a} \\ b_i &= T_{ij} a_j \end{aligned}. \tag{2.13}$$

The coefficients T_{ij} of this linear mapping:

$$
\begin{matrix}
T_{11} & T_{12} & T_{13} \\
T_{21} & T_{22} & T_{23} \\
T_{31} & T_{32} & T_{33}
\end{matrix} \; .
$$

represent the *components* of the tensor **T**.

A rather familiar example of second-order tensor is provided by the **stress tensor** σ_{ij} (see Section 2.9), which can be thought as the linear mapping that associates to each unit normal *n* the traction vector *t*, acting on the surface of outward normal *n*:

$$
t_i^{(n)} = \sigma_{ji} n_j \; .
$$

The definition of a second-order tensor as a linear mapping can be generalized to higher order tensors, by observing that the set of tensors defined on E^3, denoted as $Lin(E^3)$, is itself a (9-dimensional) vector space, since it can be proved that the axioms of addition and scalar multiple are satisfied. Then, a *fourth-order tensor* **C** is the linear mapping:

$$
\mathbf{C} := Lin(E^3) \to Lin(E^3),
$$

which means that a fourth-order tensor is a linear function whose argument is a second-order tensor and whose value is also a second-order tensor.

As an example, when dealing with constitutive equations (Chapter 3), we will introduce a *fourth-order tensor,* which relates the second-order small strain tensor ε_{hk} to the second-order stress tensor σ_{ij}:

$$
\sigma_{ij} = C_{ijhk} \varepsilon_{hk} \; .
$$

The components of the fourth-order tensor C_{ijkl} are 81 scalar quantities.

We now introduce a special product of two vectors (which is not to be confused with the scalar or vector product previously mentioned), called **dyadic product** (or **open product** or **tensor product**).

The tensor product of two vectors, represented in the form $\mathbf{a} \otimes \mathbf{b}$, or simply as **ab** (without multiplication sign), operates as a linear vector function and is therefore a *second-order tensor.*

It assigns to each vector $\mathbf{u} \in V$ the vector $\mathbf{a}\,(\mathbf{b} \cdot \mathbf{u})$, i.e.

$$
(\mathbf{a} \otimes \mathbf{b})\,\mathbf{u} = \mathbf{a}\,(\mathbf{b} \cdot \mathbf{u}) \; . \tag{2.14}
$$

The components of the tensor product are given by nine distinct product combinations of individual components of vectors **a** and **b**, namely $(\mathbf{a} \otimes \mathbf{b})_{ij} = a_i b_j$.

The dyadic product $\mathbf{a} \otimes \mathbf{b}$ is called **dyad** or **simple tensor**, because any dyad operates as a second-order tensor. However, not every tensor can be represented as a dyad, but it can always be expressed as a linear combination of n^2 dyads, formed from n base vectors of the *n-dimensional* space vector in which the tensor is defined. In fact, considering the dyadic product of the base vectors:

$$e_1 e_1 = \begin{pmatrix} 1 & 0 & 0 \\ 0 & 0 & 0 \\ 0 & 0 & 0 \end{pmatrix} \qquad \dots\dots\dots\dots\dots\dots \qquad e_1 e_3 = \begin{pmatrix} 0 & 0 & 1 \\ 0 & 0 & 0 \\ 0 & 0 & 0 \end{pmatrix}$$

$$\dots\dots \qquad \dots\dots\dots,$$

$$\dots\dots \qquad e_3 e_3 = \begin{pmatrix} 0 & 0 & 0 \\ 0 & 0 & 0 \\ 0 & 0 & 1 \end{pmatrix}$$

as the base for second-order tensors and by using the summation convention, any second-order tensor $\mathbf{T}$ can be represented as:

$$\mathbf{T} = T_{ij} e_i \otimes e_j. \tag{2.15}$$

If a second-order tensor is expanded as a dyadic, then the results of scalar product and composition of tensors can be obtained from the algebra of dyads. As a first example, consider the **scalar product of two dyads**, defined as follows:

$$(a \otimes b) \cdot (c \otimes d) = (a \cdot c)(b \cdot d), \tag{2.16}$$

which means *that the first and the second vector of both dyads are scalarly multiplied.*

Then, by using the representation (2.15), it follows that the scalar product of two second-order tensors is the following scalar quantity:

$$\mathbf{A} \cdot \mathbf{B} = \left(A_{ij} e_i \otimes e_j \right) \cdot \left(B_{kl} e_k \otimes e_l \right) = A_{ij} B_{kl} \delta_{ik} \delta_{jl} = A_{ij} B_{ij}. \tag{2.17}$$

Similarly, the **tensor product of two dyads**:

$$(a \otimes b)(c \otimes d) = (b \cdot c)(a \otimes d), \tag{2.18}$$

is formed according to the rule that *the vectors on the inner side are scalarly multiplied and the tensor product of the base vectors on the outer side forms the base.* Therefore, it follows that the tensor product of two general tensors (also called the **operational product of two tensors**) is a second-order tensor defined as follows:

$$\mathbf{AB} = A_{ij} B_{kl} \left(e_i \otimes e_j \right) (e_k \otimes e_l) = A_{ij} B_{kl} \delta_{jk} \left(e_i \otimes e_l \right) = A_{ij} B_{jl} \left(e_i \otimes e_l \right). \tag{2.19}$$

The operational product corresponds to the *composition of the two operation* $\mathbf{A}$ and $\mathbf{B}$, with $\mathbf{B}$ performed first, i.e.:

$$(\mathbf{AB})\mathbf{v} = \mathbf{A}(\mathbf{Bv}). \tag{2.20}$$

Before ending this section, we summarize a set of definitions that will be extensively used in the sequel.

The **transpose** of $\mathbf{T}$, denoted by $\mathbf{T}^T$, is defined as a tensor such that, $\forall \mathbf{a}, \mathbf{b} \in E^3$:

$$\mathbf{a} \cdot \mathbf{Tb} = \mathbf{b} \cdot \mathbf{T}^T \mathbf{a}. \tag{2.21}$$

A tensor is called **symmetric** if:

$$\begin{aligned} \mathbf{T}^T &= \mathbf{T} \\ T_{ji} &= T_{ij} \end{aligned} . \tag{2.22}$$

and is called **skew-symmetric** if:

$$\begin{aligned} \mathbf{T}^T &= -\mathbf{T} \\ T_{ji} &= -T_{ij} \end{aligned} . \tag{2.23}$$

Every tensor $\mathbf{T}$ can be expressed uniquely as the sum of a symmetric tensor $\mathbf{S}$ and a skew-symmetric tensor $\mathbf{W}$, defined as:

$$\mathbf{S} = \frac{\mathbf{T} + \mathbf{T}^T}{2}; \quad \mathbf{W} = \frac{\mathbf{T} - \mathbf{T}^T}{2}. \tag{2.24}$$

A tensor $\mathbf{Q}$ is defined **orthogonal** if it preserves the inner product:

$$\mathbf{Qa} \cdot \mathbf{Qb} = \mathbf{a} \cdot \mathbf{b}. \tag{2.25}$$

This definition implies that the angle between $\mathbf{Qa}$ and $\mathbf{Qb}$ is equal to the angle between $\mathbf{a}$ and $\mathbf{b}$ and the length $\mathbf{Qa}$ is equal to the length of $\mathbf{a}$.

A necessary and sufficient condition for $\mathbf{Q}$ to be orthogonal is that:

$$\mathbf{QQ}^T = \mathbf{Q}^T \mathbf{Q} = \mathbf{I}, \tag{2.26}$$

or equivalently $\mathbf{Q}^{-1} = \mathbf{Q}^T$. If the orthogonal tensor has a positive determinant, it represents a **rotation** and is called a **proper orthogonal tensor**.

A tensor $\mathbf{T}$ is said to be **positive definite** if, $\forall \mathbf{a} \neq 0$, it holds:

$$\mathbf{a} \cdot \mathbf{Ta} > 0. \tag{2.27}$$

2.2.1 Change of orthonormal basis

The advantage of vector notation in applied sciences is that equations describing physical problems can be formulated without reference to any particular system. However, when doing calculations we replace vector equations by an equivalent system of scalar equations, and this requires expanding vectors to a convenient *basis*.

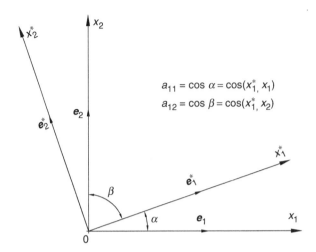

$$a_{11} = \cos \alpha = \cos(x_1^*, x_1)$$
$$a_{12} = \cos \beta = \cos(x_1^*, x_2)$$

Figure 2.2 Change of orthonormal basis.

In the three-dimensional space, a *basis* is any set of three mutually independent vectors e_1, e_2, e_3.

If the basis consists of mutually orthogonal unit vectors, it is called an **orthonormal basis,** and this will be the case in the sequel.

Once a basis has been selected, every vector **v** has a unique expansion of the form $\mathbf{v} = v_i e_i$, the numerical coefficients of the base vectors being the rectangular components.

It is now relevant to prove that the three components of a vector transform, under change of coordinates, in such a way that the new components always define the same vector.

Consider two reference systems, $0e_1 e_2 e_3$ and $0e_1^* e_2^* e_3^*$ sharing the same origin (Figure 2.2) and rotated relative to one another, so that the *direction cosines* between the the new axes (marked with star) and the old ones are given by:

$$A_{ij} = \mathbf{e}_i^* \cdot \mathbf{e}_j = \cos\left(\mathbf{e}_i^*, \mathbf{e}_j\right). \tag{2.28}$$

Each new base vector can be expressed as a linear combination of the old base vectors:

$$\mathbf{e}_i^* = A_{ij} \mathbf{e}_j, \tag{2.29}$$

and the backward relation writes:

$$\mathbf{e}_i = A_{ji} \mathbf{e}_j^*. \tag{2.30}$$

Since we are considering orthonormal bases, it must be:

$$\mathbf{e}_i^* \cdot \mathbf{e}_j^* = \delta_{ij}$$

and by using (2.29) we can also write:

$$\mathbf{e}_i^* \cdot \mathbf{e}_j^* = A_{ir}\mathbf{e}_r \cdot A_{js}\mathbf{e}_s = A_{ir}A_{js}\delta_{rs} = A_{ir}A_{jr} = \delta_{ij} \qquad (2.31)$$

which allow to obtain the *orthogonality condition* of the direction cosines:

$$A_{ir}A_{jr} = \delta_{ij}. \qquad (2.32)$$

This means that the transformation matrix A is orthogonal, i.e. $A^{-1} = A^T$.

An arbitrary vector can be expanded in the two forms $\mathbf{v} = v_i^*\mathbf{e}_i^*$ and $\mathbf{v} = v_j\mathbf{e}_j$ and by using (2.30) and (2.32) the following *forward* and *backward transformation rules* can be derived:

$$\begin{aligned} v_i^* &= A_{ij}v_j \\ v_k &= A_{ik}v_i^* \end{aligned} \qquad (2.33)$$

By applying these rules to a *dyad* **uv**, we obtain the transformation rule of a *second-order tensor*:

$$\begin{aligned} u_i^*v_j^* &= A_{im}u_m A_{jq}v_q = A_{im}A_{jq}u_m v_q \\ T_{ij}^* &= A_{im}A_{jq}T_{mq} \end{aligned} \qquad (2.34)$$

Note that the transformation rules (2.33) and (2.34) are assumed as a basic definition of vectors and tensors, i.e. *vectors and tensors are mathematical quantities, the components of which transform under change of coordinates according to the rules (2.33) and (2.34).*

2.3 Eigenvalues and eigenvectors

We introduced the tensor **T** as a linear mapping which relates the vector **a** to a vector **b**, the vector **b** being in general different from vector **a** in both direction and magnitude.

Let us now suppose to determine those vectors **a** which are not rotated by the application of **T**, i.e. all vectors such that:

$$\mathbf{Ta} = \lambda\mathbf{a} \qquad (2.35)$$

Such vectors, if they exist, are called **eigenvectors** of the tensor **T**. Their directions are called *principal directions* of the tensor **T** and the reference axes determined by these principal directions are called *principal axes*.

When referred to these principal axes, the values of the components T_{ij} are called *eigenvalues*, and these are the values of λ, for which the equation (2.35) has solutions.

We learned from linear algebra that the condition for the existence of a non-trivial solution of (2.35) is:

$$\det(\mathbf{T} - \lambda\mathbf{1}) = 0. \qquad (2.36)$$

Then, by expanding the determinant we obtain the characteristic equation:

$$-\lambda^3 + I_I \lambda^2 - I_{II} \lambda + I_{III} = 0 \tag{2.37}$$

The coefficients that appear in the characteristic equation do not depend on the choice of the basis and for this reason they are called **invariants**. These invariants can be expressed in a rather convenient way if we introduce the definition of **trace** of a second-order tensor $\mathbf{T}$:

$$tr\mathbf{T} = tr(T_{ij}\mathbf{e}_i \otimes \mathbf{e}_j) = T_{ij}\delta_{ij} = T_{ii}. \tag{2.38}$$

Then, by using this definition, the principal invariants of the tensor T can be expressed as:

$$I_I = tr\mathbf{T}$$
$$I_{II} = \frac{1}{2}\left[(tr\mathbf{T})^2 - tr(\mathbf{T}^2)\right]. \tag{2.39}$$
$$I_{III} = \det \mathbf{T}$$

Once the eigenvalues have been determined, the eigenvectors can be obtained by substituting each eigenvalue into the homogeneous equation, as it is illustrated in Example 2.1.

An important property of a symmetric tensor is the following: *the eigenvectors of a symmetric tensor are mutually orthogonal*, so that the principal directions are mutually orthogonal.

Example 2.1
Find the eigenvalues and the eigenvectors of the second-order tensor **T**:

$$\begin{pmatrix} 4 & 2 & -1 \\ 2 & 4 & 1 \\ -1 & 1 & 3 \end{pmatrix}$$

Solution The eigenvalue problem $\mathbf{Tu} = \lambda\mathbf{u}$ has non-trivial solutions only if the eigenvalue λ satisfies the characteristic equation $\det(\mathbf{T} - \lambda\mathbf{I}) = 0$. By expanding the determinant:

$$\det \begin{pmatrix} 4-\lambda & 2 & -1 \\ 2 & 4-\lambda & 1 \\ -1 & 1 & 3-\lambda \end{pmatrix} = 0,$$

we get the third-order equation in the unknown λ:

$$-\lambda^3 + 11\lambda^2 - 34\lambda + 24 = 0.$$

whose solutions are: $\lambda_1 = 1$, $\lambda_2 = 4$, $\lambda_3 = 6$.

Once the eigenvalues have been determined, the eigenvectors can be obtained by substituting each eigenvalue into the homogeneous equation. Substituting $\lambda_1 = 1$ we get the system:

$$3u_1 + 2u_2 - u_3 = 0$$
$$2u_1 + 3u_2 + u_3 = 0,$$
$$-u_1 + u_2 + 2u_3 = 0$$

and by imposing the **normalization** condition:

$$u_k u_k = 1, \qquad (2.40)$$

we obtain: $u_1 = \frac{1}{\sqrt{3}}$, $u_2 = -\frac{1}{\sqrt{3}}$, $u_3 = \frac{1}{\sqrt{3}}$.

Similarly, substitution of $\lambda = 4$ and $\lambda = 6$ gives the components of the other two eigenvectors: $\left(-1/\sqrt{6}, 1/\sqrt{6}, 2/\sqrt{6}\right), \left(1/\sqrt{2}, 1/\sqrt{2}, 0\right)$.

We leave the reader to verify that the obtained eigenvectors are mutually orthogonal.

2.4 Vector and tensor fields

A vector field is a function $\mathbf{v}(\mathbf{x})$ which assigns a vector to each point $\mathbf{x}$ in a continuous domain. Similarly a tensor field is a function that assigns to each point $\mathbf{x}$ a tensor $\mathbf{T}(\mathbf{x})$.

When dealing with vector and tensor fields, it is usual in continuum mechanics to introduce, for the sake of compactness, the subscript *comma* to denote the partial derivative with respect to the coordinate variables x_i.

As an example, we write $F_{,i}$ to indicate $\dfrac{\partial F}{\partial x_i}$ and $T_{ij,k}$ indicates $\dfrac{\partial T_{ij}}{\partial x_k}$. Similarly, $u_{i,jk}$ indicates the mixed partial derivatives $\dfrac{\partial^2 u}{\partial x_j \, \partial x_k}$. In particular, we note the identity:

$$x_{i,j} = \frac{\partial x_i}{\partial x_j} = \delta_{ij}. \qquad (2.41)$$

The operator **nabla** ∇ (or 'del') has the following representation in rectangular coordinates:

$$\nabla = \mathbf{e}_1 \frac{\partial}{\partial x_1} + \mathbf{e}_2 \frac{\partial}{\partial x_2} + \mathbf{e}_3 \frac{\partial}{\partial x_3}. \qquad (2.42)$$

Since it is an operator, by itself it has no numerical meaning but it has significance when it operates on scalar, vector and tensor fields.

The **gradient of a scalar function** $F(x,y,z)$ is a *vector*, defined as follows:

$$\nabla F = \mathbf{e}_1 \frac{\partial F}{\partial x_1} + \mathbf{e}_2 \frac{\partial F}{\partial x_2} + \mathbf{e}_3 \frac{\partial F}{\partial x_3} = \mathbf{e}_i F_{,i} . \tag{2.43}$$

Note that it is also customary to use the operator ∂_k ahead of the function, so that we have the following equivalent expressions:

$$\nabla F = \frac{\partial F}{\partial x_k} \mathbf{e}_k = \mathbf{e}_k \partial_k F . \tag{2.44}$$

A useful geometric interpretation of the gradient is related to the following example. Consider the level surface $F(x, y, z) = const$ and the directional derivative of the function given by $(dF/ds) = \nabla F \cdot \mathbf{t}$.

If the vector $\mathbf{t}$ is tangent to the surface, this derivative is zero, because F remains constant, and it follows that the vector ∇F must be normal to the level surface.

The **divergence of a vector field** is a scalar point function if the operator nabla operates on the vector as in a scalar product, i.e.:

$$\nabla \cdot \mathbf{v} = \mathbf{e}_i \frac{\partial}{\partial x_i} \cdot \mathbf{e}_j v_j = \delta_{ij} \frac{\partial v_j}{\partial x_i} = \frac{\partial v_i}{\partial x_i} = v_{i,i} . \tag{2.45}$$

If ∇ operates on the vector in a manner analogous to vector product the result is a vector, called the **curl** of the vector $\mathbf{v}$:

$$\nabla \times \mathbf{v} = \begin{vmatrix} \mathbf{e}_1 & \mathbf{e}_2 & \mathbf{e}_3 \\ \dfrac{\partial}{\partial x_1} & \dfrac{\partial}{\partial x_2} & \dfrac{\partial}{\partial x_3} \\ v_1 & v_2 & v_3 \end{vmatrix} = \varepsilon_{ijk} \frac{\partial v_k}{\partial x_j} \mathbf{e}_i = \varepsilon_{ijk} v_{k,j} \mathbf{e}_i . \tag{2.46}$$

The *divergence of a second-order tensor field* $\mathbf{T}$ is a vector field, defined as follows:

$$\nabla \cdot \mathbf{T} = (\mathbf{e}_i \frac{\partial}{x_i}) \cdot (\mathbf{e}_j \mathbf{e}_k T_{jk}) = \delta_{ij} \mathbf{e}_k \frac{\partial}{\partial x_i} T_{jk} = \frac{\partial}{\partial x_i} T_{ik} \mathbf{e}_k . \tag{2.47}$$

The *Laplacian operator* ∇^2 of a scalar field is a scalar point function:

$$\nabla^2 = \nabla \cdot \nabla F = \frac{\partial}{\partial x_i} \frac{\partial F}{\partial x_i} = F_{,ii} . \tag{2.48}$$

2.5 Kinematics

In classical continuuum mechanics the space of physical observation is the Euclidean space E^3, and to describe in mathematical terms the presence of a continuum body the following definition is introduced: *a continuum body is a set of particles* $S = \{P\}$ *which can be related into a one-to-one correspondence with the points of a region of the space* E^3. On the basis of this definition, we can say that there exists a one-to-one mapping χ, called **configuration**, which assigns to each particle a triple of real numbers, i.e.:

$$\chi : P \rightarrow \chi(P) = (x_1, x_2, x_3) \Leftrightarrow P = \chi^{-1}(x_1, x_2, x_3). \tag{2.49}$$

Then, we define the **motion** of the continuum body as *the evolution in time of its position, so that the motion is represented by the set* $K = \{\chi\}$ *of the configurations.*

In order to describe the motion of a given particle, we need to introduce a **reference configuration** $C_o \in K$:

$$C_o : P \rightarrow C_o(P) = (X_1, X_2, X_3). \tag{2.50}$$

so that the motion is described by the equations:

$$x_k = x_k(X_1, X_2, X_3, t). \tag{2.51}$$

where $t \in I$ is the time variable. The coordinates (X_1, X_2, X_3) are called **Lagrangian** or **material coordinates** and the coordinates (x_1, x_2, x_3) are called **Eulerian** or **spatial coordinates**.

Equations (2.51) are required to be one-to-one functions of class C^2. The requirement that only one **x** corresponds to each **X** assures that the body does not suffer fractures or material discontinuities; the existence of the inverse function, i.e. only one **X** corresponds to each **x**, preserves the property that two particles cannot occupy simultaneously the same position. Finally the requirement of class C^2 implies the smooth behaviour of the velocity and acceleration.

When the position **x** of a particle is described as a function of time and its position **X** in the reference configuration, i.e.:

$$\mathbf{x} := \mathbf{x}(\mathbf{X}, t), \tag{2.52}$$

the description is called *Lagrangian or material description.*

Alternatively, the position of a particle in the reference configuration can be expressed as a function of its current position **x** at time t, by means of the *inverse function*:

$$\mathbf{X} := \mathbf{X}(\mathbf{x}, t), \tag{2.53}$$

and this will provide the *Eulerian or spatial description* of the motion. Note that the spatial description focuses on a given region of space and uses the spatial coordinates and the time as independent variables. A *necessary and sufficient condition* for the existence of the inverse function is the non-vanishing of the *Jacobian determinant*:

$$J := \left| \frac{\partial x_i}{\partial X_J} \right| \neq 0. \tag{2.54}$$

The **material derivative** (represented by the differential operator d/dt or by a super-positioned dot) of *any property* of the continuum body is its time rate of change as measured by an observer attached to the particles. If this property (a scalar, vector or tensor quantity) is described by means of a *Lagrangian description*, as an example $H(\mathbf{X}, t)$, then the material derivative is simply the partial derivative with respect to time:

$$\frac{dH}{dt} := \frac{\partial H(\mathbf{X}, t)}{\partial t}, \tag{2.55}$$

since the material coordinates $\mathbf{X}$ do not change with time (they can be thought as *label* to identify particles).

However, if the same property is described by means of the *Eulerian description*, and in order to outline this description we write $h(\mathbf{x}, t)$, by substituting the motion into this spatial description we obtain:

$$h = h[\mathbf{x}(\mathbf{X}, t), t] \tag{2.56}$$

and the chain rule of calculus gives:

$$\frac{dh(\mathbf{x}, t)}{dt} = \frac{\partial h(\mathbf{x}, t)}{\partial t} + \frac{\partial h(\mathbf{x}, t)}{\partial x_k} v_k, \tag{2.57}$$

where $\mathbf{v} = \dfrac{\partial \mathbf{x}(\mathbf{X}, t)}{\partial t}$ is the velocity of the particle $\mathbf{X}$.

Note that the first term in the r.h.s. of (2.57) represents the *local change* of the property at the place $\mathbf{x}$, whereas the second term gives the rate of change as result of movement of the particle. For this reason, the second term is called *convective term*.

As a general rule, in order to obtain the material derivative of any quantity expressed in spatial coordinates we make use of the *material derivative operator*:

$$\frac{d}{dt} := \frac{\partial}{\partial t} + \mathbf{v} \cdot \nabla. \tag{2.58}$$

2.6 Finite deformation

Any arbitrary *change of configuration* is called *deformation* and usually includes a rigid-body motion as well as strains. When there is no change in the *relative position* of the material points of the continuum body, then the change of configuration is a rigid-body motion, which may be composed of a rigid translation, rigid rotation or both of them.

Strains arise because of a *relative change in the position of material points*, and this may include changes in length, changes in volume or shape.

The concept of deformation means that we compare two specific configurations without considering the sequence by which the current configuration has been reached from the reference one.

Accordingly, the mapping function (2.52) does not depend on the time as variable, so that we write:

$$\mathbf{x} := \mathbf{x}(\mathbf{X}) \qquad (2.59)$$

$$x_i := x_i(X_L),$$

The three scalar functions (2.59) and their inverse function are assumed to be of class C^1, so that the Jacobian determinant must be different than zero, and in particular, since J is equal to one in the reference configuration, it follows that:

$$J := \det\left(\frac{\partial x_i}{\partial X_L}\right) > 0. \qquad (2.60)$$

Differentiation of (2.59) gives:

$$dx_i = \frac{\partial x_i}{\partial X_L} dX_L. \qquad (2.61)$$

which defines a linear mapping which relates the infinitesimal material vector $d\mathbf{X}$, emanating from $\mathbf{X}$, onto the corresponding infinitesimal vector $d\mathbf{x}$, emanating from $\mathbf{x}$.

This mapping is the tensor $\mathbf{F}$, called the **deformation gradient tensor**:

$$d\mathbf{x} = \mathbf{F}d\mathbf{X}, \qquad (2.62)$$

whose components are the partial derivatives $\dfrac{\partial x_i}{\partial X_j}$, i.e.

$$F_{ij} = \frac{\partial x_i}{\partial X_j}. \qquad (2.63)$$

If the deformation gradient tensor F is independent of X, the deformation is called *homogeneous* (see Example 2.2). On the contrary, if $\mathbf{F}$ depends explicitly upon X, then the deformation is *inhomogeneous*.

Example 2.2
Figure 2.3 shows a volume element subjected to extension, as described by the following equations:

$$x_1 = aX_1$$
$$x_2 = bX_2.$$
$$x_3 = cX_3$$

Find the deformation gradient tensor.

Solution. By applying (2.63), the tensor $\mathbf{F}$ has the following components:

$$\left(F_{ij}\right) = \begin{pmatrix} a & 0 & 0 \\ 0 & b & 0 \\ 0 & 0 & c \end{pmatrix}.$$

Since these components do not depend upon X, the deformation is homogeneous. It can be observed that the directions of the eigenvectors are coincident with those of the edges of the element, which do not rotate during deformation, but any other material vector changes its direction, as it is the case of the diagonal OA. Moreover, the deformation is said to be volume preserving if $J = \det \mathbf{F} = 1$ (this could be the case if a, b and c are such that $abc = 1$).

In order to show how the *metric aspects* of the deformation are related to the deformation gradient tensor, let's start by considering a simple and intuitive measure of the change in length, defined by the **stretch ratio**:

$$\Lambda_{(\mathbf{N})} := \frac{ds}{dS}. \tag{2.64}$$

The scalar quantity $\Lambda_{(\mathbf{N})}$ relates the initial length dS of a fibre, which in the initial configuration has direction of the unit vector $\mathbf{N}$, to its length ds in the deformed configuration.

In order to find $\Lambda_{(\mathbf{N})}$, it is convenient to write:

$$\Lambda_{(\mathbf{N})}^2 = \frac{ds^2}{dS^2}. \tag{2.65}$$

and to observe that the change in the squared length of the material vector $d\mathbf{X}$ is given by:

$$(ds)^2 = d\mathbf{x} \cdot d\mathbf{x} = \mathbf{F}d\mathbf{X} \cdot \mathbf{F}d\mathbf{X} = d\mathbf{X} \cdot \mathbf{F}^T\mathbf{F}d\mathbf{X} = d\mathbf{X} \cdot \mathbf{C}d\mathbf{X}$$

$$(ds)^2 = dX_i C_{ij} dX_j. \tag{2.66}$$

where we have introduced the **right Cauchy-Green tensor**:

$$\mathbf{C} = \mathbf{F}^T\mathbf{F}$$

$$C_{ij} = \frac{\partial x_k}{\partial X_i}\frac{\partial x_k}{\partial X_j}. \tag{2.67}$$

By observing that $d\mathbf{X} = dS\mathbf{N}$, equation (2.65) gives the searched solution:

$$\Lambda_{(\mathbf{N})}^2 = \mathbf{N} \cdot \mathbf{C}\mathbf{N}, \tag{2.68}$$

and the worked Example 2.3 illustrates the use of this relation.

Example 2.3

The (plane) deformation shown in Figure 2.4 is called *pure shear* and is described in Cartesian coordinates by the following equations:

$$x_1 = X_1$$
$$x_2 = X_2 + kX_3 .$$
$$x_3 = X_3 + kX_2$$

Find the components of the tensors **F**, **C** and **E**, the stretch ratios of the diagonal elements and the angle ϑ_{23} in the deformed configuration.

Solution. The matrix representation of the required tensors is the following:

$$\left[F_{ij}\right] = \begin{bmatrix} 1 & 0 & 0 \\ 0 & 1 & k \\ 0 & k & 1 \end{bmatrix} ; \left[C_{ij}\right] = \begin{bmatrix} 1 & 0 & 0 \\ 0 & 1 & k \\ 0 & k & 1 \end{bmatrix}\begin{bmatrix} 1 & 0 & 0 \\ 0 & 1 & k \\ 0 & k & 1 \end{bmatrix} = \begin{bmatrix} 1 & 0 & 0 \\ 0 & 1+k^2 & 2k \\ 0 & 2k & 1+k^2 \end{bmatrix}$$

$$\left[E_{ij}\right] = \frac{1}{2}\left(\left[C_{ij}\right] - \left[\delta_{ij}\right]\right) = \begin{bmatrix} 0 & 0 & 0 \\ 0 & k^2/2 & k \\ 0 & k & k^2/2 \end{bmatrix} .$$

Since the *stretch ratio* Λ_N is obtained by using the relation: $\Lambda_N^2 = N \cdot CN$, when considering the unit vector **N** oriented as the diagonal $\overline{OB}$ we have:

$$\Lambda_{OB}^2 = \begin{bmatrix} 0, & 1/\sqrt{2}, & 1/\sqrt{2} \end{bmatrix}\begin{bmatrix} 1 & 0 & 0 \\ 0 & 1+k^2 & 2k \\ 0 & 2k & 1+k^2 \end{bmatrix}\begin{bmatrix} 0 \\ 1/\sqrt{2} \\ 1/\sqrt{2} \end{bmatrix} = (1+k)^2 .$$

Similarly, along the diagonal $\overline{CA}$

$$\Lambda_{CA}^2 = \begin{bmatrix} 0, & -1/\sqrt{2}, & 1/\sqrt{2} \end{bmatrix}\begin{bmatrix} 1 & 0 & 0 \\ 0 & 1+k^2 & 2k \\ 0 & 2k & 1+k^2 \end{bmatrix}\begin{bmatrix} 0 \\ -1/\sqrt{2} \\ 1/\sqrt{2} \end{bmatrix} = (1-k)^2 .$$

Finally, the angle ϑ_{23} in the deformed configuration is given by: $\cos\vartheta_{23} = \dfrac{C_{23}}{\sqrt{C_{22}}\sqrt{C_{33}}} = \dfrac{2k}{1+k^2}$

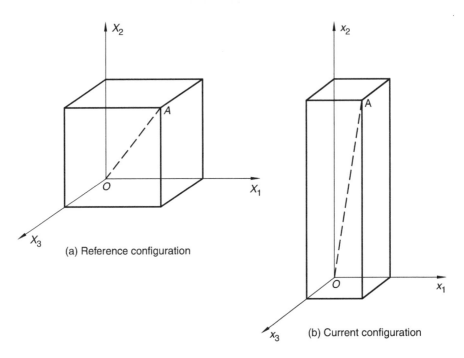

(a) Reference configuration

(b) Current configuration

Figure 2.3 Homogeneous extension.

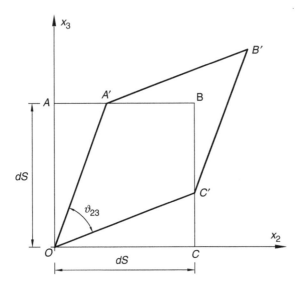

Figure 2.4 Pure shear deformation.

A similar procedure is used to find the **change in angle**, defined in the initial configuration by two unit vectors, $\mathbf{N}_1$ and $\mathbf{N}_2$. If we denote as ϑ_{12} the angle between the same elements in the deformed configuration, then:

$$\cos\vartheta_{12} = \frac{d\mathbf{x}_1 \cdot d\mathbf{x}_2}{|d\mathbf{x}_1||d\mathbf{x}_2|} = \frac{|d\mathbf{X}_1||d\mathbf{X}_2|\mathbf{N}_1 \cdot \mathbf{C}\mathbf{N}_2}{|d\mathbf{X}_1||d\mathbf{X}_2|\Lambda_{\mathbf{N}_1}\Lambda_{\mathbf{N}_2}} = \frac{\mathbf{N}_1 \cdot \mathbf{C}\mathbf{N}_2}{\Lambda_{\mathbf{N}_1}\Lambda_{\mathbf{N}_2}}. \tag{2.69}$$

As a special case, if $\mathbf{N}_1, \mathbf{N}_2$ were initially parallel to the Cartesian axes, since $\mathbf{N}_1(1,0,0)$ and $\mathbf{N}_2(0,1,0)$, the expressions (2.68) and (2.69) assume the simplified forms:

$$\Lambda^2_{(e_1)} = C_{11} \qquad \Lambda^2_{(e_2)} = C_{22} \qquad \cos\vartheta_{12} = \frac{C_{12}}{\sqrt{C_{11}}\sqrt{C_{22}}}. \tag{2.70}$$

Note that equation (2.70) also proves that, if we assume the coordinate axes in the principal directions of the tensor $\mathbf{C}$, then $C_{12} = 0$ and there is no change of the angle between the two elements.

Finally, if we consider an element of volume dV_o in the reference configuration, it can be proved that this volume transforms into the current element volume dV, given by:

$$dV = J dV_o. \tag{2.71}$$

The proof follows from the previously obtained results, if we take into consideration as a volume element a parallelepiped whose edges are in the principal directions of the Cauchy-Green tensor $\mathbf{C}$ (we observe that $\mathbf{C}$ is symmetric and positive defined second-order tensor).

Since according to equation (2.70) there is no change of the original angle between the edges, the volume element into current configuration is still a rectangular parallelepiped, so that:

$$\frac{dV}{dV_o} = \frac{dx_{(1)}dx_{(2)}dx_{(3)}}{dX_{(1)}dX_{(2)}dX_{(3)}} = \Lambda_{(1)}\Lambda_{(2)}\Lambda_{(3)} = \sqrt{C_{11}C_{22}C_{33}} =$$

$$\sqrt{III_C} = \sqrt{\det \mathbf{C}} = \sqrt{\det\left(\mathbf{F}^T\mathbf{F}\right)} = J \tag{2.72}$$

Note that, when deriving the last result, we made use of the property that:

$$\det\mathbf{A} = \det\mathbf{A}^T \tag{2.73}$$

and that:

$$\det\mathbf{AB} = \det\mathbf{A}\det\mathbf{B}. \tag{2.74}$$

Moreover, since the ratio dV/dV_o is linked to the invariant of the Cauchy-Green tensor, it is itself an invariant, so that the obtained result holds whatever are the coordinate axes.

In order to highlight the difference between finite and infinitesimal deformations (discussed in the following Sections 2.7 and 2.8), it is convenient to show how the deformation can also be described in terms of *displacement field*. To this aim, recall that the displacement vector **u** is the current position of a point relative to its position in the initial state, i.e.:

$$\mathbf{u} = \mathbf{x} - \mathbf{X} \tag{2.75}$$

and since the displacement vector can be associated to every point, we say that the displacement is a vector field.

If we introduce the **displacement gradient tensor**:

$$\nabla \mathbf{u} = \begin{vmatrix} \dfrac{\partial u_x}{\partial x} & \dfrac{\partial u_x}{\partial y} & \dfrac{\partial u_x}{\partial z} \\ \dfrac{\partial u_y}{\partial x} & \dfrac{\partial u_y}{\partial y} & \dfrac{\partial u_y}{\partial z} \\ \dfrac{\partial u_z}{\partial x} & \dfrac{\partial u_z}{\partial y} & \dfrac{\partial u_z}{\partial z} \end{vmatrix}, \tag{2.76}$$

then from (2.75) we derive:

$$\nabla \mathbf{u} = \mathbf{F} - \mathbf{I}. \tag{2.77}$$

Moreover, by using (2.67), i.e.:

$$\mathbf{C} = \mathbf{F}^T \mathbf{F} = (\nabla \mathbf{u} + \mathbf{I})^T (\nabla \mathbf{u} + \mathbf{I}),$$

we can give the right Cauchy-Green tensor the form:

$$\mathbf{C} = \mathbf{I} + 2\mathbf{E}, \tag{2.78}$$

where we have introduced the *Lagrangian finite strain tensor* in terms of gradient of displacement:

$$\mathbf{E} = \frac{1}{2} \left[\nabla \mathbf{u} + (\nabla \mathbf{u})^T + (\nabla \mathbf{u})^T \nabla \mathbf{u} \right]$$

$$E_{ij} = \frac{1}{2} \left(\frac{\partial u_i}{\partial X_j} + \frac{\partial u_j}{\partial X_i} + \frac{\partial u_k}{\partial X_i} \frac{\partial u_k}{\partial X_j} \right). \tag{2.79}$$

These latter expressions (2.78) and (2.79) will prove to be particularly useful in the subsequent section, where we analyse the infinitesimal strains.

2.7 Infinitesimal strains

Strains are defined *infinitesimal* (or simply *small*) if the components of displacement and the gradient of displacement are quantities of the first-order, so that the power of order $n \geq 2$ and the product of these quantities can be neglected.

If we introduce this assumption, the expression (2.79) of the *Lagrangian finite strain tensor* reduces to the **infinitesimal strain tensor**:

$$E_{ij} \cong \varepsilon_{ij} = \frac{1}{2} \left(\frac{\partial u_i}{\partial x_j} + \frac{\partial u_j}{\partial x_i} \right) . \tag{2.80}$$

In this section we prove how common *engineering measures of small strains* can be expressed in terms of displacement field, by using a formal mathematical approach, but we also anticipate that the beginning reader who does not have a good grasp of the concepts discussed in the previous sections can move to Section 2.8, where a more intuitive presentation is given.

The *longitudinal strain* of a linear element, collinear to the unit vector $\mathbf{N}$, is defined to be its change in length divided by its initial length, i.e.:

$$\delta_N = \frac{ds - dS}{dS} . \tag{2.81}$$

Since this measure is related to the *stretch ratio* (2.64), the use of the previous result $\Lambda_N^2 = \mathbf{N} \cdot \mathbf{CN}$ allows to obtain:

$$\Lambda_N = 1 + \delta_N = [\mathbf{N} \cdot (\mathbf{I} + 2\mathbf{E})\mathbf{N}]^{1/2} . \tag{2.82}$$

Suppose now that unit vector $\mathbf{N}$ is coincident with the base vector $\mathbf{e}_1$, then (2.82) writes:

$$\Lambda_{(1)} = 1 + \delta_{(1)} = \left[1 + 2E_{11} \right]^{1/2} ,$$

and under the assumption of infinitesimal strains we obtain $(1 + 2E_{11})^{1/2} \cong (1 + 2\varepsilon_{11})^{1/2} \cong (1 + \varepsilon_{11})$.

Therefore,

$$\delta_{(1)} = \varepsilon_{11} , \tag{2.83}$$

which proves that the diagonal terms of the infinitesimal strain tensor represent the longitudinal strain in the direction of the coordinate axes.

Engineers define as *shear strain* the change in angle between two line elements that are parallel to the coordinate axes in the reference state. If ϑ_{12} is the angle between the two coordinate axes (x_1, x_2) in the deformed configuration, then the change will be $\gamma_{12} = \dfrac{\pi}{2} - \vartheta_{12}$, and by using (2.69) we obtain:

$$\sin \gamma_{12} = \cos \vartheta_{12} = \frac{\mathbf{N}_1 \cdot (\mathbf{I} + 2\mathbf{E})\mathbf{N}_2}{\Lambda_{N_1} \Lambda_{N_2}} = \frac{2E_{12}}{(1 + 2E_{11})^{1/2} (1 + 2E_{22})^{1/2}}. \tag{2.84}$$

Here again, if we introduce the assumption of infinitesimal strains, we obtain:

$$\gamma_{12} = 2\varepsilon_{12} = \frac{\partial u_1}{\partial x_2} + \frac{\partial u_2}{\partial x_1}. \tag{2.85}$$

which proves that the engineering strear strain is twice the component of the small strain tensor component with mixed pedices.

We leave the reader to prove, as an exercise, that, under the assumption of small strains, the volume strain is equal to:

$$\varepsilon_v = \varepsilon_{xx} + \varepsilon_{yy} + \varepsilon_{zz}. \tag{2.86}$$

2.7.1 Compatibility equations.

The concept of strain compatibility refers to the physical idea that, when deformation occurs, it happens without material gaps or overlaps. To translate this concept in mathematical terms one can observe that, when the displacement field is specified, equation (2.80):

$$\varepsilon_{ij} = \frac{1}{2} \left(u_{i,j} + u_{j,i} \right)$$

allows one to compute the strain components.

The reverse process of obtaining displacement components from strains is overdetermined, because the six equations (2.80) contain only three unknowns. This means that the strain distribution cannot be assumed arbitrarily and additional conditions must be specified in order to ensure that (2.80) can be integrated to obtain the displacement field. This problem was solved by the French mathematician Barré de Saint-Venant in 1860, who showed that the strain components must satisfy six differential equations, which are both necessary and sufficient conditions to guarantee the existence of a single valued displacement field (see Malvern (1969) and Davis and Selvadurai (1996) for further insight into this subject).

2.8 Geometrical interpretation of infinitesimal strains

In the previous section we deduced the meaning of the components of the infinitesimal strain tensor by using a formal mathematical approach, which allows to capture the assumptions we introduce when moving from finite to infinitesimal strains. However, infinitesimal strains can also be thought on a more intuitive basis, as it is usually

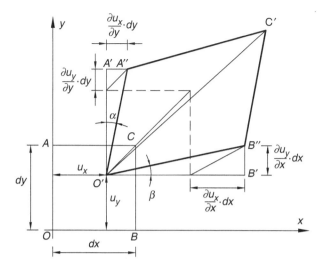

Figure 2.5 Infinitesimal strains.

done in undergraduated courses. Consider the parallelepiped shown in Figure 2.5 and assume that the face $OACB$, on the plane (x,y), be represented after the deformation by $O'B''C'A''$.

The x component of the displacement of the point O is given by u_x, and that of point B is:

$$(u_x + du_x) = \left(u_x + \frac{\partial u_x}{\partial x} dx \right).$$

The quantity

$$\frac{\partial u_x}{\partial x} dx$$

is therefore a measure of the relative displacement between B and O and represents the change in length of the original segment OB, aligned in the coordinate direction.

If we define the **longitudinal strain** ε_{xx} as the ratio between this change and the original length, then we obtain:

$$\varepsilon_{xx} = \frac{\partial u_x}{\partial x},$$ (2.87)

and similarly we have:

$$\varepsilon_{yy} = \frac{\partial u_y}{\partial y} \text{ and } \varepsilon_{zz} = \frac{\partial u_z}{\partial z}.$$

The rotation of the edge OB with respect to O is given by:

$$\tan \beta = \frac{\frac{\partial u_y}{\partial x} dx}{\left(1 + \frac{\partial u_x}{\partial x}\right) dx},$$

and because of the infinitesimal strain assumption, the quantity:

$$\frac{\partial u_x}{\partial x}$$

is negligible with respect to unity.

Moreover, we can substitute the function by its argument, so that:

$$\tan \beta \cong \beta = \frac{\partial u_y}{\partial x}.$$

Similarly for the edge OA the rotation is:

$$\alpha \cong \frac{\partial u_x}{\partial y}$$

and the **shear strain**, defined by one-half the change in the initially right angle between two material fibres aligned with the coordinate directions in the reference configuration, is given by:

$$\varepsilon_{xy} = \frac{1}{2}\left(\frac{\partial u_x}{\partial y} + \frac{\partial u_y}{\partial x}\right). \tag{2.88}$$

By considering the other couple of sides, one obtains:

$$\varepsilon_{xz} = \frac{1}{2}\left(\frac{\partial u_x}{\partial z} + \frac{\partial u_z}{\partial x}\right) \text{ and } \varepsilon_{zy} = \frac{1}{2}\left(\frac{\partial u_z}{\partial y} + \frac{\partial u_y}{\partial z}\right).$$

In order to give to these results a concise expression, let's recall the definition of the **displacement gradient tensor**:

$$\nabla u = \begin{vmatrix} \dfrac{\partial u_x}{\partial x} & \dfrac{\partial u_x}{\partial y} & \dfrac{\partial u_x}{\partial z} \\ \dfrac{\partial u_y}{\partial x} & \dfrac{\partial u_y}{\partial y} & \dfrac{\partial u_y}{\partial z} \\ \dfrac{\partial u_z}{\partial x} & \dfrac{\partial u_z}{\partial y} & \dfrac{\partial u_z}{\partial z} \end{vmatrix}, \tag{2.89}$$

and observe once again that any arbitrary deformation may include a rigid translation, a rigid rotation and strains.

Any rigid translation is characterized by the fact that the displacement vector **u** is independent of the variables (x, y, z), so that all partial derivatives would be zero. Therefore, considering the displacement gradient tensor means that we are faced with rigid rotation and strains.

If we decompose the tensor $\nabla \mathbf{u}$ into a symmetric and a skew-symmetric component:

$$\nabla \mathbf{u} = \frac{1}{2}\left[\nabla \mathbf{u} + (\nabla \mathbf{u})^{\mathrm{T}}\right] + \frac{1}{2}\left[\nabla \mathbf{u} - (\nabla \mathbf{u})^{\mathrm{T}}\right], \tag{2.90}$$

we reach the conclusion that the previous definition of strains correspond to the symmetric part of the deformation gradient tensor, i.e.:

$$\varepsilon_{ij} = \frac{1}{2}\left(\frac{\partial u_i}{\partial x_j} + \frac{\partial u_j}{\partial x_i}\right)$$

$$\varepsilon = \frac{1}{2}\left[\nabla \mathbf{u} + (\nabla \mathbf{u})^{T}\right] \tag{2.91}$$

By considering the matrix representation of the *second-order symmetric tensor*:

$$\left[\varepsilon_{ij}\right] = \begin{bmatrix} \varepsilon_{11} & \varepsilon_{12} & \varepsilon_{13} \\ \varepsilon_{21} & \varepsilon_{22} & \varepsilon_{23} \\ \varepsilon_{31} & \varepsilon_{32} & \varepsilon_{33} \end{bmatrix},$$

the diagonal terms represent the longitudinal strains and the off-diagonal terms are called shear strains.

The next step is to consider the skew-symmetric part of the displacement gradient tensor:

$$\mathbf{\Omega} = \frac{1}{2}\left[\nabla \mathbf{u} - (\nabla \mathbf{u})^{T}\right]$$

$$[\mathbf{\Omega}]_{ij} = \begin{bmatrix} 0 & \Omega_{xy} & \Omega_{xz} \\ \Omega_{yx} & 0 & \Omega_{yz} \\ \Omega_{zx} & \Omega_{zy} & 0 \end{bmatrix}, \tag{2.92}$$

and we have to prove that its components correspond to rigid rotations.

If we consider a rigid rotation around the y axis (see Figure 2.6) and assume that the angle of rotation α is small, then we can write:

$$\alpha = \frac{\partial u_x}{\partial z} = -\frac{\partial u_z}{\partial x},$$

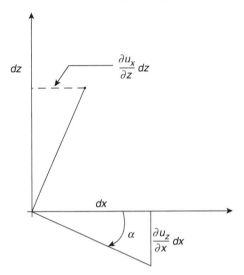

Figure 2.6 Rigid rotation around the y axis.

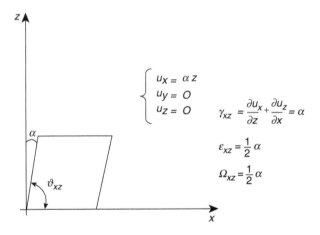

Figure 2.7 Engineering shear strain.

and therefore:

$$\Omega_{xz} = \frac{1}{2} \left(\frac{\partial u_x}{\partial z} - \frac{\partial u_z}{\partial x} \right) = \frac{1}{2} (\alpha - (-\alpha)) = \alpha,$$

Similarly we can prove that Ω_{xy} and Ω_{yz} represent a rigid rotation about the z axis and the x axis respectively.

These arguments also allow us to clarify an aspect linked to the conventional definition of *engineering shear strain*:

$$\gamma_{ij} = \left(\frac{\partial u_i}{\partial x_j} + \frac{\partial u_j}{\partial x_i} \right). \tag{2.93}$$

Only half of the engineering shear strain γ_{ij} corresponds to a measure of strain, because half of γ_{ij} corresponds to a rigid rotation of the element. This can be proved by applying the previous results to the displacement field associated to the simple shear deformation illustrated in Figure 2.7.

2.8.1 Plane strain

Many problems of engineering interest, such those related to embankments, walls, strip footings, dams, tunnels, can be analysed as plane strain problems. The relevant aspect of a plane strain problem is the following: if the longitudinal dimension (that here we assume to develop along the *y axis*) is very large if compared with the other two dimensions, and both volume and surface forces act othogonal to the *y* axis and are constant along it, then any cross-section, far from edges, exhibits a plane state of strain, since displacement components along the *y* axis must be zero.

Then, referring to the displacement field, the plane strain condition is formally expressed through the following relations:

$$u_y = 0, \quad u_{x,y} = 0, \quad u_{z,y} = 0, \tag{2.94}$$

and accordingly the following strain components vanish:

$$\varepsilon_{yy} = \varepsilon_{xy} = \varepsilon_{zy} = 0. \tag{2.95}$$

2.9 Stress

When considering the forces acting on any portion D of a continuum body C, we have to distinguish between two types of forces:

1   Those forces acting on D from the exterior of C, called **body forces**, such as *mass forces*, represented by a vector field **b** (per unit mass).
2   Those forces interior to C and acting on the boundary ∂D of the portion D, called **contact forces**.

Contact forces are associated with the specific surface we choose at a given point, so that the mathematical description of contact forces poses some problems, solved by the French mathematician Augustin Cauchy in 1823.

To introduce this subject, consider a body subjected to *external forces* P_1, P_2, P_3, as sketched in Figure 2.8. If this body is divided into two parts by an arbitrary

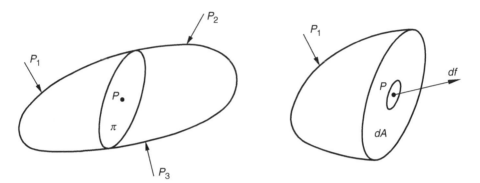

Figure 2.8 The concept of stress.

plane π through the point P and one of these parts is removed, the equilibrium of forces will no longer be satisfied, unless the system of internal forces is specified. This system represents the *contact forces* exerted by one part onto the other.

Cauchy assumed that the actions on the small element of surface dA, oriented by the unit outward normal vector n, can be reduced to the vector df, applied at the point P, and postulated that the limit:

$$\lim_{dA \to 0} \left(\frac{df}{dA} \right) = \mathbf{t} \tag{2.96}$$

exists and is finite. The vector t is called **traction** or **stress vector**. This vector depends in general on the surface on which the point P lies, so that if we consider a different surface at the same point, the traction vector will in general be different. Since the orientation of a given surface can be described by means of the unit outward vector **n** normal to the surface, Cauchy proved that the way the stress vector t depends upon the surface is expressed through the linear mapping:

$$t_i^{(n)} = \sigma_{ji} n_j, \tag{2.97}$$

and the symmetric *second-order tensor* σ_{ij} is called **stress tensor**.

Therefore, in mathematical terms we can formally state that the stress tensor is a linear mapping that associates to each unit normal **n** the traction vector **t**, acting on the surface of outward normal **n**. Its matrix representation is the following:

$$\left[\sigma_{ij} \right] = \begin{bmatrix} \sigma_{11} & \sigma_{12} & \sigma_{13} \\ \sigma_{21} & \sigma_{22} & \sigma_{23} \\ \sigma_{31} & \sigma_{32} & \sigma_{33} \end{bmatrix}. \tag{2.98}$$

Note that when considering each component σ_{ij} the first subscript indicates the coordinate plane on which the stress component acts, and the second subscript identifies the direction in which it acts. As an example, the component σ_{kl} is the stress component acting on the plane orthogonal to the x_k axis in the x_l direction. Therefore, the *diagonal components* act normal to the coordinate planes and are called **normal stresses**, while the off-diagonal components act tangential to the coordinate planes and are called **shear stresses**.

We also remark that in engineering literature the following notation is of more common use:

$$\left[\sigma_{ij}\right] = \begin{bmatrix} \sigma_x & \tau_{xy} & \tau_{xz} \\ \tau_{yx} & \sigma_y & \tau_{yz} \\ \tau_{zx} & \tau_{zy} & \sigma_z \end{bmatrix},$$
(2.99)

with the assumption that the σ symbol refers to a *normal component* and the suffix indicates the direction parallel to the coordinate axis; the symbol τ identifies the *shear stress components*, and here again the first suffix indicates the coordinate plane and the second one identifies the direction parallel to the coordinate axis (see Figure 2.9).

Therefore, the components in each row represent the components of the traction vector acting on the plane perpendicular to the coordinate axis corresponding to that row.

The above-mentioned symmetry of the stress tensor (i.e. $\tau_{xy} = \tau_{yx}$; $\tau_{xz} = \tau_{zx}$; $\tau_{yz} = \tau_{zy}$) can be proved by considering that the sum of moments about the coordinate axes must be zero (see Section 2.14).

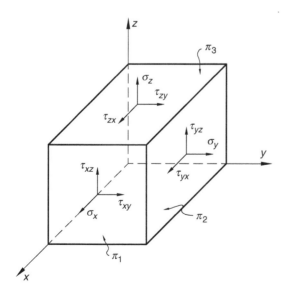

Figure 2.9 Stress components.

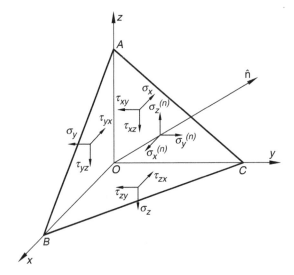

Figure 2.10 The Cauchy tetrahedron.

2.9.1 Cauchy's tetrahedron

The linear mapping (2.97) tell us that, *when the components σ_{ij} acting on any three mutually perpendicular planes through a point O are known, the stress vector on any plane through O can be determined.* To prove that this is the case, consider the equilibrium of a small portion of the continuum body in the shape of a tetrahedron, as shown in Figure 2.10. If **n** is the unit outward vector normal to the surface ABC (whose components n_x, n_y, n_z are its direction cosines with respect to the coordinate axes), the components $t_x^{(n)}, t_y^{(n)}, t_z^{(n)}$ of the stress vector acting on this surface can be obtained by considering the tetrahedron equilibrium. The force equilibrium in the x direction requires:

$$\bar{t}_x^{(n)} \cdot (area\,ABC) = \bar{\sigma}_x \cdot (area\,AOC) + \bar{\tau}_{yx} \cdot (area\,AOB) + \bar{\tau}_{zx} \cdot (area\,BOC) - \rho b_x dV$$

(2.100)

and similar equations are written in the other two directions. Note that in this equation the vector **b** is the body force per unit mass, ρ is the density and the stresses are the average stresses acting on the faces of the tetrahedron. The volume of the tetrahedron can be expressed in the form $dV = 1/3\,(h \cdot dS)$, if dS is the area of the surface ABC and h is its distance from the point O.

If we consider the limit for $h \to 0$, the term corresponding to body forces vanishes and the average stresses reduce to the value attained at the point O, as expressed by the limit (2.96). Therefore, since:

$$n_x = \frac{area\,AOC}{area\,ABC}, \quad n_y = \frac{area\,AOB}{area\,ABC} \text{ and } n_z = \frac{area\,BOC}{area\,ABC}$$

(2.101)

the equilibrium equations can be written in the following matrix form:

$$
\begin{Bmatrix} t_x^{(n)} \\ t_y^{(n)} \\ t_z^{(n)} \end{Bmatrix} = \begin{bmatrix} \sigma_x & \tau_{yx} & \tau_{zx} \\ \tau_{xy} & \sigma_y & \tau_{zy} \\ \tau_{xz} & \tau_{yz} & \sigma_z \end{bmatrix} \begin{Bmatrix} n_x \\ n_y \\ n_z \end{Bmatrix},
\tag{2.102}
$$

and in tensor notation

$$
t_i^{(n)} = \sigma_{ji} n_j.
\tag{2.103}
$$

This result, known as *Cauchy's theorem*, was first proved by Cauchy in 1827.

2.9.2 Principal stresses and invariants

Once we have established the tensorial nature of the stress state, we can also pose the problem of finding a *local reference system* in which the shear stresses vanish. In this case the coordinate axes are called **principal axes** and the coordinate planes are called **principal planes**. The stress vector acting on a principal plane is characterized by only the normal component; therefore, if **n** is the unit vector of a principal direction and λ is the modulus of the normal component, the stress vector can be expressed in the form:

$$
\mathbf{t}^{(\mathbf{n})} = \lambda \mathbf{n}.
\tag{2.104}
$$

By using (2.103) we can also write:

$$
\sigma_{ij} n_j = \lambda n_i,
\tag{2.105}
$$

so that the problem of finding the principal stresses reduces the following *eigenvalue problem*:

$$
\left(\sigma_{ij} - \lambda \delta_{ij} \right) n_j = 0.
\tag{2.106}
$$

Equations (2.106) are three homogeneous linear equations and we learned from linear algebra that the condition for the existence of a non-trivial solution is:

$$
\det(\boldsymbol{\sigma} - \lambda \mathbf{I}) = 0.
$$

This condition gives the characteristic cubic equation:

$$
\lambda^3 - I_I \lambda^2 + I_{II} \lambda - I_{III} = 0
$$

the solution of which is represented by three real eigenvalues, $\lambda_1 = \sigma_1, \lambda_2 = \sigma_2, \lambda_3 = \sigma_3$ called **principal stresses**. The substitution of each of these eigenvalues into (2.106) allows to compute the eigenvectors, representing the **principal directions**.

Note that the components of the stress tensor at a point will change in different coordinate systems, but *the three principal stresses are invariant* under coordinate transformation, i.e. they are always the same.

Furthermore, a theorem from linear algebra (see Section 2.3) states that *the eigenvectors of a symmetric tensor are mutually orthogonal*, so that the principal directions are mutually orthogonal. This implies that, if we choose a coordinate system such that the coordinate directions are parallel to the principal directions, in that system the stress tensor representation assumes the form:

$$\left[\sigma_{ij}\right] = \begin{bmatrix} \sigma_1 & 0 & 0 \\ 0 & \sigma_2 & 0 \\ 0 & 0 & \sigma_3 \end{bmatrix}.$$

Furthermore, it can be proved that the coefficients of the characteristic equations, defined by the following scalar quantities:

$$I_I = tr\mathbf{T} = \sigma_{ii} = \sigma_1 + \sigma_2 + \sigma_3$$
$$I_{II} = \tfrac{1}{2}\left[(tr\mathbf{T})^2 - tr\left(\mathbf{T}^2\right)\right] = \tfrac{1}{2}\left(\sigma_{ii}\sigma_{jj} - \sigma_{ij}\sigma_{ij}\right) = \sigma_1\sigma_2 + \sigma_2\sigma_3 + \sigma_1\sigma_3 \qquad (2.107)$$
$$I_{III} = \det\mathbf{T} = \sigma_1\sigma_2\sigma_3$$

are also invariant with respect to change of the reference axes and for this reason are called **invariant of first, second** and **third order**.

2.9.3 Spherical and deviatoric components of the stress tensor

If we define as *mean pressure* the scalar quantity:

$$p = \frac{1}{3}(\sigma_{kk}), \qquad (2.108)$$

the stress tensor can be decomposed into a spherical component, represented by the mean pressure, and a component characterized by zero trace, defined *stress deviator*, whose component will be given the symbol s_{ij} (see Section 4.14):

$$\sigma_{ij} = \delta_{ij}p + s_{ij} \qquad (2.109)$$

According to the above definition, the components with mixed indices of both the stress tensor and the stress deviator are the same, so that principal directions of the stress deviator are also coincident with the principal directions of the stress tensor. Further, by using the same arguments introduced in the previous point, we can define the invariants of the stress deviator:

$$J_I = 0$$
$$J_{II} = \frac{1}{2}tr\left(s_{ij}^2\right) = \frac{1}{2}\left[(\sigma_1 - p)^2 + (\sigma_2 - p)^2 + (\sigma_3 - p)^2\right]. \qquad (2.110)$$
$$J_{III} = \det s_{ij}$$

The interest in the above decomposition arises from the fact that there are constitutive models, within the framework of plasticity, for which the constitutive response is controlled by p and J_{II}. In this context, it is also of interest to find the stresses acting on planes equally inclined with respect to the coordinate axes. Since these planes form an octahedron, there are eight such planes and the stresses on these planes are known as **octahedral stresses**.

To find these stresses we follow the same procedure as seen for stresses on an oblique plane (see the worked Example 2.4 and Figure 2.11), so that, let's assume the coordinate axes to be the principal axes. In this case, the outward normal unit vector is defined by:

$$\mathbf{n}\left(\frac{\sqrt{3}}{3}, \frac{\sqrt{3}}{3}, \frac{\sqrt{3}}{3}\right)$$

Example 2.4
The stress tensor at point P has the following matrix expression:

$$\begin{pmatrix} 300 & -50 & 0 \\ -50 & 200 & 0 \\ 0 & 0 & 100 \end{pmatrix}.$$

Find the stress vector acting on a plane through P parallel to the plane shown in Figure 2.11.

Solution. The outward normal to this plane can be obtained by observing that it must be othogonal to the vectors $\overrightarrow{AB}$ *and* $\overrightarrow{AC}$, so that

$$\overrightarrow{AB} \times \overrightarrow{AC} = \begin{vmatrix} \mathbf{e}_1 & \mathbf{e}_2 & \mathbf{e}_3 \\ -2 & 2 & 0 \\ -2 & 0 & 3 \end{vmatrix} = 6\mathbf{e}_1 + 6\mathbf{e}_2 + 4\mathbf{e}_3.$$

Therefore the unit outward normal $\mathbf{n}$ has the following components $(3/\sqrt{22};\ 3/\sqrt{22};\ 2/\sqrt{22})$.

By applying (2.97) the required stress vector is the following:

$$t_i^{(n)} = \sigma_{ji} n_j = \begin{pmatrix} 300 & -50 & 0 \\ -50 & 200 & 0 \\ 0 & 0 & 100 \end{pmatrix} \begin{pmatrix} 3/\sqrt{22} \\ 3/\sqrt{22} \\ 2/\sqrt{22} \end{pmatrix} = \begin{pmatrix} 750/\sqrt{22} \\ 450/\sqrt{22} \\ 200/\sqrt{22} \end{pmatrix}.$$

---

and the stress vector t acting on the octahedral plane, also called π-plane, will be (by taking into account the decomposition (2.109))

$$\mathbf{t} = p\mathbf{n} + s\mathbf{n},$$

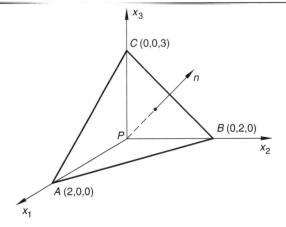

Figure 2.11 Illustrative example of computing the stress on a plane of unit outward normal **n**.

the deviator stress tensor being:

$$\mathbf{s} = \begin{bmatrix} \sigma_1 - p & 0 & 0 \\ 0 & \sigma_2 - p & 0 \\ 0 & 0 & \sigma_3 - p \end{bmatrix}.$$

It can easily be proved that the product $\mathbf{sn} \cdot \mathbf{n} = 0$, so that the conclusion is reached that the component:

$$\mathbf{sn} = \frac{\sqrt{3}}{9} \left[(2\sigma_1 - \sigma_2 - \sigma_3)\,\mathbf{e}_I + (2\sigma_2 - \sigma_3 - \sigma_1)\,\mathbf{e}_{II} + (2\sigma_3 - \sigma_1 - \sigma_2)\,\mathbf{e}_{III} \right]$$

lies on the octahedral plane. For this reason, this component is called **octahedral shear stress**, and its modulus is given by:

$$\tau_{oct} = \|\mathbf{sn}\| = \frac{\sqrt{3}}{3}\sqrt{s_1^2 + s_2^2 + s_3^2} = \sqrt{\frac{2}{3}J_{II}}. \tag{2.111}$$

2.9.4 Plane stress problem

Referring to the definition of invariants (see equations 2.107), it can be observed that if, $I_{III} = \det \sigma \neq 0$, then all three pricipal stresses must be different from zero and we are in the presence of a **triaxial stress state**.

If $I_{III} = 0$, at least one of the principal stresses must vanish and, if the additional condition $I_{II} \neq 0$ holds, then we are in presence of a **biaxial** or **plane stress**. Thus, plane stress condition refers to those circumstances in which all stresses associated to one coordinate direction are zero. An example is given by a thin plate with the face of the plate being free from stress, so that, if the $\mathbf{e}_1$ is the base vector normal to the face of the plate, the stress components $\sigma_{11}, \sigma_{12}, \sigma_{13}$ vanish, and the non-vanishing stress components will be $\sigma_{22}, \sigma_{33}, \sigma_{23} = \sigma_{32}$.

2.10 Mohr circle of stress

The representation of a state of stress, known in literature as *Mohr circle of stress*, is a geometrical representation of the linear mapping which associates to any direction n the state of stress acting on the plane oriented by the outward unit normal n.

Karl Culmann, in his book *Die graphische Statik*, published in Zurich in 1866, made a systematic introduction of graphic methods in the solution of structural problems. In particular, in his work on bending of beams, Culmann showed how stresses can be analysed graphically: known the stress components acting on the planes through a point A and perpendicular to the x and y axes, he proved that the normal and the tangential components of stress acting on any inclined plane are given by the coordinates of points on a *circle of stress* (see Timoshenko, 1983).

Cristian Otto Mohr made a more complete study of this subject in 1882, and, considering a two-dimensional case with the principal stresses σ_1 and σ_2(see Figure 2.12),

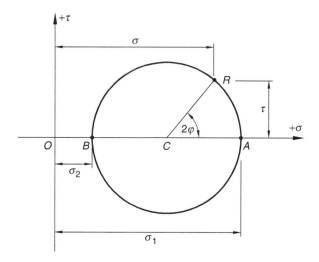

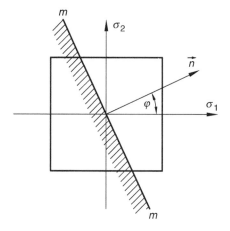

Figure 2.12 Mohr's circle.

he showed that the normal and the shear components of the stress acting on a plane mm defined by the angle φ are given by the coordinates of the point R defined by the angle 2φ on the circle.

Considering a *two-dimensional case*, the main aspects linked to the construction of the Mohr circle of stress can be summarized as follows (see Figure 2.13):

1　The *stress components*, acting on two mutually perpendicular planes, are interpreted as *coordinates* of two points *in the plane defined by the variables σ, τ.*
2　Some care must be used in order to master the *sign convention* commonly used in soil mechanics: normal components of stress are positive if they correspond to

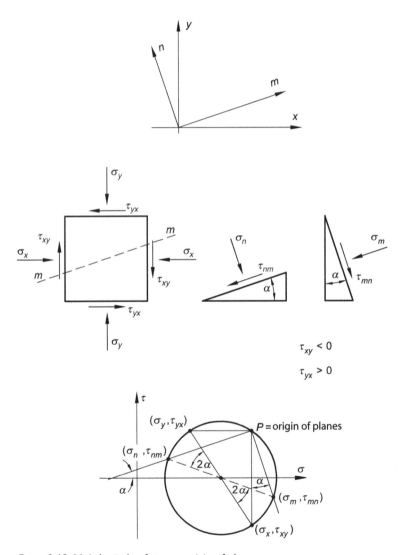

Figure 2.13 Mohr's circle of stress: origin of planes.

compression and the shear stress components are positive if they suggest a *counterclockwise rotation* (in Figure 2.13 the components τ_{yx}, acting on the horizontal planes, correspond to a counterclockwise couple and are positive; the components τ_{xy} are negative).

3 Once these two points are fixed, the *circle of stress* can be sketched, by noting that the abscissa of the centre is given by $1/2\left(\sigma_x+\sigma_y\right)$ and the radius is equal to

$$1/2\sqrt{\left(\sigma_x-\sigma_y\right)^2+4\tau_{xy}^2}.$$

4 The notion of **origin of planes** is particularly useful in order to define the state of stress acting on any other plane. It is a point P on the Mohr circle with the following property (see Figure 2.13): a line through P and parallel to a given plane intersects the Mohr circle at a point, the coordinates of which are the stress components acting on that plane.

However, an additional word of caution still refers to sign conventions: *when considering the shear stress components, the above-mentioned convention only applies to the graphical construction.*

We must revert to the mathematical convention when using the equilibrium equations (recalled for convenience in Figure 2.14): *when acting on the pair of faces of an element nearer to the origin, the positive shear stress components are in the positive directions of the parallel axis* (see Example 2.5).

2.11 Gauss and Reynolds theorems

Sections 2.12, 2.13 and 2.14 deal with some general principles of continuum mechanics: conservation of mass, balance of linear momentum and balance of angular momentum. These principles will be given and further used in local form, but will be derived from integral forms expressing fundamental postulates. The derivation will make use of two basic theorems of calculus, that, as a matter of convenience, are here briefly recalled.

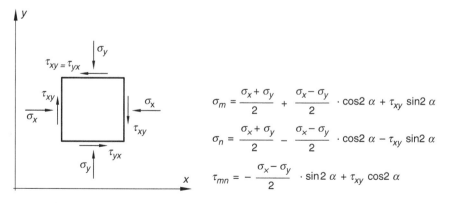

$$\sigma_m = \frac{\sigma_x+\sigma_y}{2} + \frac{\sigma_x-\sigma_y}{2} \cdot \cos2\,\alpha + \tau_{xy}\sin2\,\alpha$$

$$\sigma_n = \frac{\sigma_x+\sigma_y}{2} - \frac{\sigma_x-\sigma_y}{2} \cdot \cos2\,\alpha - \tau_{xy}\sin2\,\alpha$$

$$\tau_{mn} = -\frac{\sigma_x-\sigma_y}{2} \cdot \sin2\,\alpha + \tau_{xy}\cos2\,\alpha$$

Figure 2.14 Analytical relations between stress components and related planes on which they act.

Example 2.5

With reference to the plane stress problem sketched in Figure 2.15, find the stress components on the plane m-m.

Solution

(a) The procedure based on the Mohr circle of stress uses the notion of the *origin of planes* and the sign convention which considers as positive the shear stresses if they suggest a *counterclockwise rotation*. In this case the shear stresses on the horizontal planes, equal to 2 MPa, are positive, whereas those acting on the vertical planes are negative. Once the origin of the planes has been fixed at point P, we draw a line parallel to the plane m-m and the intersection point provides the required components ($\sigma_n = 1.77$; $\tau_{nm} = +5.33$ MPa).

(b) If we use the equilibrium equations (recalled in Figure 2.14) we must revert to the mathematical convention: *when acting on the pair of faces of an element nearer to the origin, the positive shear stress components are in the positive directions of the parallel axis.* Accordingly, the shear stresses $\tau_{xy} = \tau_{yx}$ shown in Figure 2.15 are positive, and we obtain:

$$\sigma_m = \frac{-4+6}{2} + \frac{-4-6}{2} \cdot 0.5 + 2 \cdot 0.866 = 0.23 \ MPa$$

$$\sigma_n = 1 + 2.5 - 1.73 = 1.77 \ MPa$$

$$\tau_{mn} = 5 \cdot 0.866 + 2 \cdot 0.5 = 5.33 \ MPa$$

(c) Finally we can also make use of (2.103): $t_i^{(n)} = \sigma_{ji} n_j$. Since we are dealing with a plane stress problem, the components of the stress tensor are the following:

$$[\sigma_{ij}] = \begin{bmatrix} -4 & 2 & 0 \\ 2 & 6 & 0 \\ 0 & 0 & 0 \end{bmatrix}.$$

The unit vector n normal to the plane m-m is defined by the following components $\begin{Bmatrix} -1/2 \\ \sqrt{3}/2 \\ 0 \end{Bmatrix}$, and the stress vector acting on the m-m plane is given by

$$t_i^{(n)} = \begin{bmatrix} -4 & 2 & 0 \\ 2 & 6 & 0 \\ 0 & 0 & 0 \end{bmatrix} \begin{Bmatrix} -1/2 \\ \sqrt{3}/2 \\ 0 \end{Bmatrix} = \begin{Bmatrix} 2+\sqrt{3} \\ -1+3\sqrt{3} \\ 0 \end{Bmatrix}.$$

Note that, by applying (2.103) we find the stress components along the x_i axes. The normal component is obtained through the scalar product:

$$\sigma_n = \mathbf{t}^{(n)} \cdot \mathbf{n} = \{ 2+\sqrt{3}; \ -1+3\sqrt{3}; \ 0 \} \begin{Bmatrix} -1/2 \\ \sqrt{3}/2 \\ 0 \end{Bmatrix} = 1.77 \ MPa,$$

whereas the shear component is given by the scalar product between the stress vector and the unit vector **m** normal to the face n-n: $\tau_{nm} = \mathbf{t}^{(n)} \cdot \mathbf{m} =$

$$\{2+\sqrt{3};\ -1+3\sqrt{3};\ 0\} \left\{ \begin{array}{c} \sqrt{3}/2 \\ 1/2 \\ 0 \end{array} \right\} = 5.33\,MPa\ .$$

2.11.1 Gauss's theorem or divergence theorem

Let ∂D be an oriented closed surface bounding the volume D and be **n** the unit outward normal to ∂D. Suppose that **v** is a vector field of class C^1, then the Gauss's theorem states: *the flux of v throughout the surface is equal to the divergence of the vector*

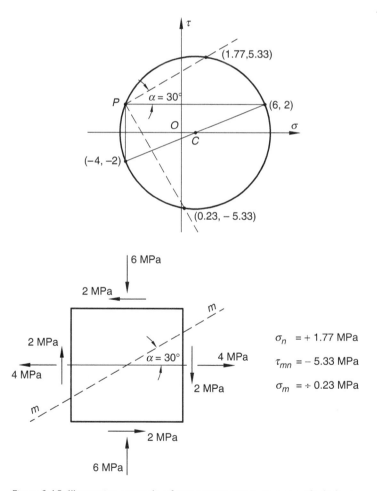

Figure 2.15 Illustrative example of computing stress on a specified plane.

over the volume bounded by the closed surface, i.e.:

$$\int_{\partial D} \mathbf{v} \cdot \mathbf{n} dS = \int_{D} \nabla \cdot \mathbf{v} dV \tag{2.112}$$

2.11.2 Reynolds's theorem or transport theorem

A volume occupied by the material as it moves and deforms is called *material volume* since it contains the same material at each time *t*. Let us now consider any continuously differentiable quantity $q = q(\mathbf{x}, t)$ and a **material volume** D bounded by a smooth (or occasionally piecewise smooth) surface ∂D, so that the outward normal is well defined. Under these assumptions, if we consider the integral:

$$\frac{d}{dt} \int_{D} q dV,$$

since the domain of integration depends on *t*, the time derivative cannot be brought under the sign of integration.

If we operate a change of variables from Eulerian to Lagrangian coordinates, then the domain of integration will be a fixed, time-independent volume, so that the derivative can be brought under the integral sign:

$$\frac{d}{dt} \int_{D} q(\mathbf{x}, t) dV = \int_{D_o} \frac{d}{dt} (Q(\mathbf{X}, t) J) dV_o. \tag{2.113}$$

If we apply to the r.h.s. the *Liouville's formula* (see Exercise 2.6):

$$\frac{dJ}{dt} = J div \mathbf{v} \tag{2.114}$$

and we revert to Eulerian coordinates, we obtain the so-called *transport theorem* or *Reynolds's theorem*:

$$\frac{d}{dt} \int_{D} q dV = \int_{D} (\frac{dq}{dt} + q div \mathbf{v}) dV = \int_{D} (\frac{\partial q}{\partial t} + div(q\mathbf{v})) dV. \tag{2.115}$$

Exercise 2.6
Proof of *Liouville's formula*: $\frac{dJ}{dt} = J div \mathbf{v}$.

We start by recalling that the *mixed product* of three vectors can be concisely expressed as:

$$\begin{vmatrix} a_1 & a_2 & a_3 \\ b_1 & b_2 & b_3 \\ c_1 & c_2 & c_3 \end{vmatrix} = \varepsilon_{ijk} a_i b_j c_k. \tag{2.116}$$

Accordingly, since $J = \det \mathbf{F}$, we can write: $J = \varepsilon_{pqr} F_{1p} F_{2q} F_{3r}$,

and it follows that $\dfrac{dJ}{dt} = \varepsilon_{pqr} \left[\dot{F}_{1p} F_{2q} F_{3r} + F_{1p} \dot{F}_{2q} F_{3r} + F_{1p} F_{2q} \dot{F}_{3r} \right]$.

Now, if we take into consideration the first term in the r.h.s.:

$$\varepsilon_{pqr} \dot{F}_{1p} F_{2q} F_{3r} = \begin{vmatrix} \dfrac{\partial v_1}{\partial X_1} & \dfrac{\partial v_1}{\partial X_2} & \dfrac{\partial v_1}{\partial X_3} \\[2mm] \dfrac{\partial x_2}{\partial X_1} & \dfrac{\partial x_2}{\partial X_2} & \dfrac{\partial x_2}{\partial X_3} \\[2mm] \dfrac{\partial x_3}{\partial X_1} & \dfrac{\partial x_3}{\partial X_2} & \dfrac{\partial x_3}{\partial X_3} \end{vmatrix} = \dfrac{D\left(v_1, x_2, x_3\right)}{D\left(X_1, X_2, X_3\right)},$$

and we observe that $\dfrac{\partial v_i}{\partial x_j} = \dfrac{\partial v_i}{\partial X_k} \dfrac{\partial X_k}{\partial x_j}$, we can write:

$$\begin{vmatrix} \dfrac{\partial v_1}{\partial x_1} & \dfrac{\partial v_1}{\partial x_2} & \dfrac{\partial v_1}{\partial x_3} \\[2mm] 0 & 1 & 0 \\[2mm] 0 & 0 & 1 \end{vmatrix} = \dfrac{D\left(v_1, x_2, x_3\right)}{D\left(x_1, x_2, x_3\right)} = \dfrac{D\left(v_1, x_2, x_3\right)}{D\left(X_1, X_2, X_3\right)} \dfrac{D\left(X_1, X_2, X_3\right)}{D\left(x_1, x_2, x_3\right)}.$$

Similar considerations can also be applied to the second and the third term in the r.h.s., and this will give the final result: $\dfrac{dJ}{dt} \dfrac{1}{J} = div\mathbf{v}$.

Incidentally, we note that, by using the properties of the permutation symbol, we can also write:

$\det F = \varepsilon_{pqr} F_{1p} F_{2q} F_{3r} = \varepsilon_{pqr} F_{p1} F_{q2} F_{r3}$, and that $\varepsilon_{pqr} \det F = \varepsilon_{ijk} F_{ip} F_{jq} F_{kr}$.

2.12 Conservation of mass

Consider a continuum body C and D be a portion of C. Let us assume that there exists a function:

$$\rho := \varphi(C) \rightarrow R^+, \tag{2.117}$$

called **density**, such that at any portion D is associated the mass:

$$m(D) = \int_D \rho dV. \tag{2.118}$$

The principle of conservation of mass states that, since a material volume contains the same material at each time, then the total mass in a material volume is constant, i.e.:

$$\frac{d}{dt} \int_D \rho(\mathbf{x}, t) dV = 0. \tag{2.119}$$

By using the transport theorem (2.115) and considering that the result (2.119) is valid for any arbitrary material volume, the integrand must vanish so that the **Eulerian** or **local expression** of the *conservation of mass* is directly obtained:

$$\frac{\partial \rho}{\partial t} + \nabla \cdot (\rho \mathbf{v}) = 0. \tag{2.120}$$

2.13 Balance of linear momentum

The principle of linear momentum states that the rate of change of the linear momentum of any portion D of a continuum body C is equal to the resultant of forces acting on it from its exterior.

When considering these forces, we have already stated that one needs to take into account two types of forces: those forces acting on D from the exterior of C, such as *mass forces* represented by vector field $\mathbf{b}$ (per unit mass); and those forces interior to C and acting on the boundary ∂D of the portion D, called *contact forces*.

In order to represent these contact forces, we introduced the **Cauchy's theorem** in Section 2.9: *the action of a continuum on the portion D is represented through the stress vectors* $\mathbf{t}$ *acting on the boundary ∂D.*

The stress vector t, also called *surface traction*, depends in general on the surface on which the point P lies and the assumption is that as long as the surfaces at P have the *same outward normal* the stress vector is the same. This assumption is known as *Cauchy's postulate*. With these assumptions, the principle of linear momentum writes:

$$\frac{d}{dt}\int_D \rho \mathbf{v} dV = \int_{\partial D} \mathbf{t} dS + \int_D \rho \mathbf{b} dV. \tag{2.121}$$

The integral in the l.h.s., by using the transport theorem and the mass conservation, can be written as:

$$\int_V \rho \frac{dv}{dt} dV.$$

Further, the stress vector can be expressed according to the Cauchy's lemma:

$$t_i = \sigma_{ji} n_j,$$

and by applying the Gauss theorem to the surface integral, equation (2.121) assumes the following Eulerian form:

$$\rho \frac{dv}{dt} = \nabla \cdot \sigma^T + \rho \mathbf{b}. \tag{2.122}$$

This equation is known as the **Cauchy's first law of motion** (1827).

If the acceleration $\dfrac{d\mathbf{v}}{dt}$ vanishes, then (2.122) reduce to the **indefinite equilibrium equations**:

$$\frac{\partial \sigma_{ji}}{\partial x_j} + \rho b_i = 0. \tag{2.123}$$

2.14 Balance of angular momentum

The term *angular momentum* (or moment of momentum) designates the moment of the linear momentum with respect to some point. Then, the principle of angular momentum states that *the time rate of change of the angular momentum with respect to a given point is equal to the moment of the body and surface forces with respect to that point.*

Accordingly we can write:

$$\frac{d}{dt}\int_D (\mathbf{r} \times \rho \mathbf{v})dV = \int_{\partial D} (\mathbf{r} \times \mathbf{t})dS + \int_D (\mathbf{r} \times \rho \mathbf{b})dV. \tag{2.124}$$

If the vector product is expressed in index notation and we make use of the Gauss theorem in order to transform the surface integral into a volume integral, then equation (2.124) can be written as:

$$\int_D e_{klm} \frac{d}{dt}(x_l v_m)\rho dV = \int_D e_{klm}\left[\frac{\partial(x_l \sigma_{jm})}{\partial x_j} + x_l \rho b_m\right]dV. \tag{2.125}$$

Because of the equation of the linear momentum: (2.122), it can be proved that (2.125) reduces to the following:

$$e_{klm}\sigma_{lm} = 0, \tag{2.126}$$

whose expansion corresponds to:

$$\sigma_{lm} = \sigma_{ml}, \tag{2.127}$$

which proves the symmetry of the stress tensor.

2.15 Summary

1   Despite the discrete nature of soils, as well as the recognition that fundamental aspects of soil behaviour cannot be properly understood without a knowledge of this discrete nature, mathematical models used in practice to predict the behaviour of geotechnical structures are based on the continuum medium idealization of soils. For this reason, learning *continuum mechanics* is an essential step.

To explain concepts such *stress* or *strain* we need to introduce more general quantities, called *tensors*, and a convenient way to introduce this concept is to consider a tensor as a *linear mapping*.

A relevant example of second-order tensor is offered by the stress tensor. The stress vector *t* acting at a point *P* on any surface through *P* oriented by its outward normal *n* is expressed through the linear mapping:

$$t_i^{(n)} = \sigma_{ji} n_j ,$$

and the symmetric *second-order tensor* σ_{ij} is called **stress tensor**. Its matrix representation is the following:

$$\left[\sigma_{ij} \right] = \begin{bmatrix} \sigma_{11} & \sigma_{12} & \sigma_{13} \\ \sigma_{21} & \sigma_{22} & \sigma_{23} \\ \sigma_{31} & \sigma_{32} & \sigma_{33} \end{bmatrix} ,$$

and we note that the *diagonal components* act normal to the coordinate planes and are called **normal stresses**, while the off-diagonal components act tangential to the coordinate planes and are called **shear stresses**.

The representation of a state of stress, known in literature as *Mohr circle of stress*, is a geometrical representation of the linear mapping which associates to any direction *n* the state of stress acting on the plane oriented by the outward unit normal *n*.

2   Any arbitrary change of configuration usually includes a rigid-body displacement as well as strains, i.e. a change in size or shape. When there is no change in the *relative position* of the material points of the continuum body, then the change of configuration is a rigid-body displacement, whereas the concept of strain is that of *relative change in the position of material points*.

Strains are defined *infinitesimal* if the components of displacement and the gradient of displacement are quantities of the first-order, so that the power of order $n \geq 2$ and the product of these quantities can be neglected. Then, the *infinitesimal strain tensor* in terms of displacement field writes:

$$\varepsilon_{ij} = \frac{1}{2} \left(\frac{\partial u_i}{\partial x_j} + \frac{\partial u_j}{\partial x_i} \right) .$$

Referring to its matrix representation:

$$\left[\varepsilon_{ij} \right] = \begin{bmatrix} \varepsilon_{11} & \varepsilon_{12} & \varepsilon_{13} \\ \varepsilon_{21} & \varepsilon_{22} & \varepsilon_{23} \\ \varepsilon_{31} & \varepsilon_{32} & \varepsilon_{33} \end{bmatrix} ,$$

the diagonal terms represent the longitudinal strains, i.e. the ratio between the change in length and the original length of a segment aligned in the coordinate direction.

The off-diagonal terms are called shear strains and represent one-half the change of the initially right angle between two material fibres aligned with the coordinate directions in the reference configuration.

3 The principle of conservation of mass states that, since a material volume contains the same material at each time, then the total mass in a material volume is constant. The *Eulerian* or *local expression* of the *conservation of mass* writes:

$$\frac{\partial \rho}{\partial t} + \nabla \cdot (\rho \mathbf{v}) = 0 .$$

4 The principle of linear momentum states that the rate of change of the linear momentum of any portion D of a continuum body C is equal to the resultant of forces acting on it from its exterior.

When considering these forces, we must take into account those forces acting on D from the exterior of C, such as *mass forces* represented by the vector field $\mathbf{b}$ (per unit mass). In addition, we have to consider those forces interior to C and acting on the boundary ∂D of the portion D (*contact forces*), represented through the *the stress vectors* $\mathbf{t}$.

The Eulerian form of the principle of linear momentum writes:

$$\rho \frac{d\boldsymbol{v}}{dt} = \nabla \cdot \boldsymbol{\sigma}^T + \rho \boldsymbol{b},$$

and if the acceleration $\dfrac{d\mathbf{v}}{dt}$ vanishes, then these equations reduce to the *indefinite equilibrium equations*:

$$\frac{\partial \sigma_{ji}}{\partial x_j} + \rho b_i = 0 .$$

2.16 Further reading

Many books offer a unified presentation of principles and concepts of continuum mechanics, so that quite often the choice is a matter of personal preference. However some classical references are:

Y.C. Fung (1965) *Foundations of Solid Mechanics*. Prentice-Hall.
A.C. Eringen (1967) *Mechanics of Continua*. Wiley.
L.E. Malvern (1969) *Introduction to the Mechanics of a Continuum Medium*. Prentice-Hall.
P. Germain (1973) *Cours de mécanique des milieux continues*. Masson.
A.J.M. Spencer (1980) *Continuum Mechanics*. Longman (reprinted by Dover, 2004).
G.T. Mase and G.E. Mase (1999) *Continuum Mechanics for Engineers*. CRC Press.
C.A. Truesdell (1991) *A First Course in Rational Continuum Mechanics*. Academic Press.

The book by A. Romano, R. Lancellotta and A. Marasco (2006) *Continuum Mechanics Using Mathematica*, Birkhäuser, focuses on mathematical aspects and provides a Mathematica-based software to solve complex and time-consuming problems.

A clear and concise presentation of tensor calculus is given in the book by I.E. Borisenko and I.E. Tarapov (1979) *Vector and Tensor Analysis with Applications*, Dover.

A reference for upper level studies is the classical book by T. Levi Civita (1926) *The Absolute Differential Calculus*, Blakie & Son Limited (reprinted by Dover, 1977).

Finally, we suggest two books dealing with the history of structural mechanics:

S.P. Timoshenko (1983) *History of the Strength of Materials*. Dover (originally published by McGraw-Hill in 1953).
E. Benvenuto (1991) *An Introduction to the History of Structural Mechanics*. Springer-Verlag.

Chapter 3

Basic constitutive models

The continuum mechanics principles, discussed in Chapter 2, have to be considered general relations, because their validity does not depend on the properties of the body under consideration. However, the experience also proves that the way a continuum deformable body reacts to changes of actions depends on its *internal constitution*. For this reason, in order to predict its deformation process, we need to introduce equations describing the response of the material, called *constitutive equations*.

From a mathematical point of view, the same conclusion is reached by observing that the balance equations do not form a closed set of *field equations*, so that we need to introduce an additional set of relations, linking the stress tensor to the strain tensor, in order to *close* the problem.

A first tempting procedure of deriving constitutive laws could be to make reference to experimental evidences and to try to match these results in a rather empirical and simple way. But the need to predict how a deformable body reacts when subjected to more complex loading histories requires a rather general theoretical framework.

The approach suggested in Rational Continuum Mechanics is to introduce some general rules, called *constitutive axioms*, representing constraints for the structure of constitutive equations, and to show how these latter can be derived within a robust general framework. There is no doubt that this is an elegant and unifying way to proceed, that can only be appreciated once the reader has already mastered the subject. However, alternatively, we can also build a general framework by gradually introducing a hierarchy of models, with emphasis on advantages and limitations of their use in practice. This is the approach followed in this chapter, because we consider it more appropriate to the beginning reader, leaving to specialized books any more complex treatment of the subject.

3.1 Elasticity

Theory of elasticity is attractive because it provides closed form solutions for the analysis of some boundary problems of geotechnical interest. Examples include the stress distribution induced within the soil mass by loads applied at the surface, as well as the displacement of loaded areas of finite dimension.

There are certainly limitations as far as the reliability of elasticity based predictions is concerned, and this aspect will be discussed in Chapter 9. However, simple calculations based on closed form solutions still offer the advantage of focusing on the most important parameters to be considered, as well as on their relative influence.

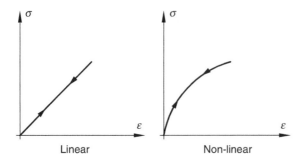

Figure 3.1 Elastic behaviour.

The behaviour of a given body is defined elastic if there is a *one-to-one correspondence between stress and strain* (see Figure 3.1). Such an assumption implies that the *work input* during the deformation process is *fully recovered* when the perturbation actions are removed, which means the process is completely reversible. As a consequence, it is possible to introduce a *state function* Ψ, which depends on the extremes of the loading path and not on the path itself, defined **elastic potential energy** (Green, 1839).

If there is any virtual strain increment $d\varepsilon_{ij}$, the work $d\Psi$ per unit volume of the body done by the stresses σ_{ij} is represented by:

$$d\Psi = \sigma_{ij} d\varepsilon_{ij}. \tag{3.1}$$

If Ψ is a function of independent variables ε_{ij}, then:

$$d\Psi = \frac{\partial \Psi}{\partial \varepsilon_{ij}} d\varepsilon_{ij}$$

and therefore we can assign to the function Ψ the property that:

$$\sigma_{ij} = \frac{\partial \Psi}{\partial \varepsilon_{ij}}. \tag{3.2}$$

This implies that the function Ψ plays the role of **strain energy** per unit volume of the elastic body or the role of *stress potential*. As a consequence, equation (3.2) allows one to derive, given the state function, the relationship between stress and strain, thus representing the *elastic constitutive law*. Furthermore, assuming that the stress–strain relationship is linear, the initial state is undeformed and the stresses are equal to zero in the absence of deformation, then the state function can be written in the form:

$$\Psi\left(\varepsilon_{ij}\right) = \frac{1}{2} C_{ijhk} \varepsilon_{ij} \varepsilon_{hk}, \tag{3.3}$$

and, by using the property that the function Ψ is a stress potential (3.2), we derive:

$$\sigma_{ij} = C_{ijhk} \varepsilon_{hk}. \tag{3.4}$$

In the above equations the 81 scalar quantities C_{ijhk}, called *elastic constants*, are the components of a *fourth-order tensor*, defined *stiffness tensor*. If the elastic constants are the same for all points of the medium, then the body is said to be elastically *homogeneous*; on the contrary, if they vary from point to point, the body is said to be *inhomogeneous*.

Note that the symmetry of the stress tensor requires that $C_{ijhk} = C_{jihk}$ and the symmetry of the strain tensor implies $C_{ijhk} = C_{ijkh}$. Further, if a strain energy exists, from which we can compute the stress by differentiating with respect to strain, i.e.:

$$\sigma_{ij} = \frac{\partial \Psi}{\partial \varepsilon_{ij}}, \quad \text{then} \quad C_{ijhk} = \frac{\partial \sigma_{ij}}{\partial \varepsilon_{hk}} = \frac{\partial^2 \Psi}{\partial \varepsilon_{hk} \partial \varepsilon_{ij}} = C_{hkij}.$$

It follows that only 21 constants are really independent. Usually, the further assumption of **isotropy** is introduced, i.e. in a given point the elastic constants are independent of the choice of reference axes. Then, the constants C_{ijhk} must be the components of a fourth-order isotropic tensor, symmetric in either ij or hk:

$$C_{ijhk} = \lambda \delta_{ij} \delta_{hk} + \mu \left(\delta_{ih} \delta_{jk} + \delta_{ik} \delta_{jh} \right), \tag{3.5}$$

and the constitutive law can be written as:

$$\sigma_{ij} = \lambda \varepsilon_{kk} \delta_{ij} + 2\mu \varepsilon_{ij}, \tag{3.6}$$

where the two elastic constants λ and μ are known as *Lamé's constants* (1852).

The inverse relationship is commonly presented in the form:

$$\varepsilon_{ij} = -\frac{\upsilon}{E} \sigma_{kk} \delta_{ij} + \frac{1+\upsilon}{E} \sigma_{ij}, \tag{3.7}$$

where E is the *Young's modulus* (1807) and υ is the *Poisson's ratio* (1829).

These two constants are the most used elastic constants in engineering, because of the direct experimental determination of them. If we consider an elastic bar, subjected to uniaxial tension σ_{zz} (the ends are subjected to uniform tension and the lateral surface is stress free, $\sigma_{xx} = \sigma_{yy} = 0$) and measurements are made of the extensional strain ε_{zz}, then Young's modulus is given by $E = \sigma_{zz}/\varepsilon_{zz}$ and has dimensions of stress.

The same experimental test also shows that, in response to the bar elongation, there is a lateral contraction, leading to lateral strains. Suppose that the bar is isotropic, then $\varepsilon_{xx} = \varepsilon_{yy} = -\upsilon \varepsilon_{zz}$, and this relation allows to obtain the Poisson's ratio, that is a dimensionless quantity.

However, it is more usual and appropriate in soil mechanics to refer to alternative constants: the **shear modulus** G (describing change in shape at constant volume) and the **bulk modulus** K (describing change in volume), because, if, as already seen in Chapter 2, we introduce the definition of *stress deviator*:

$$s_{ij} = \sigma_{ij} - p\delta_{ij}, \tag{3.8}$$

and *strain deviator*:

$$e_{ij} = \varepsilon_{ij} - \frac{1}{3}\varepsilon_v \delta_{ij},$$

(3.9)

then, the constitutive law assumes the significative form:

$$p = K\varepsilon_v$$
$$s_{ij} = 2Ge_{ij}.$$

(3.10)

This form proves the *uncoupled response* of an isotropic material: the increments of shear strains only depend on increments of the corresponding deviator stress and the volume changes only depend on increments of the corresponding spherical component of the stress tensor.

As only two constants are really independent, there must be a link between the introduced constants, and in fact the following relationships hold:

$$\mu = G = \frac{E}{2(1+v)}$$

$$\lambda = \frac{vE}{(1+v)(1-2v)}$$

$$K = \lambda + \frac{2}{3}G = \frac{E}{3(1-2v)}$$

$$E = \frac{G(3\lambda + 2G)}{\lambda + G}$$

(3.11)

Furthermore, it is relevant to recall that the elastic constants cannot take any values, but there are constraints that they have to satisfy. To show this we need to recall that the stored strain energy Ψ is a positive definite function of its arguments σ_{ij} and ε_{ij}. If we think of the stored energy as the sum of two parts, that associated to volumetric deformation and that associated to distortion, then it can be proved that the requirement for the strain energy to be a positive definite function implies:

$$G > 0 \quad (\lambda + 2/3G) > 0,$$

and, because of the relations (3.11), we also obtain:

$$K > 0 \quad E > 0 \quad -1 < v < 0.5.$$

3.1.1 Cross anisotropy

During deposition soils usually experience one-dimensional deformation, so that there is a tendency for displacements to only occur vertically without any lateral component. The so-called *inherent anisotropy* reflects this depositional history and a rather realistic model is, in this case, the *cross-anisotropic medium*: the soil response is different if the

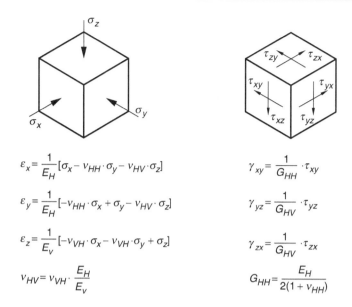

$$\varepsilon_x = \frac{1}{E_H}[\sigma_x - v_{HH} \cdot \sigma_y - v_{HV} \cdot \sigma_z]$$

$$\varepsilon_y = \frac{1}{E_H}[-v_{HH} \cdot \sigma_x + \sigma_y - v_{HV} \cdot \sigma_z]$$

$$\varepsilon_z = \frac{1}{E_v}[-v_{VH} \cdot \sigma_x - v_{VH} \cdot \sigma_y + \sigma_z]$$

$$v_{HV} = v_{VH} \cdot \frac{E_H}{E_v}$$

$$\gamma_{xy} = \frac{1}{G_{HH}} \cdot \tau_{xy}$$

$$\gamma_{yz} = \frac{1}{G_{HV}} \cdot \tau_{yz}$$

$$\gamma_{zx} = \frac{1}{G_{HV}} \cdot \tau_{zx}$$

$$G_{HH} = \frac{E_H}{2(1 + v_{HH})}$$

Figure 3.2 Cross anisotropy.

loading direction changes from vertical to horizontal, but it is the same when changes occur in the horizontal plane.

The relationships between stress and strains are shown in Figure 3.2, but we have to consider that, because for an elastic material the constitutive matrix must be symmetric, this implies:

$$\frac{v_{hv}}{E_H} = \frac{v_{vh}}{E_v},$$

so that this special case of anisotropy requires the introduction of only five independent constants:

$$E_V, E_H, G_{VH}, v_{VH}, v_{HH}$$

E_V and E_H are Young's moduli for unconfined compression in the vertical and horizontal direction; G_{VH} is the shear modulus for shearing in the vertical plane; v_{VH} is the Poisson's ratio for horizontal strain due to vertical strain and v_{HH} is the Poisson's ratio for horizontal strain due to horizontal strain at right angles.

Here again, the requirement that the strain energy must be a positive definite function of its arguments implies bounds on the values that the elastic constants can take, as discussed by Pickering (1970) and Lings *et al.* (2000).

Reference to this model will be made when interpreting field tests, in particular when dealing with seismic methods (see Section 7.11), in order to clarify if the measured shear modulus is the one relevant to shearing in the horizontal plane or to shearing in the vertical plane.

3.2 Cylindrical coordinates

Cylindrical coordinates are most appropriate when dealing with axi-symmetric problems. Seeing that we are faced with these problems when using common laboratory tests, it is relevant to have an overview of strain components and equilibrium equations as expressed in cylindrical coordinates.

Cartesian rectangular coordinates (x, y, z) are mapped into cylindrical orthogonal coordinates (r, ϑ, z) through the relations:

$$x = r \cos \vartheta$$

$$y = r \sin \vartheta \,,$$

$$z = z$$

and the displacement gradient tensor now assumes the form:

$$\nabla \mathbf{u} = \begin{bmatrix} \dfrac{\partial u_r}{\partial r} & \dfrac{1}{r}\dfrac{\partial u_r}{\partial \vartheta} - \dfrac{u_\vartheta}{r} & \dfrac{\partial u_r}{\partial z} \\[2ex] \dfrac{\partial u_\vartheta}{\partial r} & \dfrac{1}{r}\dfrac{\partial u_\vartheta}{\partial \vartheta} + \dfrac{u_r}{r} & \dfrac{\partial u_\vartheta}{\partial z} \\[2ex] \dfrac{\partial u_z}{\partial r} & \dfrac{1}{r}\dfrac{\partial u_z}{\partial \vartheta} & \dfrac{\partial u_z}{\partial z} \end{bmatrix}. \tag{3.12}$$

Therefore, the *components* of the infinitesimal strain tensor in cylindrical coordinates write:

$$\varepsilon_r = \frac{\partial u_r}{\partial r}; \quad \varepsilon_\vartheta = \frac{1}{r}\frac{\partial u_\vartheta}{\partial \vartheta} + \frac{u_r}{r}; \quad \varepsilon_z = \frac{\partial u_z}{\partial z}$$

$$\varepsilon_{r\vartheta} = \frac{1}{2}\left(\frac{1}{r}\frac{\partial u_r}{\partial \vartheta} + \frac{\partial u_\vartheta}{\partial r} - \frac{u_\vartheta}{r}\right); \quad \varepsilon_{\vartheta z} = \frac{1}{2}\left(\frac{\partial u_\vartheta}{\partial z} + \frac{1}{r}\frac{\partial u_z}{\partial \vartheta}\right);$$

$$\varepsilon_{zr} = \frac{1}{2}\left(\frac{\partial u_z}{\partial r} + \frac{\partial u_r}{\partial z}\right). \tag{3.13}$$

Further, if b_r, b_ϑ, b_z are the components of the body force, the equilibrium equations assume the following form:

$$\frac{\partial \sigma_r}{\partial r} + \frac{1}{r}\frac{\partial \tau_{\vartheta r}}{\partial \vartheta} + \frac{\partial \tau_{zr}}{\partial z} + \frac{\sigma_r - \sigma_\vartheta}{r} + \rho b_r = 0$$

$$\frac{\partial \tau_{r\vartheta}}{\partial r} + \frac{1}{r}\frac{\partial \sigma_\vartheta}{\partial \vartheta} + \frac{\partial \tau_{z\vartheta}}{\partial z} + \frac{2\tau_{r\vartheta}}{r} + \rho b_\vartheta = 0 \quad . \tag{3.14}$$

$$\frac{\partial \tau_{rz}}{\partial r} + \frac{1}{r}\frac{\partial \tau_{\vartheta z}}{\partial \vartheta} + \frac{\partial \sigma_z}{\partial z} + \frac{\tau_{rz}}{r} + \rho b_z = 0$$

The **axi-symmetric condition** imposes the following contraints to the displacement field:

$$u_\vartheta = 0$$

$$\frac{\partial u_r}{\partial \vartheta} = \frac{\partial u_z}{\partial \vartheta} = 0 \,, \tag{3.15}$$

so that the non-vanishing strain components are:

$$\varepsilon_r = \frac{\partial u_r}{\partial r}; \quad \varepsilon_\vartheta = \frac{u_r}{r}; \quad \varepsilon_z = \frac{\partial u_z}{\partial z}; \quad \varepsilon_{zr} = \frac{1}{2}\left(\frac{\partial u_z}{\partial r} + \frac{\partial u_r}{\partial z}\right). \tag{3.16}$$

Moreover, when considering the state of stress, $\tau_{r\vartheta} = \tau_{z\vartheta} = 0$, and the non-vanishing components must be independent on ϑ, so that the equilibrium equations reduce to the following:

$$\frac{\partial \sigma_r}{\partial r} + \frac{\partial \tau_{zr}}{\partial z} + \frac{\sigma_r - \sigma_\vartheta}{r} + \rho b_r = 0$$
$$\frac{\partial \tau_{rz}}{\partial r} + \frac{\partial \sigma_z}{\partial z} + \frac{\tau_{rz}}{r} + \rho b_z = 0 \tag{3.17}$$

A further remark concerns the fact that, when dealing with orthogonal curvilinear coordinates, the components of a tensor in a point are coincident with the Cartesian components referred to a local system, defined by axes tangent to the coordinate curves y^i (Figure 3.3). Therefore, any algebraic relation involving the components preserves

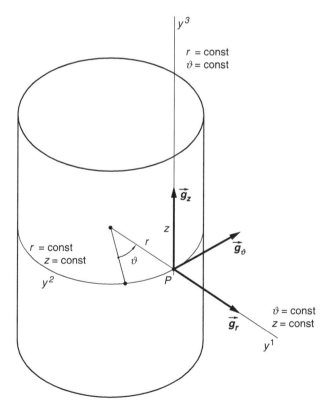

Figure 3.3 Cylindrical coordinates.

the same form that it assumes when written in Cartesian coordinates, so that the constitutive relationships still write in the form:

$$\sigma_r = \lambda\varepsilon_v + 2\mu\varepsilon_r; \quad \sigma_\vartheta = \lambda\varepsilon_v + 2\mu\varepsilon_\vartheta; \quad \sigma_z = \lambda\varepsilon_v + 2\mu\varepsilon_z$$
$$\tau_{zr} = 2\mu\varepsilon_{zr}; \quad \tau_{r\vartheta} = 2\mu\varepsilon_{r\vartheta}; \quad \tau_{\vartheta z} = 2\mu\varepsilon_{\vartheta z}$$

(3.18)

and the inverse relations (3.7) also apply.

Example 3.1
A cylindrical sample, whose initial dimension are $h_o = 200\,\text{mm}$ and $2R_o = 100\,\text{mm}$, is composed of a material characterized by the following values of Young's modulus and Poisson's ratio: $E = 28,000\,\text{MN/m}^2$, $v = 0.15$. Find the height and the diameter values, when the sample is subjected to uniform uniaxial compressive stress $\sigma_z = 5\,\text{MN/m}^2$ (see Figure 3.4).

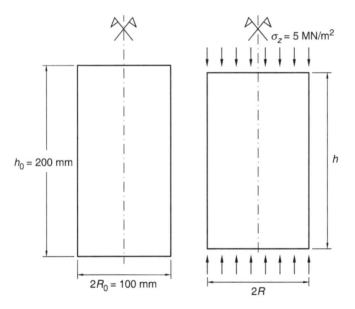

Figure 3.4 Uniform axial compression.

Solution. When dealing with experimental tests, any laboratory apparatus should induce within the tested sample a uniform state of stress and strain, because only if this basic requirement is satisfied can the sample be considered a volume element.

The boundary conditions for the test under consideration are specified as stress boundary conditions:

$$\sigma_z = 5\,\text{MN/m}^2$$

$$\sigma_r = 0\,\text{MN/m}^2.$$

These are principal stresses, so that the matrix representation of the stress tensor is the following:

$$\left[\sigma_{ij}\right] = \begin{bmatrix} \sigma_z & 0 & 0 \\ 0 & \sigma_r & 0 \\ 0 & 0 & \sigma_\vartheta \end{bmatrix}.$$

Moreover, the assumption of uniform state of stress requires that $\dfrac{\partial \sigma_r}{\partial r} = 0$, and the indefinite equilibrium equation in the radial direction (3.9) then implies $\sigma_r = \sigma_\vartheta$.

By applying the constitutive law:

$$\varepsilon_{ij} = \frac{1+\upsilon}{E}\sigma_{ij} - \frac{\upsilon}{E}\sigma_{kk}\delta_{ij},$$

we obtain:

$$\varepsilon_z = \frac{\sigma_z}{E} - \frac{\upsilon}{E}[\sigma_r + \sigma_\vartheta] = \frac{\sigma_z}{E}$$
$$\varepsilon_r = \varepsilon_\vartheta = -\frac{\upsilon}{E}\sigma_z = -\upsilon\varepsilon_z$$

By using the the material parameters we have (we retain the sign convention adopted in solid mechanics):

$$\left[\varepsilon_{ij}\right] = \begin{bmatrix} -1.79 \cdot 10^{-4} & 0 & 0 \\ 0 & 2.69 \cdot 10^{-5} & 0 \\ 0 & 0 & 2.69 \cdot 10^{-5} \end{bmatrix},$$

so that the required values of the height and the diameter in the deformed configuration are:

$$h = h_0(1 + \varepsilon_z) = 199.96\,\text{mm}$$
$$R = R_0(1 + \varepsilon_\vartheta) = 100.003\,\text{mm}$$

Example 3.2
A cylindrical sample, confined by a frictionless rigid ring, is subjected to a compressive uniform vertical stress (see Figure 3.5). *Find* the radial and the circumferential stress, assuming the sample exhibits a linear isotropic behaviour.

Solution. The confined test sketched in Figure 3.5 is an idealization of one of the most common tests in soil mechanics, known as *oedometer test* (this test will be extensively discussed in Chapter 4). In this case, the prescribed conditions are mixed boundary conditions, since we know the applied uniform vertical stress σ_z and we prescribe the radial displacement $u_r = 0$.

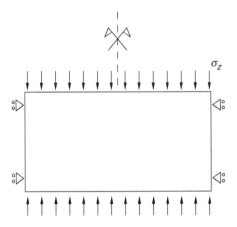

Figure 3.5 Idealized one-dimensional compression (no lateral displacement).

The axi-symmetric conditions $u_\vartheta = 0$ and $\partial u_r/\partial\vartheta = \partial u_z/\partial\vartheta = 0$ and the boundary condition $u_r = 0$ imply $\varepsilon_r = \varepsilon_\vartheta = 0$. In addition, since u_z does not depend on r, i.e. $\partial u_z/\partial r = 0$, the strain tensor has the following matrix representation:

$$\left[\varepsilon_{ij}\right] = \begin{bmatrix} 0 & 0 & 0 \\ 0 & 0 & 0 \\ 0 & 0 & \varepsilon_z \end{bmatrix}.$$

The axi-symmetric condition imposes $\tau_{\vartheta r} = \tau_{\vartheta z} = 0$, and since $\tau_{rz} = 0$ the stress tensor has the following matrix representation:

$$\left[\sigma_{ij}\right] = \begin{bmatrix} \sigma_r & 0 & 0 \\ 0 & \sigma_\vartheta & 0 \\ 0 & 0 & \sigma_z \end{bmatrix}.$$

Moreover, the assumption of uniform state of stress requires that $\dfrac{\partial\sigma_r}{\partial r} = 0$, and the indefinite equilibrium equation in the radial direction (3.17) then implies $\sigma_r = \sigma_\vartheta$.

In order to compute the radial stress, we observe that the constitutive equation allows to express the linear strain in the radial direction in the following form:

$$\varepsilon_r = \frac{1}{E}\left[\sigma_r - \upsilon\left(\sigma_z + \sigma_r\right)\right],$$

and because this is zero, due to the constraint of the test, it follows:

$$\sigma_r = \frac{\upsilon}{1-\upsilon}\sigma_z.$$

Note that this relation is only valid under the presumed linear elastic behaviour.

Example 3.3

Consider the *torsion test* on a cylindrical sample, as shown in Figure 3.6, the base of the sample being fixed and the top subjected to the torque M_t. Assuming that the torsion angle is small, the circular shape of any cross-section is preserved, as is the dimension of the initial diameter and the distance between two arbitrary cross-sections. It is required to find the strain and stress tensors.

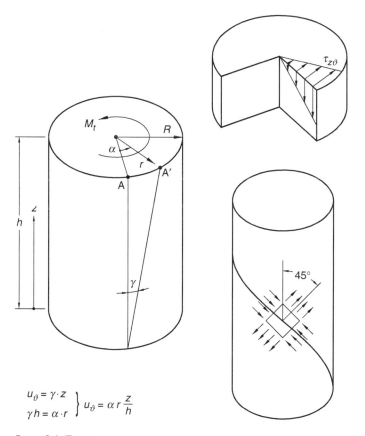

$$u_\vartheta = \gamma \cdot z \atop \gamma h = \alpha \cdot r \Bigg\} \; u_\vartheta = \alpha r \frac{z}{h}$$

Figure 3.6 Torsion test.

Solution. The basic assumptions allow to define the displacement field:

$$u_z = 0; \qquad u_r = 0; \qquad u_\vartheta = r\alpha \frac{z}{h},$$

and by applying (3.13) we obtain:

$$\varepsilon_z = \varepsilon_r = \varepsilon_\vartheta = 0$$

$$\varepsilon_{r\vartheta} = 0; \quad \varepsilon_{rz} = 0; \quad \varepsilon_{z\vartheta} = \frac{1}{2}\alpha\frac{r}{h}.$$

The imposed stress state is such that the only non-vanishing stress component is $\tau_{z\vartheta}$. If the sample behaves as a linear elastic isotropic medium, then the constitutive relationship gives $\tau_{z\vartheta} = G\alpha\frac{r}{h}$. At the top of the sample, the boundary condition requires:

$$M_t = 2\pi \int_0^R r^2 \tau_{z\vartheta}\, dr = \pi\frac{\alpha R^4}{2h}G$$

so that the shear stress can be expressed in terms of the applied torque:

$$\tau_{z\vartheta} = \frac{2M_t r}{\pi R^4}.$$

The Mohr circle of stress shows that in this case the principal stresses act on planes whose normal is inclined 45° with respect to the vertical, and when the material is unable to withstand tension, this will produce helicoidal cracks.

3.3 Plasticity

Experimental data support the assumption that many materials exhibit a reversibile behaviour at very small strains and that irrecoverable, permanent deformations occur beyond a threshold strain.

This kind of behaviour, described by means of plasticity theory, applies particularly to metals, because of their crystalline structure. In this case, plastic flow arises from dislocations in crystal lattices.

Another example of great practical relevance is provided by soils, where the irreversible nature of deformations is mainly the result of relative movements between particles.

Anyway, these microscopic processes are disregarded in the phoenomenological representation given by plasticity theory, that only reflects the most relevant effects we can observe at a macroscale. In particular, one of these relevant effects is the quantitative definition of the **yield stress**, below which no dislocation occurs.

The aim of this section is to introduce the reader to fundamental ideas of the plasticity theory. The attention will not be focused on a particular model, rather the basic ingredients needed to build a model are the object of discussion. In Chapter 5 we will concentrate on the so-called *Cam Clay Model*, a model developed at Cambridge University in the 1960s (Schofield and Wroth, 1968), which pioneered the application of plasticity to soils, in order to predict their mechanical behaviour in a consistent framework.

3.3.1 Basic assumptions of plasticity theory

A fundamental assumption of plasticity is that the total strain can be additively decomposed into the elastic and plastic component:

$$d\varepsilon_{ij} = d\varepsilon_{ij}^e + d\varepsilon_{ij}^p. \tag{3.19}$$

The elastic component of strain can be computed according to the elastic constitutive relations already seen in Section 3.1. In addition, we note that, because of the assumption of isotropy, the principal directions of the strain tensor are coincident with those of the stress tensor, so that the elastic relationships also define the strain directions.

On the contrary, to identify the plastic deformations it is required to introduce:

- a yield criterion
- a plastic flow rule
- a hardening rule

which allow one to specify *when plastic deformations occur, their direction and their magnitude.*

In order to highlight the above-mentioned features of an elasto-plastic behaviour, let's start by considering the simple *uniaxial case*, depicted in Figure 3.7. In this case, defining a *yield criterion* is equivalent to assuming that the material exhibits an elastic response if the applied stress is below a threshold value σ_p, called **yield stress**. Beyond this stress, plastic deformations also occur, and when passing from point A to point B we have

$$\delta\varepsilon_a = \lambda\delta\sigma_a. \tag{3.20}$$

Once point B has been reached, if an unloading-reloading cycle is performed, the sample behaviour is assumed to be elastic if the applied stress is below the new yield stress, from now on represented by point B. This is a particularly relevant feature of the constitutive law we are dealing with: because of plastic deformations occurring in the previous path $A-B$, the material seems to have gained **memory** of the *past stress (or deformation) history*, and it reveals its memory through an increase of the yield stress. *The evolution of the yield stress with the occurrence of plastic deformations is known in literature as* **hardening**. If, according to

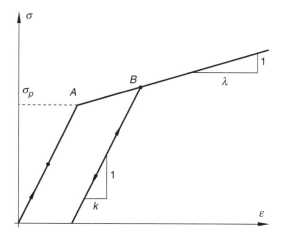

Figure 3.7 Idealized elastic-plastic behaviour.

the meaning of the symbols in Figure 3.7, the increment of elastic deformation is given by:

$$\delta\varepsilon_a^e = k\delta\sigma_a,$$ (3.21)

the plastic component is:

$$\delta\varepsilon_a^p = \delta\varepsilon_a - \delta\varepsilon_a^e = (\lambda - k)\delta\sigma_a.$$ (3.22)

Then, the direct relationship writes:

$$\delta\sigma_a = \frac{1}{\lambda - k}\delta\varepsilon_a^p = H\delta\varepsilon_a^p$$ (3.23)

and the modulus H is named **hardening modulus**.

This example offers the advantage of illustrating in a rather simple way the main features to be taken into account in our constitutive model.

First, in order to predict deformations under a given applied stress, we need to account in some way for the previous **stress history**: there is no more a one-to-one relationship between stress and strain, as it was in elasticity. As a consequence, constitutive laws of plasticity are not equations, but *functional relations*. However, the common approach to reduce the relations to equations is to account for the history throughout the introduction of *hidden variables*, whose meaning will be clarified in the sequel.

Second, only an incremental formulation can be given to elasto-plastic relationships. In this respect we note that it is usual in plasticity theory to denote the increments $\delta\varepsilon^p$ by using the superimposed dot, i.e. $\dot\varepsilon^p$. The symbol would suggest velocity, rather then increments, but we outline that in plasticity theory the time is an independent variable introduced just to order the events, and could be replaced by any other parameter, that is a monotonic function of time. The stress–strain relationships are rate independent.

Finally, referring to Figure 3.8, where both cases of *positive* and *negative hardening* are shown (the second case is also called *softening*), in order to compute strains we need to specify the alternative, loading or unloading, i.e.:

$$
\begin{aligned}
\delta\sigma &= \frac{1}{\lambda}\delta\varepsilon \quad \textit{if } \delta\varepsilon \geq 0 \\
\delta\sigma &= \frac{1}{k}\delta\varepsilon \quad \textit{if } \delta\varepsilon \leq 0
\end{aligned}
$$ (3.24)

and reverting to the inverse relationships, that only apply to the case of positive hardening:

$$
\begin{aligned}
\delta\varepsilon &= \lambda\delta\sigma \quad \textit{if } \delta\sigma \geq 0 \\
\delta\varepsilon &= k\delta\sigma \quad \textit{if } \delta\sigma \leq 0
\end{aligned}
$$ (3.25)

The need to specify the above alternatives proves that the elasto-plastic relationships are *incrementally non-linear*. For this reason, a demarcation of domains arises and each possibility generates field equations of one kind or another. As a consequence, it is not possible to write the general initial-boundary value problem in one explicit formulation, as already done in elasticity.

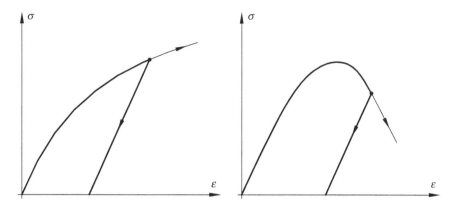

Figure 3.8 Positive and negative hardening.

3.3.2 Hardening plasticity

When the above considerations are translated into a more general state of stress, the first step is to define analytically a **yield function**, in order to establish a criterion, which allows one to specify when plastic deformations occur.

In general, we can assign to this yield function a structure such as:

$$f = f\left(\sigma_{ij}, h\right) \leq 0, \tag{3.26}$$

where h is a vector of variables dependent on the history of the process through the plastic deformations, i.e.:

$$h = h\left(\varepsilon_{ij}^{p}\right). \tag{3.27}$$

It is usual to refer these variables as *hidden variables*, as they do not appear in balance equations.

A geometrical meaning can be associated to the equation (3.26), if we assume that the stress components are coordinates of a special space, the stress space. Because of the six independent stress components, we would need a six-dimensional stress space, in which to represent a stress point. An alternative is to make reference to the three principal stresses, plus information related to the principal directions. In addition, if we introduce the assumption of isotropy, i.e. there is no dependence of material behaviour on a given direction, then the six independent components σ_{ij} can be replaced, with no loss of generality, by the three principal stresses. Therefore, if the principal stresses are used as coordinates to define a three-dimensional space called the *principal stress space*, in this space the yield function represents a **yield surface** (Figure 3.9).

Since for a state of stress inside this surface only elastic deformations occur, whereas, if the state of stress lies on it, plastic deformations also occur, the yield surface can be interpreted as the *instantaneous boundary of the elastic domain*. The adjective *instantaneous* is here introduced to recall that the yield surface evolves and in particular, if the yield surface expands uniformly, the process is named **isotropic hardening**; if

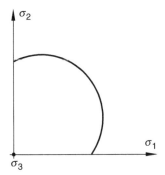

Figure 3.9 Yield surface.

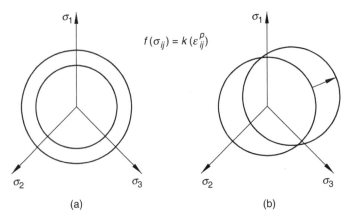

Figure 3.10 (a) isotropic hardening; (b) kinematic hardening.

the surface simply translates without modifying its shape and size, then the process is named **kinematic hardening** (Figure 3.10).

By assuming that strains can be additively decomposed into elastic and plastic components, i.e.:

$$\varepsilon_{ij} = \varepsilon_{ij}^e + \varepsilon_{ij}^p, \tag{3.28}$$

we can distinguish among the following cases:

(a) if $f < 0$, the stress state is represented by a point internal to the yield surface, so that the elastic relationship applies, i.e. $\delta\varepsilon_{ij} = C^e_{ijhk}\delta\sigma_{hk}$;

(b) if $f = 0, df < 0$, the image of the stress state lies on the yield surface, but the material is subjected to unloading and the elastic relation continues to apply;

(c) if $f = 0, df = 0$, plastic deformations occur and in this case, according to the basic assumption (3.28), we can write $\delta\varepsilon_{ij} = \delta\varepsilon_{ij}^e + \delta\varepsilon_{ij}^p = \left(C^e_{ijhk} + C^p_{ijhk}\right)\delta\sigma_{hk}$.

In order to derive the expression of the plastic components of the constitutive tensor, the adopted approach relies on experimental evidences initially provided by Taylor and Quinney (1931). These authors conducted experiments on metal hollow tubes subjected to tension and torsion and they first showed that the direction of the plastic strain increment vector does not depend on the stress increment vector, but it depends on the stress state, at which the yield surface was reached.

More specifically, the mechanism of elastic deformation depends on stress increments, while the mechanism of plastic deformation depends on stresses (see Wood, 1990, 57–64). According to such evidences, a scalar function $g = g(\sigma_{ij})$ is then introduced, playing the role of a **plastic potential**, so that we can write:

$$\delta\varepsilon_{ij}^p = \Lambda \frac{\partial g}{\partial \sigma_{ij}}. \tag{3.29}$$

This assumption can be visualized in the following way: suppose that the space of principal strain increments is superimposed on the space of principal stresses, in a way that the principal axes coincide (this assumption is known as **coaxiality hypothesis of Saint Venant**), then the plastic strain increment vector is directed in the direction of the gradient of the plastic potential (see Figure 3.11).

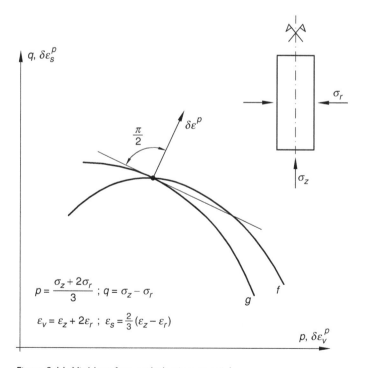

Figure 3.11 Yield surface and plastic potential.

The scalar Λ can be derived by imposing the **consistency condition**:

$$df = \frac{\partial f}{\partial \sigma_{hk}} \delta \sigma_{hk} + \frac{\partial f}{\partial h} \frac{\partial h}{\partial \varepsilon_{ij}^p} \delta \varepsilon_{ij}^p = 0, \tag{3.30}$$

which ensures that the image point of the stress state lies on the yield surface. In fact, should the stress state be outside the yield surface, plastic deformations would occur during unloading, in contradiction to the definition of yield surface. By substituting (3.29) into (3.30), one obtains:

$$\Lambda = -\frac{\dfrac{\partial f}{\partial \sigma_{hk}} \delta \sigma_{hk}}{\dfrac{\partial f}{\partial h} \dfrac{\partial h}{\partial \varepsilon_{ij}^p} \dfrac{\partial g}{\partial \sigma_{ij}}} = \frac{1}{H} \frac{\partial f}{\sigma_{hk}} \delta \sigma_{hk}, \tag{3.31}$$

where the **hardening modulus** assumes the expression:

$$H = -\frac{\partial f}{\partial h} \frac{\partial h}{\partial \varepsilon_{ij}^p} \frac{\partial g}{\partial \sigma_{ij}}. \tag{3.32}$$

Accordingly, the constitutive plastic tensor is given by

$$C_{ijhk}^p = \frac{1}{H} \frac{\partial g}{\partial \sigma_{ij}} \frac{\partial f}{\partial \sigma_{hk}}. \tag{3.33}$$

3.3.3 Associated plastic flow

Let now assume that, in addition to the above considerations, the material is required to be **stable,** in the sense specified by Drucker (1959): *a material is defined* **stable in the small** *if any perturbation produces a non-negative work*:

$$\delta \sigma_{ij} \delta \varepsilon_{ij} \geq 0;$$

a material is defined **stable in the large** *if, given a stress history* $\sigma_{ij}(t)$, *then*:

$$\int_{t_1}^{t_2} \sigma_{ij} \delta \varepsilon_{ij} dt \geq 0.$$

Because in a loading cycle the elastic component of the deformation is reversibile, the above two requirements can be summarized in the following one:

$$\left(\sigma_{ij} - \sigma_{ij}^{amm}\right)\delta \varepsilon_{ij} \geq 0 \quad \forall \sigma_{ij}^{amm}/f\left(\sigma_{ij}^{amm}\right) \leq 0. \tag{3.34}$$

This requirement is of great importance, because it has the following two implications:

- the yield surface must be convex;
- the yield surface also plays the role of plastic potential.

To prove the first implication, let us consider a stress increment from a state of stress, represented by a point A internal to the yield surface, to a point P on the surface $f(\sigma_{ij}) = 0$. The requirement (3.34) suggests that the angle between the stress increment vector and the vector of the plastic deformation must be acute. Then, if we consider a plane which contains the point P and is perpendicular to the plastic deformation vector, the (3.34) is satisfied by all points A located below such a plane only if the yield surface is **convex** (recall that a closed surface is defined convex if through every point on it there is a plane such that all points interior to the surface lie on the same side of the plane).

Once the convexity of the yield surface has been proved, we observe that (3.34) is fulfilled by any arbitrary stress increment vector (i.e. including a tangent stress vector) only if the vector representing the strain increment is directed like *gradf*. It follows that the yield surface also plays the role of plastic potential, and in this case the flow rule is referred as **associated flow rule**.

We shall return on this point in Chapter 8, as the above requirements have relevant implications when dealing with upper and lower bound theorems of plastic collapse.

3.4 Visco-elasticity

Soils exhibit a time-dependent response. The potential for this phenomenon depends on compositional factors (see Mitchell, 1976) and, in general, the greater the organic content and the higher the plasticity, the more pronounced is the viscous behaviour.

Different rheological models have been suggested to analyse the time-dependent behaviour of soils. In general they are a combination of the following basic models.

3.4.1 Maxwell model

Let us consider a dashpot, whose constitutive law is expressed as:

$$\sigma = \eta \dot{\varepsilon} . \tag{3.35}$$

Assuming that the **coefficient of viscosity** η is constant, the relationship (3.35) describes the behaviour of a **Newtonian medium**. When the Newton dashpot is combined in series with a Hooke spring (see Figure 3.12) the obtained model is known as **Maxwell model** (1868).

If a time $t = 0$ an instantaneous stress σ_o is applied and then it is mantained constant with time, the strain will be the sum of an instantaneous elastic component and a time-varying component, i.e.:

$$\varepsilon = \frac{\sigma_o}{K} + \frac{\sigma_o}{\eta} t . \tag{3.36}$$

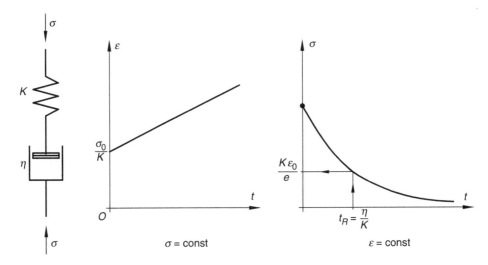

Figure 3.12 Maxwell model.

Alternatively, assuming that an impressed strain ε_o is maintained constant over time, since:

$$\dot{\varepsilon} = \frac{\dot{\sigma}}{K} + \frac{\sigma}{\eta}, \tag{3.37}$$

we obtain

$$\sigma = \left(K\varepsilon_o \right) e^{-t/t_R} . \tag{3.38}$$

The stress decay with time, expressed by (3.38), is called **relaxation** and the characteristic time $t_R = \eta/K$, which appears as a normalizing factor in equation (3.38), is known as **relaxation time**. It represents the time required for the stress to reduce to $1/e$ its initial value.

3.4.2 Kelvin model

In the Kelvin (1837) model the spring and the dashpot are connected in parallel (Figure 3.13), so that the response is:

$$\sigma = K\varepsilon + \eta\dot{\varepsilon} . \tag{3.39}$$

In this case, when we explore the behaviour under a constant stress σ_o, the integration of (3.39) provides the evolution of strain with time:

$$\varepsilon = \frac{\sigma_o}{K}\left(1 - e^{-t/t_R}\right), \tag{3.40}$$

and this phenomenon is named **transient creep**, because the strain has the tendency towards the asymptotic value σ_o/K. The characteristic time $t_R = \eta/K$ is now called **delay time**.

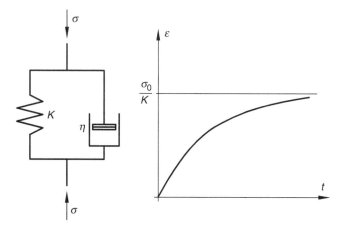

Figure 3.13 Kelvin model.

3.4.3 Burgers model

Both the previous models have shortcomings when applied to soils: Maxwell's model does not consider that the actual rate of deformation decreases with time under sustained load, and the Kelvin model does not consider the initial deformation independent of time. For this reason there is a need to use more complex models, obtained from a combination of these basic ones.

The Burgers (1935) model consists of a Maxwell element in series with a Kelvin element (Figure 3.14). If ε_1 and ε_2 are the strains of these two elements, the total strain will be:

$$\varepsilon = \varepsilon_1 + \varepsilon_2. \tag{3.41}$$

As the Maxwell model gives:

$$\dot{\varepsilon}_1 = \frac{\dot{\sigma}}{K_1} + \frac{\sigma}{\eta_1} \tag{3.42}$$

and the Kelvin model:

$$\sigma = K_2 \varepsilon_2 + \eta_2 \dot{\varepsilon}_2, \tag{3.43}$$

the following relation can be derived:

$$\ddot{\varepsilon} = \ddot{\varepsilon}_1 + \ddot{\varepsilon}_2 = \frac{\ddot{\sigma}}{K_1} + \frac{\dot{\sigma}}{\eta_1} + \frac{\dot{\sigma}}{\eta_2} - \frac{K_2}{\eta_2}(\dot{\varepsilon} - \dot{\varepsilon}_1). \tag{3.44}$$

If this latter is substituted into (3.42), one gets:

$$\eta_2 \ddot{\varepsilon} + K_2 \dot{\varepsilon} = \frac{\eta_2}{K_1} \ddot{\sigma} + \left(1 + \frac{K_2}{K_1} + \frac{\eta_2}{\eta_1}\right) \dot{\sigma} + \frac{K_2}{\eta_1} \sigma. \tag{3.45}$$

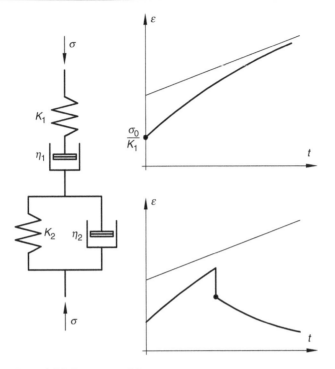

Figure 3.14 Burgers model.

Here again, if we consider the case of a constant with time stress σ_o, the integration of (3.45) allows to compute the evolution of strain:

$$\varepsilon = \frac{\sigma_o}{K_1} + \frac{\sigma_o}{K_2}\left(1 - e^{-t/t_{R_2}}\right) + \frac{\sigma_o}{\eta_1}t. \tag{3.46}$$

It can be observed that the model can simulate the presence of an instantaneous strain, a transient creep and a steady state creep. Unloading reveals the inelastic character of the response.

Example 3.4

By using the Maxwell model, we require to explore the stress–strain relationship, when the tests are performed at constant strain rate.

Solution. If we impose the condition $\dot{\varepsilon} = \text{const} = D$ into $\dot{\varepsilon} = \dfrac{\dot{\sigma}}{K} + \dfrac{\sigma}{\eta}$, then we obtain:

$$\dot{\sigma} + \frac{\sigma}{t_R} = KD. \tag{3.47}$$

Equation (3.47) is a first-order differential equation, whose integral writes:

$$\sigma = \sigma_o e^{-t/t_R} + e^{-t/t_R} \int_{t=0}^{t} K D e^{t/t_R} \, dt.$$ (3.48)

If the stress at the initial instant of time is zero, then:

$$\sigma = \eta D \left(1 - e^{-\varepsilon/D t_R} \right).$$ (3.49)

Since

$$\frac{d\sigma}{d\varepsilon} = \frac{\eta}{e^{\varepsilon/D t_R}} \cdot \frac{1}{t_R},$$

the obtained solution proves that the apparent stiffness increases if the rate of strain increases (Figure 3.15) and merges into the value of the elastic spring when the rate tends to be infinite.

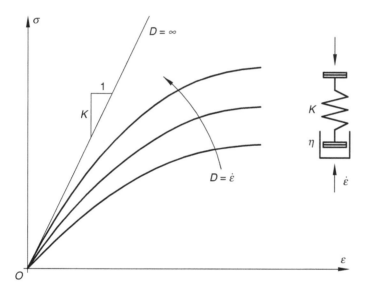

Figure 3.15 Example of rate effect on stress–strain behaviour.

3.5 Internal constraints

When formulating constitutive relationships, it must be considered that some materials could not exhibit any arbitrary deformation if they are subjected to *internal constraints*. As an example, if we consider an incompressible material, deformations that change the volume are prevented. It follows that, when considering a material subjected to *mechanical internal constraints*, the stress tensor assumes the following structure:

$$\sigma = \mathbf{P} + \sigma(\mathbf{F}),\qquad(3.50)$$

the tensor $\mathbf{P}$ being the reaction to the internal constraint and $\sigma(\mathbf{F})$ the constitutive component. The reaction $\mathbf{P}$, defined through the equations of motion, must satisfy the requirement that *there is no work done due to the reaction stress*, according to the *principle of determinism for constrained materials* (Truesdell and Noll, 1965): *the stress for a material body with mechanical constraint is determined by the history of motion only within an arbitrary tensor* $\mathbf{P}$ *that does no work*, i.e:

$$\mathbf{P} \cdot \mathbf{D} = 0,\qquad(3.51)$$

where $\mathbf{D}$ is the symmetric part of the velocity gradient tensor:

$$\mathbf{D} = Symm(\nabla \mathbf{v}).\qquad(3.52)$$

If, as an example, we consider the constraint of incompressibility, i.e. the deformation must be volume preserving, this condition can be written as $\det \mathbf{F} = 1$, or, equivalently as:

$$\nabla \cdot \mathbf{v} = 0\qquad(3.53)$$

and, since

$$\nabla \cdot \mathbf{v} = tr(\nabla \mathbf{v}) = tr(\mathbf{D}) = \mathbf{I} \cdot \mathbf{D},\qquad(3.54)$$

we reach the conclusion that the reaction must assume the following form:

$$\mathbf{P} = \lambda \mathbf{I}.\qquad(3.55)$$

The reaction to the incompressibility constraint is any arbitrary pressure λ and the stress tensor must have the form:

$$\sigma = \lambda \mathbf{I} + \sigma',\qquad(3.56)$$

where σ' is the constitutive component.

3.6 Summary

(1) The experience proves that the way a continuum deformable body reacts to changes of actions depends on its *internal constitution*. For this reason, in order to predict its deformation process, we need to introduce equations describing the response of the material, called *constitutive equations*.

    The most common example of constitutive laws is provided by the elasticity theory. The behaviour of a given body is defined elastic if there is a *one-to-one correspondence between stress and strain*. Such an assumption implies that the *work input* during the deformation process is *fully recovered* when the perturbation actions are removed, which means the process is completely reversible.

By introducing the assumption of *isotropy*, i.e. in a given point the elastic constants are independent of the choice of reference axes, the constitutive law can be written as:

$$\sigma_{ij} = \lambda \varepsilon_{kk} \delta_{ij} + 2\mu \varepsilon_{ij},$$

where the two elastic constants λ and μ are known as *Lamé's constants* (1852).

The inverse relationship is commonly presented in the form:

$$\varepsilon_{ij} = -\frac{\upsilon}{E} \sigma_{kk} \delta_{ij} + \frac{1+\upsilon}{E} \sigma_{ij},$$

where E is the *Young's modulus* (1807) and υ is the *Poisson's ratio* (1829).

However, it is usually more appropriate to refer to alternative constants: the *shear modulus G* (describing change in shape at constant volume) and the *bulk modulus K* (describing change in volume), because, if we introduce the definition of *stress deviator*:

$$s_{ij} = \sigma_{ij} - p\delta_{ij},$$

and *strain deviator*:

$$e_{ij} = \varepsilon_{ij} - \frac{1}{3}\varepsilon_{v}\delta_{ij},$$

then, the constitutive law assumes the particularly significant form:

$$p = K\varepsilon_{v}$$
$$s_{ij} = 2Ge_{ij}.$$

This form proves the *uncoupled response* of an isotropic material: the increments of shear strains only depend on increments of the corresponding deviator stress and the volume changes only depend on increments of the corresponding spherical component of the stress tensor.

(2) Experimental data support the assumption that many materials exhibit a reversible behaviour at very small strains and that irrecoverable, permanent deformations occur beyond a threshold strain. This kind of behaviour, described by means of plasticity theory, applies particularly to soils, where the irreversible nature of deformations is mainly the result of relative movements between particles.

A fundamental assumption of plasticity is that the total strain can be additively decomposed into the elastic and plastic component, i.e.:

$$d\varepsilon_{ij} = d\varepsilon_{ij}^{e} + d\varepsilon_{ij}^{p}.$$

The elastic components of strain are computed according to the elastic constitutive relations. To identify the plastic deformations it is required to introduce a *yield criterion*, a *plastic flow rule* and a *hardening rule*, which allow one to specify *when plastic deformations occur, their direction and their magnitude*.

(3) Soils exhibit a time-dependent response. The potential for this phenomenon depends on compositional factors and, in general, the greater the organic content and the higher the plasticity, the more pronounced is the viscous behaviour.

Different rheological models have been suggested to analyse the time-dependent behaviour of soils. In general they are a combination of the basic Maxwell and Kelvin models.

3.7 Further reading

Further reading on elasticity and plasticity theory for upper-level undergraduated students are:

R.O. Davis and A.P.S. Selvadurai (1996) *Elasticity and Geomechanics*. Cambridge University Press.

R.O. Davis and A.P.S. Selvadurai (2002) *Plasticity and Geomechanics*. Cambridge University Press.

More specialized references for reaserch students are:

A.E.H. Love (1927) *A Treatise on the Mathematical Theory of Elasticity*. Cambridge University Press (Dover Edition, 1944).

A.E. Green and W. Zerna (1954) *Theoretical Elasticity*. Oxford University Press.

R. Hill (1950) *The Mathematical Theory of Plasticity*. Oxford University Press.

D.R. Bland (1960) *The Theory of Linear Viscoelasticity*. Pergamon Press.

Major developments in the theory of plasticity as applied to geotechnical analyses and design are presented in a comprehensive manner in the book by H.S. Yu (2006) *Plasticity and Geotechnics*, Springer.

The last decades have seen a great development in the theory of constitutive equations, both in the formulation of general principles restricting the forms that constitutive equations can take, as well as in the formulation of specific models. The book by I-Shih Liu (2002) *Continuum Mechanics*, Springer, is a reference related to the first aspect. The book by D. Muir Wood (2004) *Geotechnical Modelling*, Spon Press, discusses the most recent developments in soil mechanics.

Chapter 4

The porous medium

Soils are multi-phase media and therefore they are idealized as *superimposed continua*. In particular, a recurrent example in this book is a fully saturated soil: the behaviour of this body will be analysed as a *continuum medium*, representing the solid phase, superimposed to another continuum medium representing the fluid phase. This assumption allows us to apply to soils stress and strain tensor concepts, as well as the continuum mechanics laws we discussed in previous chapters.

However, these two phases exhibit a different behaviour, as it can be realized by observing that the fluid phase cannot withstand shear stress, has a compressibility that significantly differs from that of the solid phase and can flow through the interconnected voids; on the contrary, the solid phase can resist shear and provides the strength and the stiffness of the porous medium. Therefore, the analysis of problems involving volume change, distortion and strength then requires to realize that it is the change of effective stresses that affects the response of the soil. The dependence of soil behaviour on effective stresses is a basic postulate of soil mechanics and may be considered as a fundamental step in the development of a continuum theory of soils.

Moreover, because of its discrete (particulate) nature and depositional conditions, soil behaviour is inelastic and anisotropic, so that the response of a soil element to a given increment of stress depends not only on the increment in itself but also on the overall *loading history* and on the *loading path*.

To give a convenient representation of loading paths, a methodological approach, known in soil mechanics as the '*stress path method*', is extensively used when analysing the behaviour of geotechnical structures and when programming or interpreting experimental tests.

The aim of this chapter is to introduce the principle of effective stress, to extensively discuss the geological history of a soil deposit, to show how this history can be quantified through the yield stress and the overconsolidation ratio, and to illustrate how one-dimensional compression processes can be simulated by means of a one-dimensional compression test (oedometer test).

4.1 The principle of effective stress

A convenient way of introducing the *principle of effective stress* may be to refer to the description given by Terzaghi at the First International Conference of Soil

Mechanics and Foundation Engineering (1936) (actually Terzaghi published in 1923 the paper: Die Berechnung der Durchlassigkeitsziffer des Tones aus dem Verlauf der hydrodynamischen Spannungserscheinungen, Sitzber. Akad. Wiss. Wien, Abt. IIa, Vol. 132, in which he already introduced the decomposition of the total stress into the pore pressure and the effective stress):

> The stresses acting in any point of a section through a mass of soil can be computed from the total principal stresses $\sigma_I, \sigma_{II}, \sigma_{III}$, which act in this point. If the voids of the soil are filled with water under a pressure u, the total principal stresses consist of two parts. One part, u, acts in the water and in the solid in every direction with equal intensity. It is called neutral stress (or pore pressure). The balance $\sigma_I' = \sigma_I - u, \sigma_{II}' = \sigma_{II} - u, \sigma_{III}' = \sigma_{III} - u$, represents an excess over the neutral stress u, and it has its seat exclusively in the solid phase of the soil. This fraction of the total principal stresses will be called effective principal stresses.
>
> [...]
>
> All the measurable effects of a change of stress, such as compression, distortion and a change of shearing resistance, are exclusively due to changes in the effective stresses $\sigma_I', \sigma_{II}', \sigma_{III}'$. Hence every investigation of the stability of a saturated body of soil requires the knowledge of both the total and neutral stresses.

This sentence can be conveniently subdivided into two parts. The first part is a definition of the effective stress, that we write in the more general form:

$$\sigma_{ij}' = \sigma_{ij} - u\delta_{ij} .$$

(4.1)

The total stress, denoted as σ_{ij}, is the stress that has a global meaning, in the sense that it satisfies the requirements for equilibrium, and is carried partly by the solid phase and partly by the fluid phase.

Since in the context of soil mechanics the ability of water to carry shear stress is negligible, all shearing stress components are only supported by the solid phase (as highlighted by equation 4.1), while the normal components of stress are decomposed into the water pressure and the remaining normal stress components, supported by the solid phase, called effective stresses.

The effective stresses are traditionally referred by using the stress symbol with a prime, i.e. σ_{ij}', and the pore pressure is denoted by the scalar u.

The second part of the sentence is the basic postulate of soil mechanics: *the behaviour of soil is related only to the effective stresses*. Since this concept is the first basic step for the development of a continuum theory of soils, it can be easily imagined that, due to its relevance, various attempts have been made to prove the validity of the above postulate, referred in geotechnical literature as the **principle of effective stress** (see Mitchell, 1993). But there is no need to open the door to this debate, since within the context of this book it is enough to observe that, when investigating the behaviour of saturated soils in terms of causes and effects, this principle has never been falsified, provided that the compressibility of soil particles is negligible in comparison with that of the soil skeleton.

4.2 Geostatic stresses

Stresses within soils are due to external loads applied to the soil and to the weight of the material itself. Stresses due to own weight are indicated as **geostatic stresses** and are the object of this section.

The pattern of stresses can be quite complicated, but a common situation in which the weight of the soil originates a simple pattern of stresses is when the soil surface (without surface loading) is horizontal and there is no variation of the nature of the soil in the horizontal direction.

Suppose that the soil deposit in Figure 4.1 satisfies the above assumptions, so that any vertical section can be thought a section of symmetry, and therefore on the vertical and the horizontal planes the shear stresses must vanish. This implies that the non-vanishing independent component of the stress tensor are:

$$\sigma_{11} = \sigma_{vo}$$
$$\sigma_{22} = \sigma_{33} = \sigma_{ho},$$

where σ_{vo} and σ_{ho} are respectively called the total vertical (or overburden) stress and the total horizontal stress.

Throughout this book, the standard convention is to consider compressive stresses as positive, as the tradition in soil mechanics, which is just opposite to that used in continuum mechanics. Accordingly, if we assume the vertical axis positive downward, the indefinite equilibrium equation in the vertical direction writes:

$$\frac{\partial \sigma_{vo}}{\partial z} - \gamma = 0, \tag{4.2}$$

and its integration, with the boundary condition of traction free surface, gives the total vertical stress at any depth:

$$\sigma_{vo} = \gamma z. \tag{4.3}$$

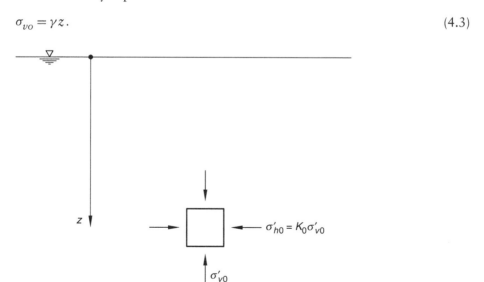

Figure 4.1 Geostatic stresses.

Assuming the groundwater table at surface and hydrostatic conditions, the pore water pressure is given by:

$$u_o = \gamma_w z \, .$$
(4.4)

so that the effective vertical stress (effective overburden stress) can be computed as:

$$\sigma'_{vo} = \sigma_{vo} - u_o \, .$$
(4.5)

Despite the ease with which we compute the vertical overburden stress, the assessment of the horizontal stress is a complex task, because its actual value depends on the stress history of the deposits.

The reason is immediately apparent if we consider the indefinite equilibrium equation in the horizontal direction:

$$\frac{\partial \sigma_{ho}}{\partial x} = 0 \, .$$
(4.6)

The horizontal stress cannot be determinated univocally from this equation, which only tells us that we can give the horizontal stress an expression, such as:

$$\sigma_{ho} = F(z) + u_o \, .$$

The function $F(z)$ remains arbitrary and does not need to be continuous, since we can predict discontinuities in presence of stratigrafical discontinuities of the soil deposit (see the worked Example 4.1).

As a matter of practical convenience we can assign to the function $F(z)$ a structure of this kind: $F(z) = K_o \sigma'_{vo}$, so that the horizontal effective stress writes:

$$\sigma'_{ho} = K_o \sigma'_{vo} \, ,$$
(4.7)

where the unknown coefficient K_o, which may depend on the z coordinate, is known as **coefficient of earth pressure at rest.**

In Section 4.4, it will be shown that this coefficient is linked to the history of the deposit, so that in order to determine its actual value we need direct measurements in the field (see Chapter 7) or empirical correlations, that, in some way, take into account this history.

Typically, K_o ranges from 0.5 to 2, with usual values of 0.45 to 0.55 for coarse-grained soils and values of 0.55 to 0.70 for soft fine-grained soils.

The reader should be aware that the K_o coefficient is a ratio between effective stresses, so that when defining the initial stress state the stress components should be computed in the following order, to avoid any mistake:

$$\sigma_{vo} = \gamma z$$
$$\sigma'_{vo} = \sigma_{vo} - u_o$$
$$\sigma'_{ho} = K_o \sigma'_{vo}$$
$$\sigma_{ho} = \sigma'_{ho} + u_o$$

Example 4.1

Given the soil profile in Figure 4.2, it is required to compute the pattern of geostatic stresses with depth.

Solution. The water table is located at the interface between the first and the second layer, so that within the first layer (in which we can neglect any capillary rise, see Section 4.3), total and effective stresses are coincident and at point A:

$$\sigma_{vo} = \sigma'_{vo} = \gamma_d z = 1.5 \cdot 9.81 \cdot 3 = 44.15 \text{ kN/m}^2 \,.$$

At point B we need to account for the hydrostatic pore pressure regime, so that the total and the effective overburden stress will be:

$$\sigma_{vo} = \sum \gamma_i \cdot \Delta z_i = 14.72 \cdot 3 + 18.64 \cdot 5 = 137.36 \text{ kN/m}^2$$
$$u_o = \gamma_w \cdot z = 1 \cdot 9.81 \cdot 5 = 49.05 \text{ kN/m}^2$$
$$\sigma'_{vo} = \sigma_{vo} - u_o = 137.36 - 49.05 = 88.31 \text{ kN/m}^2$$

The same procedure, when considering the depth corresponding to point C, gives:

$$\sigma_{vo} = 290.39 \quad u_o = 127.53 \quad \sigma'_{vo} = 162.86 \quad (\text{kN/m}^2)\,.$$

Finally, by using the relationship (4.7) and the values of K_o coefficient reported in Figure 4.2a, we obtain the horizontal stress value, as indicated in Figure 4.2b.

Remarks. (1) Across the interface, there is no jump in the value of the vertical stress, although its derivative can be discontinuous if there is change of the bulk density across the discontinuity. On the contrary the horizontal stress exhibits discontinuity across the interface, if the value of the earth pressure coefficient above the interface is different from that below the interface.
(2) The term pore pressure indicates the pressure measured using a gauge pressure, i.e. the pressure above the atmospheric pressure.

Example 4.2

With reference to the stratigraphical sequence depicted in Figure 4.2, it is required to prove that the rise of water table to the ground surface will reduce the effective stress, while an increase of effective stress is produced when lowering the water table.

The reader should also prove that a further rise of the water table above the ground surface produces an increase of both the total and the pore pressure, but the effective stress remains unchanged.

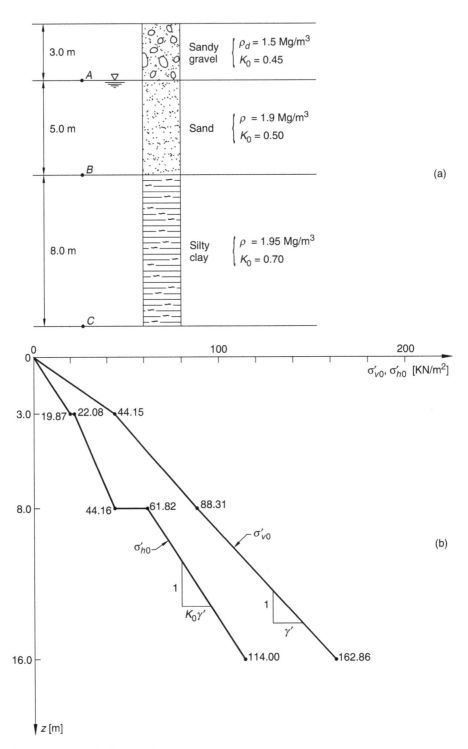

Figure 4.2 Example of computing geostatic stresses.

4.3 Capillarity

We learned from Physics courses that when a gas-liquid interface meets a solid surface, the interface curves up or down near the solid surface. If the molecules of the liquid are attracted to each other less strongly than to the solid surface (an example is water in a glass container), we say that the liquid *wets* or adheres to the solid surface. In this case, the gas-liquid interface curves up and the angle at which it meets the surface (called **contact angle**) is less than 90°. On the reverse, if the attraction between liquid molecules is stronger than the adhesion to the surface (as with mercury and glass), the interface curves down and we call the liquid a *non-wetting liquid*. In both cases the curved liquid surface is called *meniscus*.

When the contact angle is less than 90°, a surface tension force acts upward along the line of contact with the solid surface and the liquid in a narrow tube rises until it reaches an equilibrium height at which the tension force balances the weight of the liquid in the tube.

The ability of a fluid to rise in a narrow tube against gravity is called **capillarity**.

Surface tension T is a property of water and depends on temperature (it decreases if the temperature rises): at temperature of 10°C, T is about $7 \cdot 10^{-5}$ kN/m.

Related to the surface tension is the *negative pore pressure*. We know that pore pressure is usually compressive, but we will see in the sequel that water can sustain negative pore pressure.

If we consider the water rise in a narrow capillary tube (Figure 4.3), the upward force is the vertical component of the surface tension and the equilibrium requires that this force must balance the weight of the water in the tube:

$$\frac{\pi}{4}d^2\gamma_w h_c = \pi d \cdot T \cos\alpha . \tag{4.8}$$

where α is the contact angle.

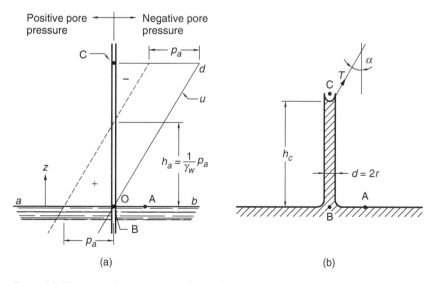

Figure 4.3 The rise of water in a capillary tube.

Table 4.1 Approximate capillarity height

Soil	h(m)
Gravel	0.05–0.30
Coarse sand	0.03–0.80
Medium sand	0.12–2.40
Fine sand	0.30–3.50
Silt	1.50–12.00
Clay	>10.00

Solving for capillary height in the case of pure water and clean glass tube ($\alpha \cong 0; \cos\alpha \cong 1$), we obtain a formula which allows us to give an estimation of the capillary rise, depending on the diameter of the capillary tube:

$$h_c = \frac{4T}{\gamma_w d} = \frac{0.03}{d\,(\text{mm})} \,. \tag{4.9}$$

Now we observe that, since the water is in equilibrium, the pressure in the water is the same in every point of a horizontal section. In particular, at the level of the free water surface the pressure is zero, whereas the pressure u, acting at an elevation z above this level, can be obtained from the equilibrium of the column in the tube (Figure 4.3):

$$\begin{aligned} u + \gamma_w z &= 0 \\ u &= -\gamma_w z \end{aligned} \,. \tag{4.10}$$

The obtained result proves that water, in a capillary rise, sustains negative pore pressure. Soils are random aggregates of particles and the resulting void distribution hardly bears resemblance to a capillary tube. However, this analogy still explains capillary phenomena observed in soils, and the approximate values of the height of capillary rise, reported in Table 4.1, are also of practical utility.

One consequence of the values quoted in Table 4.1 is that coarse-grained soils above the phreatic surface (on which the pore pressure is zero) tend to be unsaturated and little water is contained in the pores. On the contrary, fine-grained soils may remain saturated for many metres above the phreatic surface, with negative pore pressure.

Note that the negative pore pressure increases the effective stresses, so that moist sands can exhibit an unconfined compressive strength. This strength accounts for the possibility to build sand castles and also makes apparently stable unsupported sides of excavations in silts and fine sands. But flooding of the excavation or rain and slight vibrations will suddenly prove the unstable nature of these materials.

This strength, arising from negative pore pressure, cannot be considered a soil property.

4.4 A mechanistic picture of geological processes: yield stress and overconsolidation ratio

There are some terms we reserve in geotechnical engineering for a specific meaning, so that it is convenient to introduce here a brief glossary.

The **structure of a soil** is the combination of both the particle arrangements and the interparticle forces (Mitchell, 1993), whereas the term **fabric** is used to indicate the particle association and arrangements.

A soil sample can be considered an **undisturbed soil sample** if it has been so carefully sampled to preserve both the structure and the water content of the soil in the field. However, the process of sampling (see Chapter 7), trimming and mounting the soil sample in the test equipment can have some influence on the structure of the soil. Those changes associated to the sampling of soil are called **disturbance.**

A **reconstituted** clay is a clay paste remoulded at a water content of between w_L and $1.5w_L$ and successively compressed (usually or preferably one-dimensionally compressed).

Since the structure of the reconstituted sample is certainly different from the one in its natural state, being independent from all depositional and post-depositional events occurring in a natural soil deposit, properties associated to reconstituted soils can be seen as basic or *inherent properties* (as suggested by Burland (1990) in his Rankine Lecture). For this reason, it will be recurrent the following definition: **intrinsic** properties are those basic or inherent properties of a reconstituted clay, which are independent of its natural state.

The term **compression** is commonly used when describing the change in void ratio due to the change in effective stresses, without any reference to the time interval over which this compression occurs. The time dependent phenomenon which describes the evolution of pore pressure and the deformation of the porous medium is defined as **consolidation**, and will be analysed in detail in Chapter 6.

Having these definitions in mind, we can now start the description of the **geological history** of a soil deposit; that means the history in precise sequence (hopefully) of all events experienced by a soil deposit since its formation to its actual configuration.

And in order to appreciate the importance of the stress history and its consequence on the mechanical behaviour of soils, we will try to give a mechanistic picture of main events.

During **deposition**, a soil element compresses by gravitational compaction and, by assuming that this process can be thought of as one-dimensional processes, a convenient way to describe this phase is to make use of two state variables: the void ratio and the effective vertical stress.

With reference to Figure 4.4, a clay element, recently deposited, is represented by the point A, characterized by a high void ratio (at depth of 10 cm from the surface, the void ratio can attain values of 6 or higher). Any further increases of effective stresses, due to additional deposition of sediments, will reduce the void ratio, and the representative point will move from point A to B and, successively, to C, along the curve called **sedimentation compression curve** by gravitational compaction .

Experimental evidences (Skempton, 1970; Burland, 1990) show that at a given σ'_{vo} the corresponding void ratio depends on the nature and the amount of clay minerals as reflected by the liquid limit, so that the higher the liquid limit the higher will be the void ratio. Additional important factors affecting the fabric of the sediment are the rate of deposition (see Table 4.2) and the stillness of water, since a slow deposition in fresh water gives higher value of void ratio than a rapid deposition in the presence of currents.

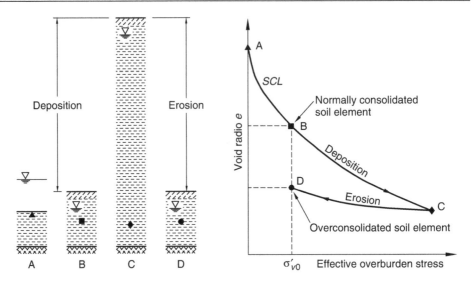

Figure 4.4 Stress history of a soil element (adapted from Skempton, 1970).

Table 4.2 Rate of sedimentation of clay deposits

Deposition environment	Rate of sedimentation, expressed as mm/year
Deltaic	from 8 to 17
Deltaic (Mississippi)	120
Estuarine	from 2.0 to 2.5
Marine (shallow water)	from 0.8 to 1.2
Marine (deep sea)	0.03

Source: Adapted from Skempton, 1970.

A soil element is defined **normally-consolidated** (NC) if it has never been compressed by a vertical effective stress greater than the existing effective overburden stress.

As already seen in Section 4.2, the effective horizontal stress can be expressed as:

$$\sigma'_{ho} = K_o(NC) \cdot \sigma'_{vo} , \qquad (4.11)$$

and when dealing with a NC soil deposit the earth pressure coefficient $K_o(NC)$ can be estimated through the expression suggested by Jaky (1944):

$$K_o(NC) = \left(1 + \frac{2\sin\varphi'}{3}\right) \frac{1-\sin\varphi'}{1+\sin\varphi'} \cong 1-\sin\varphi' , \qquad (4.12)$$

where φ' is the angle of shearing resistance.

However, most natural soil deposits, after deposition and gravitational compaction, have experienced processes of erosion. The removal of sediments by erosion corresponds to a reduction of effective stresses, so that the actual state of soil elements moves from point C to D in Figure 4.4.

We can now observe that two different values of void ratio may correspond to the same vertical effective stress, depending on the stress history. And it is also of relevance to observe that unloading from C to D reveals that soil deformations are essentially irreversibile (or plastic).

If we consider the compression curve within the context of the plasticity theory, we can attribute to the vertical effective stress corresponding to point C the meaning of a **yield stress**, and soil behaviour along the path CD can be expected to be reversible. Accordingly, deformations will be elastic until the applied vertical effective stress will not exceed point D, wheareas for greater loads plastic deformations will play the major role.

This conceptual model allows us to highlight practical consequences of the stress history. As an example, if we are dealing with the design of a shallow footing, depending on the loading path we impose to a soil element in relation to its previous history, we can expect tolerable settlement if the amount of load does not exceed the yield stress or, on the reverse, quite large settlements if the loading path develops along the virgin compression curve. For this reason it is of relevance to distinguish between normally consolidated and overconsolidated soils. For these latter, the following definition applies: a soil element is defined **overconsolidated** if, at present time, it is subjected to an effective stress lower than the value which acted in its past history.

The maximum vertical effective stress, which acted over its geological and the most recent stress history (see point C in Figure 4.4), is defined **preconsolidation stress**, if its magnitude can be inferred from the geological history, or alternatively and with greater generality, is defined **yield stress**. In the following the yield (or preconsolidation) stress will be denoted by the symbol σ'_p.

Further, in order to quantify the entity of the overconsolidation process, reference is made to the **OverConsolidation Ratio** (OCR):

$$OCR = \frac{\sigma'_p}{\sigma'_{vo}} .\tag{4.13}$$

Note that the actual vertical effective stress is computed as already shown in the Example 4.1; in natural soil deposits the preconsolidation stress is assessed experimentally by using the procedure described in Section 4.6.

When the soil is being unloaded, that is when it is moving from a normally consolidated state to an overconsolidated one, the coefficient of earth pressure at rest can be evaluated through the following empirical relation (Schmidt, 1966; Alpan, 1967):

$$K_o(OC) = K_o(NC) \cdot OCR^\alpha ,\tag{4.14}$$

where the exponent α is of the order of 0.42, for low plasticity clays, and 0.32, for high plasticity clays (Ladd *et al.*, 1977).

It should be noticed that during unloading the $K_o(OC)$ coefficient cannot increase indefinitely, the upper limit being established by the passive failure condition (see Section 8.3).

Example 4.3
Figure 4.5 shows a cycle of events producing overconsolidation in a soil deposit:

(1) During deposition under water, a soil element compresses along the compression curve 'a' in Figure 4.5b.
(2) After deposition, suppose that the water table will be drawdown to a depth z_1, so that the soil element will experience further compression along the curve 'b', due to an increase of the vertical effective stress, which now attains its maximum value σ'_p.
(3) Finally the water table rises to z_0 and the soil element is unloaded along the curve 'c'.

It is required to define the pattern with depth of the overconsolidation ratio (OCR).

Solution. The value of the maximum vertical effective stress (the preconsolidation stress), at any depth below z_1, is given by:

$$\sigma'_p = \gamma \cdot z - \gamma_w \cdot (z - z_1) \quad \text{if} \quad z > z_1$$

and the actual value of the effective vertical stress

$$\sigma'_{vo} = \gamma \cdot z - \gamma_w \cdot (z - z_0) \quad \text{if} \quad z > z_0 .$$

It follows that the required pattern of OCR with z will be given by the ratio:

$$OCR = \frac{\gamma'z + \gamma_w z_1}{\gamma'z + \gamma_w z_o} .$$

Note that the above relationship applies to any depth below the maximum depth of the water table z_1, because above this depth desiccation effects produce scattered and unpredictable values of OCR (broken curve in Figure 4.5d).

Also note that, below the depth z_1, the overconsolidation produced by groundwater changes will be characterized by a constant difference $\sigma'_p - \sigma'_{vo} = \gamma_w \cdot (z_1 - z_o)$ with depth (Figure 4.5c).

4.5 Modelling one-dimensional compression: the oedometer test

A one-dimensional compression process is simulated in the laboratory by compressing a soil specimen in a special testing apparatus called **oedometer**. Figure 4.6 shows the main components of this apparatus: an undisturbed soil specimen is carefully trimmed and placed into a rigid and (supposed) frictionless ring, that does not allow any radial displacement or radial flow of water during compression.

Porous stones at the top and bottom of the sample allow drainage during consolidation. Dry stones should be used in both soft and stiff clays and the sample should be flooded only after a vertical effective stress equal to σ'_{vo} has been reached. It is necessary

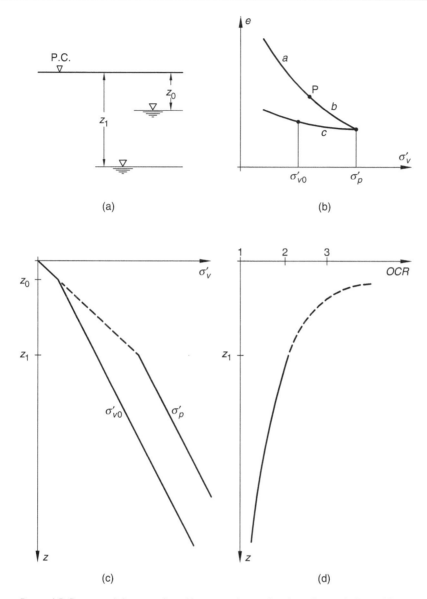

Figure 4.5 Preconsolidation induced by groundwater level oscillation (adapted from Parry, 1970).

to flood the soil sample in order to prevent dessiccation during the test or to allow the sample to absorb water during swelling.

The dimensions of the soil sample depend on different requirements: in general the sample should be representative of the soil structural features, a factor which requires large dimensions. Since this prevents the possibility of obtaining good quality specimens from conventional tube samplers, the usual diameters are of the order of 50 mm.

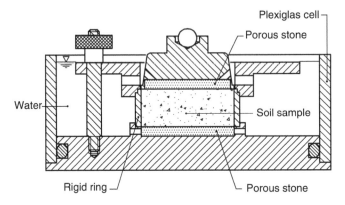

Figure 4.6 Oedometer testing apparatus.

To reduce the influence of friction along the lateral surface, the ratio of the diameter to the height should be greater than 2.5, but less than 6 to avoid disturbance during trimming. Often, a floating ring apparatus is used for this reason (as suggested by Lambe, 1951), rather than the more common fixed ring.

Experimental evidences show that a load increment from 20 to 40 kPa produces almost the same compression as produced by a load increment from 40 to 80 kPa, i.e. there is a progressive increase of stiffness, due to the dominating effect of increasing the mean stress level.

This observation justifies the conventional sequence of vertical loads, each of which is usually double the previous value (i.e. $\Delta N/N = 1$, but this ratio should be reduced to 0.5 to obtain a better defined curve in the proximity of the yield stress, mainly when testing sensitive clays or soft clays with high liquidity index).

It also justifies the conventional use of a semilog plane (void ratio versus logarithm of the vertical effective stress) when presenting the experimental data.

When we investigate the one-dimensional behaviour of fine-grained soils, we have to realize that fine-grained soils do not react instantaneously to the applied load. In fact, the decrease of void ratio of saturated soils requires that water must flow from the pores, and, since the rate of flow depends on soil permeability and drainage path (this will be discussed in detail in Chapter 6), in low permeability soils neither void ratio nor effective stress can change instantaneously. For this reason, after each load the sample is allowed to consolidate until the excess pore pressure is equal to zero, the load being usually maintained for 24 hours and the compression readings observed at convenient intervals, as shown in Figure 4.7.

Because this procedure requires a lot of time and offers only discontinuous compression data, new apparatuses have been developed, such as the controlled gradient test (Lowe *et al.*, 1969; Evangelista, 1973) and the constant rate of strain test (Wissa *et al.*, 1971).

When dealing with the conventional incremental loading test, since each increment is maintained for one day, but the end of consolidation occurs in less than one hour, the final reading D in Figure 4.7 also includes secondary compression due to creep. Point B in the same figure corresponds to the end of consolidation and it is usually obtained by the intersection of the two linearized portions of the curve.

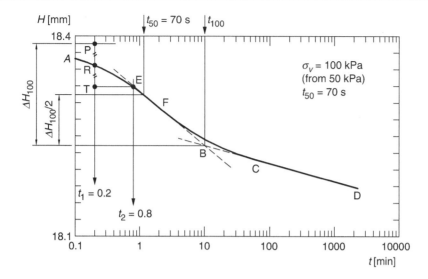

Figure 4.7 Compression readings versus time.

It will be here shown that only this component must be considered in analysing the soil response, rather than the final compression obtained at 24 hours, and will be referred to as *primary compression*.

Once the end of primary compression has been defined for a given loading step, the corresponding value of void ratio e is plotted versus vertical effective stress, as shown in Figure 4.8a, but we said it is more common to present data by plotting the void ratio versus the logarithm of effective vertical stress (see Figure 4.8b).

The void ratio is accurately computed from the following data: the weight W_s of the solid matter, obtained by drying and weighting the sample at the end of the test; the cross-section A of the container; the thickness of the sample, as given by dial readings; the unit weight of solids γ_S and by using the following relations:

$$V = A \cdot H \quad ; \quad V_S = \frac{W_S}{\gamma_S} \quad ; \quad e = \frac{V}{V_S} - 1 .$$

Alternative possibilities of presenting data make use of variables such as (v, σ'_v) or $(\varepsilon_z, \sigma'_v)$, being $v = (1 + e)$ the specific volume and $\varepsilon_z = \delta H/H_o$ the vertical strain.

Once the compression test has been performed, we are faced with the interpretation of experimental data. Note that, when deriving from experiments any sort of parameters characterizing the material response, we assume, deliberately or tacitly, a constitutive model. Let us give an interpretation within the framework of an elasto-plastic model.

1 The break point B in Figure 4.8a or 4.8b separates the small strain elastic behaviour from large irreversibile strains. The corresponding vertical effective stress σ'_p is therefore a yield stress in terms of mechanical behaviour.

2 Accordingly, the first portion of the compression curve, from point A to B, is defined **recompression curve,** and soil behaviour is assumed to be non-linear but

σ'_v [kPa]	$\varepsilon_z = \Delta H / H_0$ [%]	e [-]
25	0.92	1.477
50	1.91	1.452
100	4.39	1.390
200	12.98	1.176
400	20.39	0.990
800	26.99	0.825
1600	32.88	0.678
400	31.50	0.713
100	29.50	0.763
25	27.50	0.813
100	29.00	0.775
400	30.50	0.738
1600	34.00	0.650
3200	38.95	0.526
6400	43.33	0.417

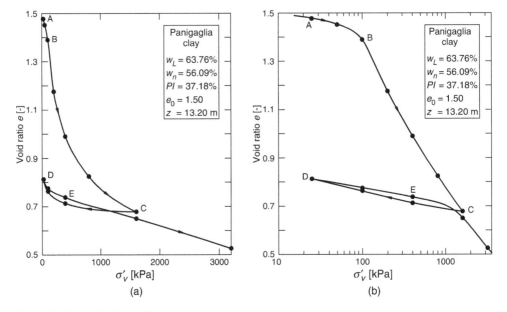

Figure 4.8 Example of one-dimensional compression curves.

almost completely reversible along this path. The second portion, from B to C, is called **compression curve**, and is characterized by largely irreversibile strains, as can be proved by unloading the specimen from C to D.

3   The soil has *memory* of its previous loading history, as it is revealed from the fact that, when the specimen is further reloaded starting from D, its behaviour appears to be reversible until the current yield stress σ'_{vC} has been attained. The yield stress evolves with the history of plastic strains and this evolution is called *hardening*.

4.6 Experimental determination of yield stress

There are many procedures suggested for determining the yield stress (Casagrande, 1936; Burmister, 1951; Schmertmann, 1955; Janbu, 1969; Butterfield, 1979).

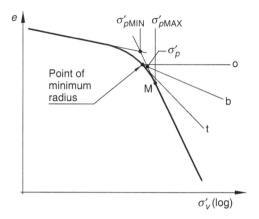

Figure 4.9 Casagrande's procedure of determining the yield stress.

The most common is that suggested by Casagrande (1936), that involves the following steps (Figure 4.9):

(a) consider the point of maximum curvature on the $(e, \log \sigma'_v)$ plot;
(b) from this point draw the tangent line t, the horizontal line o and the bisector line b;
(c) extend the compression line until this latter intersects the bisector;
(d) the point of intersection is assumed to be the yield stress σ'_p.

By considering the subjectivity in the choice of the maximum curvature point, it is also suggested to compare the obtained value with its bounds: the lower bound value is given by the intersection of the recompression line with the compression line and the upper bound value at M, the starting point of the compression straight line portion.

As already outlined in the previous section, the trend of yield stress and overburden effective stress with depth represents a concise picture of the stress history of the deposit (Ladd, 1971; Jamiolkowski *et al.*, 1985), and inspection of this trend can also help to understand the basic mechanisms that may have affected the soil deposit during its geological history:

(a) for example, a *mechanical overconsolidation* is characterized by a constant difference $\left(\sigma'_p - \sigma'_{vo} \right)$ with depth;
(b) if the preconsolidation is ascribed to ageing, a constant ratio σ'_p / σ'_{vo} with depth should be expected;
(c) desiccation due to drying and freeze-thaw cycles produce scattered values of σ'_p;
(d) physio-chemical phenomena can produce an increase of yield stress, with a quite variable profile.

Different mechanisms can obviously act together and quite often the stress history profiles are not so simple. But despite the fact that in principle the preconsolidation pressure could have a different meaning when related to its cause, the engineer can refer to its mechanical implications, and in this optic the σ'_p is regarded as a yield stress, independently of its nature.

To highlight these points, we now consider two examples. The first one is related to Panigaglia soft clay and a representative compression curve is that shown in Figure 4.8. From this curve we can infer a yield stress $\sigma'_p = 105\,\mathrm{kPa}$ and, the sample being from a depth of 13.20 m, the effective overburden stress is equal to $\sigma'_{vo} = 82\,\mathrm{kPa}$. Then:

$$OCR = \frac{\sigma'_p}{\sigma'_{vo}} = \frac{105}{82} = 1.28 \, .$$

Additional values of the yield stress are reported in Figure 4.10 and from the trend with depth it can be concluded that Panigaglia clay is a soft clay, only slightly overconsolidated ($OCR = 1.1 \pm 0.1$), due to ageing.

A more complex example is the one related to the stress history of Pisa clay. In the subsoil profile of the Piazza dei Miracoli (see Figure 4.11), the 'horizon C', found between depths of 40 m and at least 120 m, is composed of marine sands deposited during the Flandrian transgression. Above these dense sands is the 'horizon B', a 30 m thick marine clay deposit, formed at a time of rapid eustatic rise. Between the so-called 'lower clay' and the 'upper clay', known locally as 'Pancone Clay', there is a 2 m thick layer of intermediate sand and a 4 m thick layer of stiff 'intermediate clay'. The rate of eustatic rise decreased during the last 10,000 years, so that the upper sediments (the 'horizon A' in Figure 4.11) became increasingly estuarine in character. These sediments are in fact characterized by pronounced spatial variability, inclined layering and they differ over short horizontal distances. It will be further discussed in Chapter 9 that, by considering Cone Prenetration Test profiles (see Figure 9.14), the material to the south of the Pisa Tower appears to be finer-grained than that to the north, and the sand layer

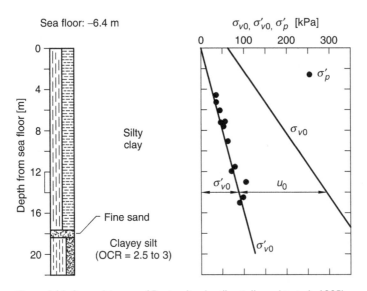

Figure 4.10 Stress history of Panigaglia clay (Jamiolkowski et al., 1985).

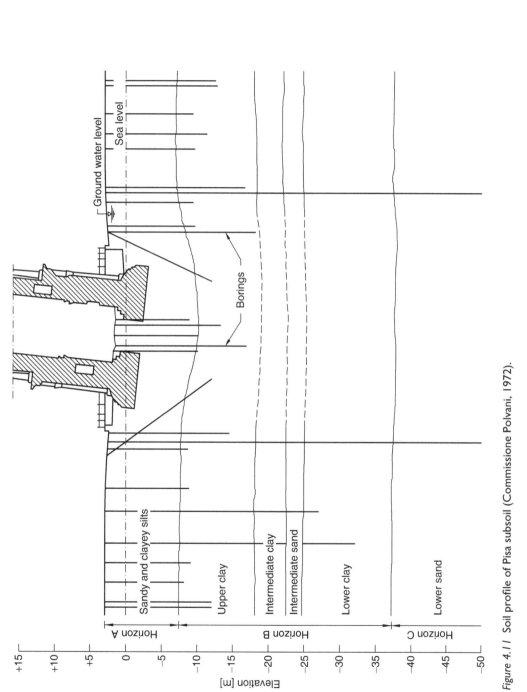

Figure 4.11 Soil profile of Pisa subsoil (Commissione Polvani, 1972).

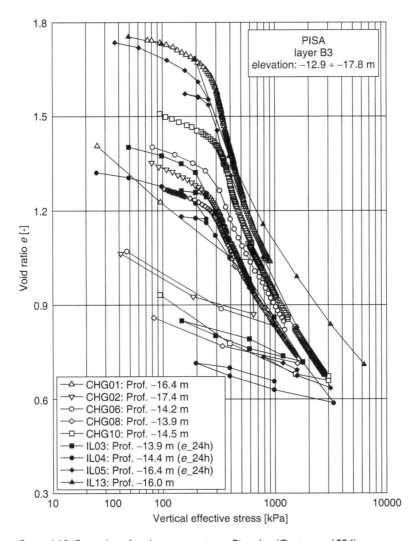

Figure 4.12 Examples of oedometer tests on Pisa clay (Costanzo, 1994).

is locally thinner. This soil variability was probably responsible for the initial tilting of the Pisa Tower.

Figure 4.12 gives nine examples of compression tests performed on undisturbed samples from different depths and Figure 4.13 summarizes the trend with depth of the effective overburden stress and OCR.

These profiles are more complex, if contrasted to the previous case, and cannot be linked to a single mechanism: the overconsolidation can be ascribed to aging, to groundwater oscillations as well as to erosion not exceeding 50 to 60 kPa. Moreover, OCR values in horizon A and layer B5 can be related to desiccation during groundwater fluctuations.

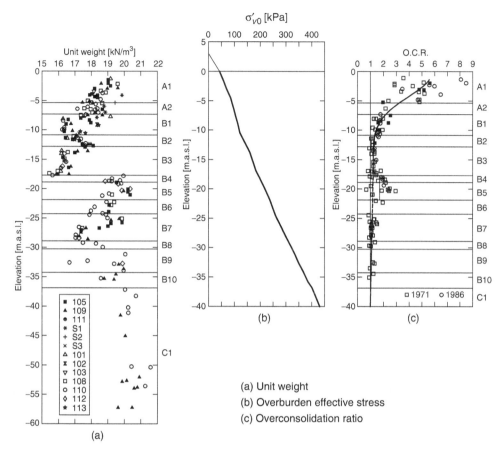

Figure 4.13 Stress history of Pisa clay. (Calabresi *et al.*, 1993).

4.7 Soil compressibility

By plotting the void ratio versus the logarithm of effective vertical stress (see Figure 4.14), it is generally argued that this plot can be represented by two straight linear portions: the first one, from A to B, is defined *recompression curve*, and soil behaviour is assumed to be non-linear but almost completely reversible along this path; the second portion, from C to D, is called *compression curve*, and is characterized by largely irreversible strains, as can be proved by unloading.

The most intuitive way of describing the stress–strain relationship is to make reference to the slope of this idealized behaviour, so that the **compression index** C_c is the slope of the compression portion, i.e.:

$$C_c = \frac{-\delta e}{\delta \log \sigma'_v},$$

(4.15)

and the same definition applies to the **recompression index** C_r or to the **swelling index** C_s, the only difference being that they refer to different portions of the curve.

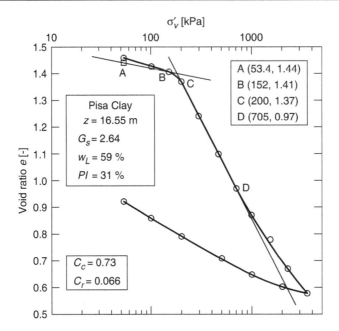

Figure 4.14 Soil compressibility.

As an alternative, by considering any point $(\varepsilon_z, \sigma_v')$ the current value of soil compressibility can be expressed by the **coefficient of volume change** m_v:

$$m_v = \frac{-\delta e}{1+e} \frac{1}{\delta \sigma_v'} \tag{4.16}$$

or, by the ratio:

$$E_{oed} = \frac{\delta \sigma_v'}{-\delta e}(1+e) \tag{4.17}$$

referred to as **constrained modulus**.

Note that (4.17) gives the tangent to the compressibility curve on the point of coordinates $(\varepsilon_z, \sigma_v')$. Interestingly we observe that by substituting (4.15) into (4.17), the constrained modulus can be expressed as:

$$E_{oed} = 2.3 \frac{1+e}{C_c} \sigma_v'. \tag{4.18}$$

which proves that this parameter is not a constant material property, but it depends on the current state, as represented by the current value of void ratio and effective stress.

The above defined compressibility parameters are of practical use when predicting settlement of structures on clay deposits (see Section 4.8 and Chapter 9), so that the reader should have familiarity with their order of magnitude.

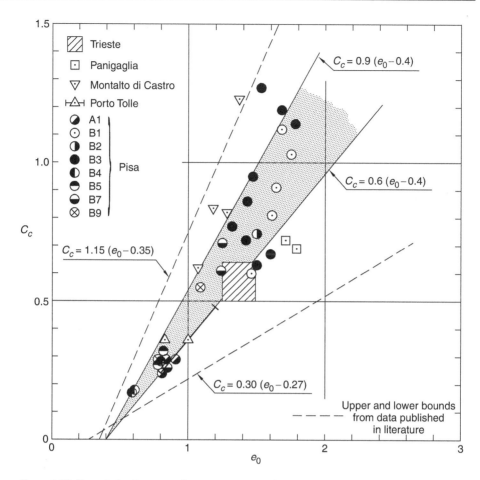

Figure 4.15 Correlation between the compression index and the void ratio.

In this respect, we have already discussed that soil composition and depositional environment affect the potential range of soil parameters, but their actual value also depends on the structure the soil exhibits as a result of post-depositional changes.

Accordingly, we can expect that the compressibility of a reconstituted soil sample may be related to its liquid limit; but if a soil element reaches an equilibrium stage represented by high values of void ratio, under a given overburden effective stress, it will also show high compressibility. Therefore, we can predict that the compressibility of a natural soil can be related to the void ratio, as shown in Figure 4.15.

Natural clays of low sensitivity exhibit values of the compression index lower than 1.0; Swedish and Canadian sensitive clays have values of C_c from 1 to 3 and, in some circumstancies (as is the case for Canadian clays), up to 5. Finally, when dealing with organic soils and peat, showing a water content from 200% to 1500%, the compression index can reach unusually higher values from 2 to 12 (Terzaghi et al., 1996).

When dealing with stiff clays, Leonards (1976) suggested that, in order to detect a representative value of the recompression index, an unloading-reloading cycle should be performed in the range of σ'_{vo} to $(\sigma'_{vo} + \Delta\sigma'_v)$, the increment being equal to the expected value in the field, because it is presumed that the slope of the recompression index depends on the starting point and on the cycle amplitude. Typical values should be in the range of 0.015 to 0.035.

Finally, sandy soils, in the stress range of practical interest $(\sigma'_v \leq 500\,\text{kPa})$, may exhibit values of the constrained modulus ranging from 15 to 50 MPa, in a loose packing, and from 40 to 150 MPa if densified.

4.8 Prediction of one-dimensional compression settlement

One-dimensional compression occurs when soils form by sedimentation from a suspension, a circumstance we have discussed when dealing with the geological history of a soil deposit. It is also of engineering relevance because it occurs when a fill of large lateral extent is superimposed on a soil stratum. Further, one-dimensional compression can be considered an acceptable approximation of the situation occurring beneath the centre line of a large foundation, as an embankment. In all these cases, the settlement occurring in the field may be computed by using the one-dimensional approximation, which involves the following steps:

1   Divide the soil stratum into a convenient number of horizontal layers.
2   Estimate the initial and final vertical effective stress in each layer.
3   Compute the change of thickness of each layer, by using the results of oedometer tests.
4   Compute the settlement by summing up the results for all layers.

As an example, consider the compression curve in Figure 4.14, performed on a Pisa clay specimen sampled at a depth of 16.55 m. By using the results of such a test, it is required to compute the one-dimensional settlement of a soil layer, 1 m thick, given an increment of vertical effective stress equal to 227 kPa.

First we observe that the quality of the tested sample is so good that the yield stress can be inferred by simple inspection from the plot in Figure 4.14 and this gives a value of 200 kPa. Being the vertical effective stress equal to 153 kPa, the overconsolidation ratio is 1.3.

The compression index can be computed by using the coordinates of points C and D, so that:

$$C_c = \frac{-\delta e}{\delta \log \sigma'_v} = \frac{0.40}{\log \frac{705}{200}} = 0.73 ,$$

whereas the coordinates of points A and B allow to compute the recompression index:

$$C_r = \frac{-\delta e}{\delta \log \sigma'_v} = \frac{0.03}{\log \frac{152}{53.4}} = 0.066 .$$

The initial conditions are defined by the values:

$$e_o = 1.41 \, ; \, \sigma'_{vo} = 153 \, kPa$$

and in the final stage the vertical effective stress will reach the value $\sigma'_{vf} = 153 + 227 = 380 \, kPa$.

Now it is relevant to observe that, by neglecting particle compressibility, the reduction of thickness of a soil element, laterally constrained, is related to the change in void ratio by:

$$\frac{\Delta H}{H_o} = \frac{-\Delta e}{1 + e_o} \, . \tag{4.19}$$

The reduction of void ratio can be computed by taking into account the soil compressibility and the stress changes, so that the reduction of thickness of the layer assumes the expression:

$$\Delta H = \frac{H_o}{1 + e_o} \left(C_r \log \frac{\sigma'_p}{\sigma'_{vo}} + C_c \log \frac{\sigma'_{vf}}{\sigma'_p} \right) , \tag{4.20}$$

and specifically we have:

$$\Delta H = \frac{1.00}{2.41} \left(0.066 \log \frac{200}{153} + 0.73 \log \frac{380}{200} \right) = 0.088 \, m \, .$$

Note that, in addition to computing the magnitude of the settlement, an estimate is often required of the rate at which the settlement occurs. When dealing with coarse-grained soils, characterized by high values of hydraulic conductivity (the ease with which water can flow), it can be assumed that the settlement occurs during the construction time. On the contrary, with fine-grained soils the time required for the final settlement to develop may be of the order of months or even years. For this reason, it is of relevance to study how this delayed settlement can affect the performance of the construction. This question will be answered in Section 6.10, where the theory of consolidation will be studied in detail.

Further aspects related to settlement prediction as well as to the reliability of the one-dimensional approach will be discussed in Chapter 9 (Section 9.4).

4.9 Secondary compression

Other phenomena, in addition to those described in Section 4.4, also take part in the geological history of a soil deposit. *Creep*, *leaching*, *exchange of cations* and *cementation* deserve special attention and are referred to as post-depositional processes.

Bjerrum and Rosenqvist (1956) and Leonards and Altschaeffl (1964) attempted to reproduce through laboratory tests the natural sedimentation process. Figure 4.16, adapted from Leonards and Altschaeffl (1964), is one of these compression tests, performed by using the oedometer cell.

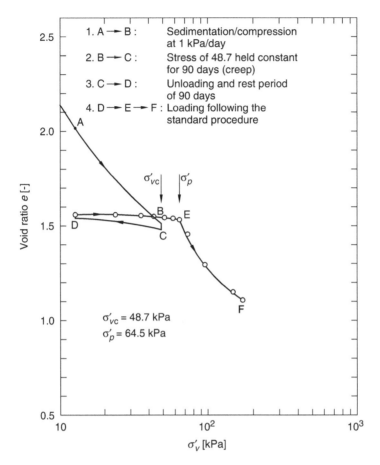

Figure 4.16 Preconsolidation induced by creep (adapted from Leonards and Altschaeffl (1964) and Burland (1990)).

During the first stage of the test (from point A to B in Figure 4.16), the load was applied so slowly (1 kPa/day) in order to simulate the consolidation taking place by gravitational compaction.

Once the effective vertical stress of $\sigma'_{vc} = 48.7$ kPa was reached, the soil specimen was left under constant applied load for a time interval as long as 90 days, during which the soil experienced further compression (as shown by the reduction of void ratio from point B to C).

This phenomenon of delayed compression, which continues under constant effective stress, is called **delayed** or **secondary compression** (or simply **creep**) (Leonards and Altschaeffl, 1964; Bjerrum, 1967).

Note that creep contributes to further modify the structure of the material, giving rise to a sort of preconsolidation. In fact, once the soil specimen was subjected to an unloading–reloading cycle, it attained a yield stress of $\sigma'_p = 64.5$ kPa, greater than the maximum stress of 48.7 kPa, experienced during the previous loading path.

This experimental evidence suggests that, by considering the age of natural deposits, we can expect that most natural clays are at least slightly overconsolidated.

The amount of preconsolidation due to creep depends on the clay plasticity, since the OCR increases if the plasticity index increases. Should this phenomenon be the only one producing the preconsolidation, the OCR would be constant with depth, and the maximum value of OCR can be expected to be of the order of 1.6, when considering post-glacial clays (see Example 4.4).

Prediction of secondary compression is usually performed by using the results of oedometer tests. By referring to Figure 4.7, it is well accepted in geotechnical literature to define as *primary compression* the volume change associated to the change of effective stresses (portion AB of the experimental curve) and as *secondary compression* the volume change at constant effective stress (portion BD). The second contribution depends on the viscous nature of the soil structure and its explanation would require investigations at a microscale. Implicit in this definition is the assumption that creep effects do not play a relevant role during the primary process of excess pore pressure dissipation. This assumption is difficult to accept mainly in the field, especially

---

Example 4.4
Try to find an estimate of the OCR due to creep.

Let us assume, on the basis of experimental evidence, a linear change of void ratio with the log of time, i.e.

$$\delta e = -C_\alpha \delta \log t ,$$

where the coefficient C_α is known as coefficient of secondary compression.

If the compression due to creep is represented in Figure 4.18 by the path from point A to C (i.e. a path at constant effective stress), and the point B represents the yield stress produced by creep, we can write:

$$e_A - e_C = (e_A - e_B) - (e_C - e_B) .$$

Compression and unloading curves plot in the $(\log \sigma', e)$ plane as straight lines, defined by their slope, respectively C_c and C_s, so that the previous relation can also be expressed in the form:

$$C_\alpha \log \frac{t_C}{t_A} = C_c \log \frac{\sigma'_{vB}}{\sigma'_{vA}} - C_s \log \frac{\sigma'_{vB}}{\sigma'_{vC}} ,$$

from which it follows:

$$OCR = \frac{\sigma'_{vB}}{\sigma'_{vA}} = \left(\frac{t_C}{t_A} \right)^{\frac{C_\alpha}{C_c - C_s}} .$$

Since the ratio C_s/C_c usually ranges from 0.1 to 0.2 and the ratio C_α/C_c is of the order of 0.04, by assuming the age of most soft clays from 4000 to 10,000 years, we reach the conclusion that OCR due to creep may reach values up to 1.6.

with relatively thick deposits, and must simply be regarded as a working engineering hypotesis. According to such an assumption, the experimental data (Figure 4.7) allows us to define a **secondary compression index**:

$$C_\alpha = -\frac{\delta e}{\delta \log t},$$
(4.21)

corresponding to the slope CD.

When using the above definition, it is implicitly accepted that the secondary compression index does not depend on the stratum thickness, nor on the stress increment, but caution must be taken in selecting its value over the relevant stress level, as the stress history is of paramount influence for this parameter (see Figure 4.17).

In addition, it must also be observed that the deformation conditions can differ in the field from the one-dimensional case analysed here and the secondary rate depends on the relevant stress ratio.

In order to have an estimate of the secondary compression index, we can note that Mesri and Choi (1985) have found that, despite the fact that both primary and secondary compression indices change over the considered stress range, their ratio remains constant (see Table 4.3).

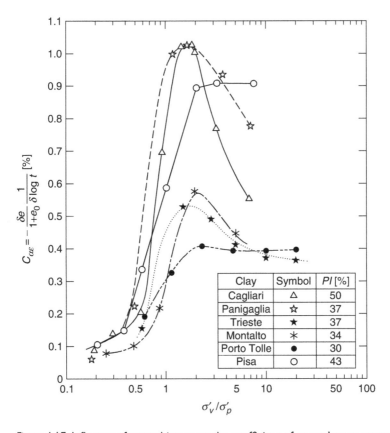

Figure 4.17 Influence of stress history on the coefficient of secondary compression.

Table 4.3 Typical values of the ratio C_α/C_c

Clay type	C_α/C_c
Soft organic clays	0.05 ± 0.01
Soft inorganic clays	0.04 ± 0.01
Sands	from 0.015 to 0.03

Source: Mesri and Choi (1985).

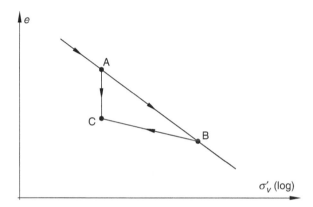

Figure 4.18 Estimating the preconsolidation induced by creep.

4.10 Further post-depositional phenomena: leaching, exchange of cations, cementation

Further phenomena, which can produce significant changes in properties from an engineering point of view, have been discussed by Bjerrum (1967) in its Rankine Lecture. These phenomena are *leaching*, *exchange of cations* and *cementation*.

1 *Leaching* is caused by a slow flow of fresh water through marine sediments. The flow results in a gradual exchange of the salt water in the pore voids with fresh water, and this leaching produces a reduction of the liquid limit. Since the water content remains unaltered, at the end of the leaching process the water content may result higher (or even considerably higher) than the liquid limit, so that the clay will have the consistency of a liquid in its remoulded state. This peculiar behaviour has given the leached clays the name of **sensitive** or **quick clays**. The **sensitivity** of a clay is defined as the ratio of undrained strength of undisturbed soil to undrained strength of the soil after it has been remoulded.

2 The description of microstructural features given in Section 1.2 highlights that the seat of positive charges and the seat of negative charges do not coincide, so that a clay particle exhibits a surface negatively charged. In addition, because Al^{3+} can partly substitute Si^{4+} in the tedrahedral and Mg^{2+} can substitute Al^{3+} in the octahedral sheet, a net unit charge deficiency results.

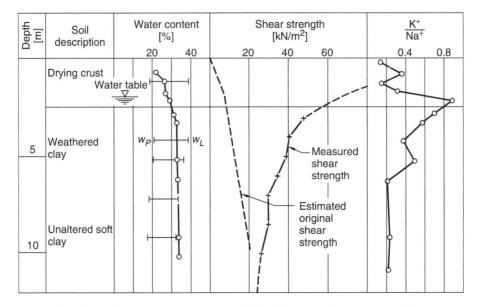

Figure 4.19 Effect of weathering on properties of clay (Moum and Rosenquist, 1957).

In marine clays, the net charge deficiency is neutralized by cations Na^+, but if the type and concentration of cations in the pore water changes, the cations on the surface of clay particles will exchange with those of the solution. This exchange will produce a change in plasticity, compressibility and shear strength.

As an example, Figure 4.19 shows the effect of cation exchange on shear strength. When the normally consolidated clay deposit rose above the seawater level (as a consequence of the withdrawal of glaciers and the rebound of the previously glaciated areas), the upper two or three metres of clay have been subjected to desiccation. Below the desiccated crust up to a depth of 6 to 7 metres, the clay developed additional strength, although there is no reduction of water content. This increase of shear strength is the result of cation exchange, proved by measurements of the type of adsorbed cations and by the ratio K^+/Na^+. This ratio is about 1 just below the desiccated crust and decreases towards the value of 0.2 for the unaltered clay.

3   Finally, in natural soils there are soluble chemical agents such as carbonates, gypsum, aluminium and iron compounds and organic matter. In given conditions these chemical agents can precipitate, giving rise to chemically stable cements or gels, strengthening the clay structure.

4.11 Importance of detailed geological history

The stress history of a soil deposit is not always sufficiently simple to be depicted as a phase of compression and subsequent unloading so that, although we give to

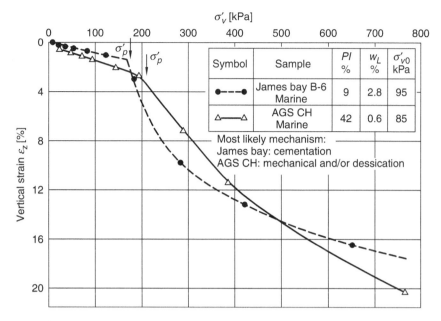

Figure 4.20 Compression curves for clays having different preconsolidation mechanisms (Jamiolkowski *et al*, 1985).

the preconsolidation stress the meaning of a yield stress, independently from the nature of the events, the experimental data in Figure 4.20 show that preconsolidation due to cementation will produce a more brittle soil structure than the mechanical overconsolidation does.

Moreover, more often the soil has been unloaded and reloaded, during several cycles, therefore it can be observed in Figure 4.21 that for a given OCR there is more than one possibility of $K_o(OC)$, depending on the precise loading sequence. The important conclusion therefore follows that if the soil deposit has been subjected to more than one cycle of loads during its history, as well as to ageing, cementation and other phenomena, its actual effective horizontal stress cannot be predicted. The previous empirical relations should be restricted to truly normally consolidated soils, or to the case of overconsolidation due to simple unloading. In more complex situations there is the need to refer to *in situ* measurements of the horizontal stress, as discussed in Chapter 7. Such measurements include the use of total pressure cells, the hydraulic fracturing technique, the self-boring pressuremeter and the flat dilatometer.

The previous considerations, developed for clay deposits, also apply for coarse-grained soils. In this case the $K_o(NC)$ can also be estimated by means of equation (4.12) and it is therefore possible to postulate a dependency of $K_o(NC)$ from the relative density. During unloading, relation (4.14) also applies, the exponent α ranging from 0.4 to 0.5, with higher values for higher relative density.

However, it must be outlined that in a sand deposit there is no possibility of obtaining undisturbed soil samples by means of common sampling techniques, so that the yield stress cannot be measured and the overconsolidation ratio cannot be quantified in absence of detailed geological information.

4.12 Importance of sample quality and testing procedures

As far as the reliability of the yield stress determined through oedometer tests is concerned, there are several factors to be taken into consideration. The quality of the sample is by far the most important requirement. *Disturbance* of a soil sample (Figure 4.22) tends to modify the original soil structure and all information related

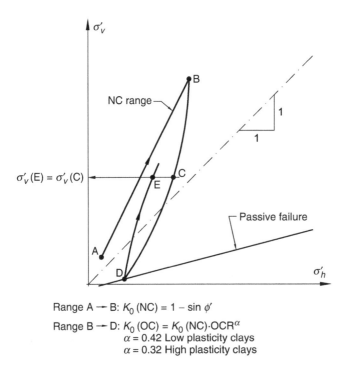

Range A → B: K_0 (NC) = 1 – sin ϕ'

Range B → D: K_0 (OC) = K_0 (NC)·OCR$^\alpha$
α = 0.42 Low plasticity clays
α = 0.32 High plasticity clays

Figure 4.21 The influence of the sequence of events on the effective horizontal stress.

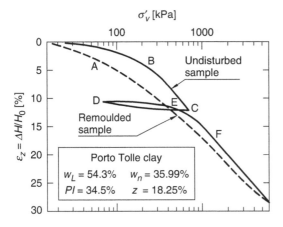

Figure 4.22 Influence of disturbance on oedometer test results.

to this structure, therefore the soil partially loses its memory and the laboratory compression curve appears more and more gently smoothed, without a well-defined break point.

On some occasions the soil can be exposed by excavation and a *block* can be cut by hand, a process resulting in high quality samples. But usual situations require the samples of soil to be extracted by a *sampler* lowered into the soil through a *borehole* (see Chapter 7). Even if the sampling process is done in a rather perfect manner, the quality of sample tends to be inferior to that of a sample obtained by blocks hand-cutted, as can be inferred from the data of Figure 4.23, where compression curves of specimens from an *Osterberg piston sampler* are compared with those from block samples.

In addition, it must be observed that in conventional loading tests the end of primary compression occurs in less than 1 hour. Therefore, data referred to 24 hours include one or more cycles of secondary compression, so that the compression portion tends to be displaced downward, giving a yield stress generally lower than that obtained from the end of primary curve (Figure 4.24).

4.13 Stress paths

Soils cannot be treated as elastic or isotropic materials, but they must be considered inelastic, anisotropic and history dependent materials, so that the response of a soil element to a given increment of stress depends not only on the increment in itself but also on the overall *loading history* and on the *loading path*. For this reason the need arises to consider stress paths, and in order to give a proper and convenient representation of these loading paths a methodological approach, known as '**stress path method**', is extensively used in soil mechanics to analyse both the behaviour of geotechnical structures and data of experimental tests.

The stress path method was first introduced by Lambe (1967); it was further extended to a wide range of geotechnical problems by Wood (1984) and it is discussed in detail by Lambe and Whitman (1969) and Wood (1990).

However, before considering the main aspect of this topic, it is convenient to address the problem of an appropriate choice of stress and strain variable. First we observe that volume changes are recognized to be one of the most important features of the mechanical response of soils. For this reason it is convenient to divide soil deformation into compression (change of volume) and distortion (or change of shape). Then we can argue that, when dealing with constitutive modelling of soil behaviour, we need to specify increments of work done when deforming soil elements, so that the idea arises to conjugate stress and strain variables.

If we consider volumetric effects, the natural choice is to consider the mean effective stress p' such that the work done in changing the volume of a unit element is given by:

$$W = p' \delta \varepsilon_v ,$$

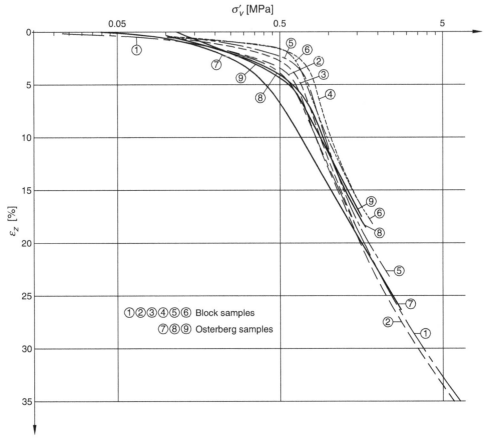

Figure 4.23 Influence of sample quality on oedometer test results (Holtz et al., 1986.)

Montalto di Castro clay

Sample	test	Depth [m]	w_L [%]	PI [%]	w_n [%]	G_s [-]
1	IL	13.00	73.5	54.0	46.4	2.76
2	IL	13.00	79.1	58.2	50.3	2.76
3	CHG	13.00	79.1	58.2	46.8	2.76
4	CRS	13.00	79.1	58.2	45.8	2.76
5	CRS	13.00	79.1	58.2	49.8	2.76
6	CHG	13.00	73.5	54.0	47.2	2.76
7	IL	14.50	64.6	50.2	43.6	
8	CHG	13.30	68.3	46.3	43.9	2.71
9	CHG	16.00	59.8	39.3	40.0	2.70

so that the volumetric stress and strain variable are:

$$p' = \frac{1}{3}(\sigma_1' + \sigma_2' + \sigma_3')$$

$$\varepsilon_v = \varepsilon_1 + \varepsilon_2 + \varepsilon_3$$

(4.22)

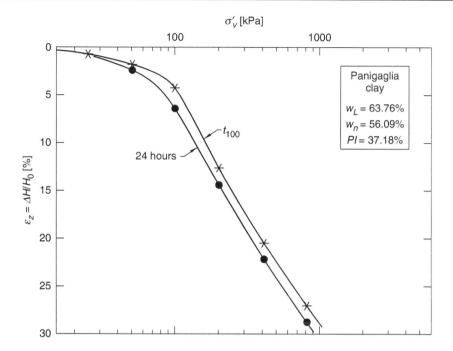

Figure 4.24 Influence of secondary compression on the assessment of yield stress.

It follows that it is convenient to decompose the stress and strain tensor into the above-defined spherical component and in a deviator component (the meaning of the term 'deviator' is clarified in Section 4.14):

$$\sigma_{ij} = p\delta_{ij} + \sigma_{ij}^d$$
$$\varepsilon_{ij} = \frac{1}{3}\varepsilon_v\delta_{ij} + \varepsilon_{ij}^d \,, \tag{4.23}$$

Note that the pore pressure affects only the spherical component of the stress tensor and not the deviator component, so that:

$$p' = p - u \,, \tag{4.24}$$

$$(\sigma_{ij}')^d = \sigma_{ij}^d \,. \tag{4.25}$$

For this reason there is no need to introduce the apex when dealing with the deviator component. Most of the data discussed in this book are from triaxial tests, in which we impose conditions of axial symmetry. In this test we have two degrees of freedom, i.e. the axial principal stress σ_a and the radial principal stress σ_r, because the assumption of homogeneous state of stress implies $\sigma_\vartheta = \sigma_3 = \sigma_r = \sigma_2$ (see Chapter 5). It is then convenient to represent the stress path in the plane (p', q), the mean effective stress p'

and the deviator stress q being:

$$p' = \frac{1}{3}\left(\sigma_1' + 2\sigma_3'\right)$$

$$q = \sigma_1 - \sigma_3$$

(4.26)

Note that, for an axisymmetric state of stress, a generalized definition of the above stress variable is:

$$p = \frac{I_1}{3},$$

$$q = \sqrt{3J_2}$$

(4.27)

where I_1 and J_2 are respectively the first invariant of the stress tensor and the second invariant of the stress deviator, as defined in Section 2.9.

Once the deviator (or distortional) stress variable has been selected, the choice of the distortional strain variable (whose symbol will be ε_s) is dictated by the requirement that the total work done per unit volume must be:

$$W = \sigma_1' \delta\varepsilon_1 + \sigma_2' \delta\varepsilon_2 + \sigma_3' \delta\varepsilon_3 ,$$

(4.28)

or, in terms of (p', q) variables:

$$W = p' \delta\varepsilon_v + q \delta\varepsilon_s ,$$

(4.29)

and it can easily be proved that the above requirements is satisfied if the following definition is assumed for the strain variables:

$$\varepsilon_v = \varepsilon_1 + \varepsilon_2 + \varepsilon_3$$

$$\varepsilon_s = \frac{2}{3}\left(\varepsilon_1 - \varepsilon_3\right)$$

(4.30)

The volumetric strain variable ε_v and the distortional strain variable ε_s (often also called *triaxial shear strain*) are the strain variables conjugate to the stress variables p' and q.

In spite of the fact that most of the data presented in this book are from axisymmetric tests (oedometer and triaxial tests), conditions of **plane strain** are of practical occurrence in geotechnical problems (footings, walls, embankments, whose length is large compared with the cross-section, are examples of plane strain problems).

In all these cases, the behaviour of any arbitrary cross-section, perpendicular to the longitudinal axis y and far enough from the extremities of the structure, will be independent of its location, so that by imposing conditions of symmetry we reach the conclusion that displacements along the longitudinal axis and the strain components $\varepsilon_y, \gamma_{yx}, \gamma_{yz}$ are zero, and the non-vanishing components of the strain tensor reduce to:

$$\varepsilon_x, \varepsilon_z, \gamma_{xz} .$$

The stress component σ'_{yy}, which from symmetry is a principal stress, can be considered to be a dependent variable, because its value must satisfy the imposed condition:

$$\varepsilon_y = 0 . \tag{4.31}$$

The work done per unit volume assumes the expression:

$$W = \sigma'_x \delta\varepsilon_x + \sigma'_z \delta\varepsilon_z + \tau'_{xz} \delta\gamma_{xz} . \tag{4.32}$$

and with coaxiality of principal axes:

$$W = \sigma'_1 \delta\varepsilon_1 + \sigma'_3 \delta\varepsilon_3 . \tag{4.33}$$

Therefore, it follows that the most appropriate stress and strain variables in the plane strain case are:

$$
\begin{aligned}
s' &= \frac{\sigma'_1 + \sigma'_3}{2} \\
t &= \frac{\sigma_1 - \sigma_3}{2} \\
\varepsilon_v &= \varepsilon_1 + \varepsilon_3 \\
\varepsilon_\gamma &= \varepsilon_1 - \varepsilon_3
\end{aligned}
\tag{4.34}
$$

and the unit work can be written as:

$$W = s' \delta\varepsilon_v + t \delta\varepsilon_\gamma . \tag{4.35}$$

Once the stress and strain variables have been selected, we can illustrate the main steps of the **stress path method** (see Wood, 1990):

(a) given a geotechnical structure the first step is to identify representative soil elements;
(b) the second step is to estimate the stress path of these elements as the structure is being loaded;
(c) then the final step is to perform laboratory tests by simulating this stress path and to use the results of these tests to estimate the response of the structure.

To illustrate how stress paths can be represented, let us now refer to the plane strain structure in Figure 4.25a (a smooth vertical wall). In its initial undisturbed state, a soil element has a stress state represented by the overburden effective stress σ'_{vo} and the horizontal effective stress $\sigma'_{ho} = K_o \sigma'_{vo}$. If the wall is moving forward, so that soil deforms towards a final condition of active failure, the vertical stress will remain unchanged, whereas the horizontal stress will reduce. This stress path can be depicted through a succession of Mohr circles in a (σ', τ) plane (the passage from circle A to circle B in Figure 4.25b), or, more conveniently, in terms of stress variables (s', t),

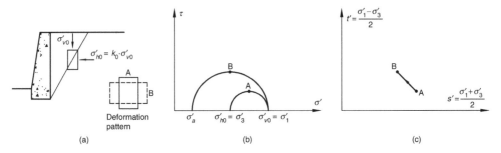

Figure 4.25 The stress-path concept.

so that the stress path will be represented by the path AB in Figure 4.25c, being in this case:

$$\delta s = \frac{\delta \sigma_b}{2} < 0$$

$$\delta t = \frac{-\delta \sigma_b}{2} > 0$$

If, on the contrary, the smooth wall is pushed against the soil (Figure 4.26), the soil element near the wall experiences an increment of the horizontal stress, without change of the vertical overburden stress. This case is representative for the problem of passive earth pressure, and the stress path will be represented by the path AB for which:

$$\delta s = \frac{\delta \sigma_b}{2} > 0$$

$$\delta t = \frac{-\delta \sigma_b}{2} < 0$$

Note that in both the previous examples the soil has been assumed to be dry, for sake of simplicity. In the presence of pore pressure, we need to make a clear distinction between the total stress path (TSP) and the effective stress path (ESP). As illustrated in Figure 4.27, since the pore pressure does not affect the deviator stress component, the effective stress path will be simply translated with respect to the total stress path by an amount equal to the current value of the pore pressure.

4.14 On the meaning of the term 'deviator'

Any second-order tensor $\mathbf{T}$ can always be decomposed into a spherical tensor and a **deviator** tensor, whose trace is zero:

$$\mathbf{T} = \frac{1}{3}(tr\mathbf{T})\mathbf{I} + \mathbf{T}^d \, . \tag{4.36}$$

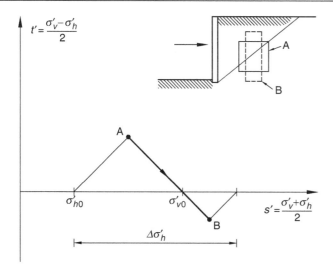

Figure 4.26 Stress path of a soil element simulating a passive failure of conditions.

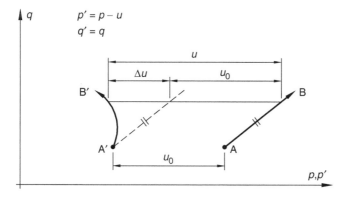

u = Pore pressure measured during loading
u_0 = Stationary pore pressure
Δu = Excess pore pressure

Figure 4.27 Total (A→B) and effective (A′ → B′) stress paths.

As extensively shown in Section 2.2, a second-order tensor can be thought of as a linear mapping of the Euclidean vector space into itself, so that, given:

$$p = \frac{1}{3} tr\mathbf{T} \,,$$

if we apply the tensor $\mathbf{T}$ to any arbitrary vector $\mathbf{a}$, it follows:

$$\mathbf{b} = \mathbf{Ta} = p\mathbf{a} + \mathbf{T}^d\mathbf{a} \,. \tag{4.37}$$

The vector **b**, in which the vector **a** is mapped, differs from **a** both in terms of modulus as well as its orientation. But (4.37) also proves that in the absence of the deviator component T^d, the vector **b** will be collinear to **a**, because the application of a spherical tensor to **a** modifies its modulus but not its orientation.

In order to *deviate* the vector from its original direction we need to apply the deviator component of the tensor **T**, and this fact clarifies the definition of *deviator* given to this component (having zero trace).

4.15 Summary

The principle of effective stress postulates that any measurable effect, such as volume change, distortion and changes in shear strength, depends on the effective stress component, which is the difference between the total stress and the pore pressure.

The stress history of a soil element is quantified by the yield stress and the Over Consolidation Ratio (OCR), the ratio between the yield stress and the actual vertical effective stress. On the basis of the OCR, soils are divided into normally consolidated and overconsolidated materials.

To define the actual initial stress state, we need to introduce the coefficient of earth pressure at rest K_o. This coefficient depends on the history of the events in precise sequence, therefore empirical correlations can only be applied to virtually normally consolidated deposits or when the overconsolidation is produced by simply unloading. In more complex cases, the horizontal stress must be measured in the field.

Because of inelastic and anisotropic soil behaviour, it is important to visualize the history and the direction of loading. This can be conveniently done with the use of the stress path concept.

One-dimensional compression is of engineering relevance when a fill of large lateral extent is superimposed on a soil stratum. Further, one-dimensional compression can be considered an acceptable approximation of the situation occurring beneath the centre line of a large foundation, as an embankment. In all these cases, the settlement occurring in the field may be computed by using the one-dimensional approach and compressibility parameters derived from oedometer tests.

We have introduced in this chapter quantities that are soil properties, state variables and soil parameters. Soil properties are material constants, i.e. quantities independent of the current state of the soil sample. State variables are quantities that describe changes of state of a soil element as it deforms. The most relevant state variables are the specific volume, the effective stress and the yield stress. Soil parameters (examples include soil stiffness and peak strength) are quantities depending on the state of a soil element.

4.16 Further reading

A complete discussion of the sedimentation compression curves, with examples of clays of different geological ages, is given in the paper by A.W. Skempton (1970) *The Consolidation of Clays by Gravitational Compaction*, Q.J. Geo. Soc. London, 125, 373–412.

A review of basic compressibility and shear strength properties of natural sedimentary clays, compared with the corresponding properties of the reconstituted material, is given in the thirtieth Rankine Lecture by J.B. Burland (1990) *On the Compressibility and Shear Strength of Natural Clays*, Rankine Lecture, Géotechnique, 40, 3, 329–378.

Changes in properties which occur after deposition are discussed in the Rankine Lecture by L. Bjerrum (1967) *Engineering Geology of Norwegian Normally Consolidated Marine Clays as Related to Settlements of Buildings*, Rankine Lecture, Géotechnique, 17, 214–235.

For futher insight into the subject of capillarity, shrinkage, swelling and frost action, see Chapter 6 of R.D. Holtz and W.D. Kovaks (1981) *An Introduction to Geotechnical Engineering*, Prentice-Hall.

Additional contributions related to natural soil deposits can be found in the following papers:

R.J. Chandler (2002) Clay sediments in depositional basins: the geotechnical cycle. The Third Glossop Lecture, *Quarterly J. of Engineering Geology and Hydrogeology*, 33, 7–39.

A.W. Skempton (1961) Horizontal stresses in an overconsolidated Eocene clay, *Fifth Int. Conf. Soil Mech. Found. Eng.*, Paris, 1, 351–357.

S. Leroueil and P.R. Vaughan (1990) The general and congruent effects of structure in natural soils and weak rocks, *Géotechnique*, 3, 467–488.

M.E. Barton and S.N. Palmer (1989) The relative density of geologically aged, British fine-medium sands, *Quarterly J. of Engineering Geology*, 22, 49–58.

M. Jamiolkowski, C.C. Ladd, J.T. Germaine and R. Lancellotta (1985) New developments in field and laboratory testing of soils, *Theme Lecture, XI Int. Conf. Soil Mech. Found. Eng.*, San Fancisco, 1, 57–153.

S. Leroueil and D.W. Hight (2003) Behaviour and properties of natural soils and soft rocks, in *Characterisation and Engineering Properties of Natural Soils*, Tan *et al.* (eds), Swets & Zeitlinger, Lisse, 1, 29–254.

Chapter 5

Stress-strain behaviour of soils

When analysing experimental results related to soil behaviour, undergraduated students are faced with a common difficulty of establishing links between many aspects, such as drained and undrained paths, the response of normally consolidated and over-consolidated soil samples, the meaning of undrained strength and many others. One possibility to circumvent this difficulty is to provide the reader, from the beginning, with a simple but robust conceptual framework, and for this reason reference is made to the Critical State Theory.

What do we mean by Critical State Theory? The central idea relies on the consideration that volume changes play a role so important as the effective stresses and it is the relation between the initial state and the critical state that has a major influence on soil behaviour (Wood, 1990; Atkinson, 1993).

This chapter (in particular Sections 5.1 and 5.4 to 5.10) essentially deals with this central idea, but also offers a presentation of other aspects usually thought in upper level courses.

One aspect which deserves attention is the possibility, provided by elastic-plastic models, to establish a link between behaviour and modelling of soils, and for this reason Section 5.11 provides a concise presentation of the original Cam-Clay model.

The objective of this presentation is intentionally didactic, so that it is aimed at capturing essential aspects of soil behaviour. However, care has been taken to suggest selected references, where more advanced soil models are discussed, especially those aimed at capturing specific aspects of natural soils.

5.1 The ability of soils to carry stresses: soil strength and Coulomb's failure criterion

When the term *failure* is applied to soils, it indicates a relative slip occurring on any plane within the soil mass if the ratio of the shear stress and the effective normal stress, acting on this plane, reaches a critical value. The reason is that soil strength is controlled by friction, so that the *failure criterion* (first introduced by Coulomb, 1773) assumes the form:

$$|\tau| = \mu \sigma' . \tag{5.1}$$

When plotted on the (σ', τ) plane (Figure 5.1), equation (5.1) defines two straight lines and, since the stress at a point in the soil is represented by a Mohr circle, the

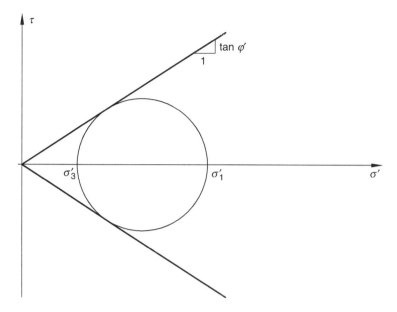

Figure 5.1 Coulomb failure criterion.

failure condition is attained if the Mohr circle is tangent to these lines. In fact, the coordinates of the contact points satisfy the criterion (5.1), and the stress vector acting on the failure plane reaches the maximum possible *obliquity*, given by the ratio τ/σ'.

This obliquity can be expressed by an angle, so that we can give to the constant of proportionality μ the form $\mu = \tan\varphi'$ and for this reason the angle φ' is called *angle of shear resistance*. Accordingly, the failure criterion is usually expressed in the form:

$$|\tau| = \sigma' \tan\varphi' . \tag{5.2}$$

For reasons explained in Section 5.3, the most widely used test is the triaxial test, which would be more properly defined as a cylindrical compression test, in that a cylindrical soil sample is loaded by applying the axial stress σ_z and the radial stress σ_r. The applied stresses are principal stresses, so that it is convenient to express the failure criterion in terms of principal stresses. This can be done by observing (see Figure 5.1) that the abscissa s' of the centre of the Mohr circle is equal to $(\sigma_1' + \sigma_3')/2$, the radius R is given by $(\sigma_1' - \sigma_3')/2$ and, since $R = s' \sin\varphi'$, it follows that the failure criterion can be written in the form:

$$\sigma_1' = \sigma_3' \frac{1 + \sin\varphi'}{1 - \sin\varphi'} . \tag{5.3}$$

Note that, by assuming that in the triaxial apparatus the stress state is uniform, i.e. $\dfrac{\partial\sigma_r}{\partial r} = 0$, the indefinite equilibrium equation:

$$\frac{\partial\sigma_r}{\partial r} + \frac{\sigma_r - \sigma_\theta}{r} = 0$$

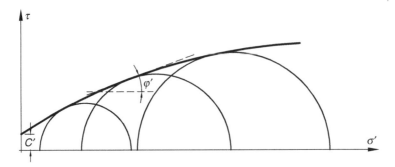

Figure 5.2 Failure envelope.

allows us to reach the conclusion that $\sigma_\theta = \sigma_r$. The state of stress is then represented by a single Mohr circle, i.e. that corresponding to couple (σ'_1, σ'_3). But even in a more general case, since the failure criterion (5.3) is not affected by the value of the intermediate principal stress, it is only the largest circle which dictates the attainment of a limiting state of stress.

Determining the strength parameters requires to perform a series of experimental tests, so that if the Mohr circles at failure are plotted on the (σ', τ) plane, the envelope of these circles represents the locus of points at failure (Figure 5.2) and is referred to as *failure envelope*.

Often this envelope is linearized over the stress range of interest, and is represented by the equation:

$$\tau = c' + \sigma' \tan\varphi'. \tag{5.4}$$

The first component c' is usually referred as *cohesion*, but this term should be thought of as an *intercept* which only gives the position of the linearized failure envelope. In addition, it must be noticed that the linearized failure envelope may be unsafe when the stress level in the field is smaller than the effective stress range applied in laboratory tests, because the actual failure envelope, representative of stiff soils, shows a significant curvature near the origin.

These introductory notes must be considered as a short and preliminary presentation of the subject, because understanding the real meaning of strength parameters, as well as their use in stability analyses, requires a deeper scrutiny. This is the subject of subsequent sections.

5.2 Drained and undrained conditions: relative rate of loading

When considering coarse-grained soils, the hydraulic conductivity reaches such high values that excess pore pressure due to external loads rapidly dissipates during the same loading stage. In this case the transient phase of pore pressure decay can be ignored and soil behaviour can be analysed by referring to the so-called **drained conditions**.

Formally, the term *drained condition* indicates the circumstance in which changes of effective stresses coincide with changes of total stesses in any soil element, i.e. locally in the domain. On the contrary, the hydraulic conductivity of fine-grained soils is so low that a soil element tested in the laboratory can attain drained conditions only if free draining boundary conditions are provided, so that pore water can flow out of the sample, and the rate of loading is so slow as to be comparable to the rate of excess pore pressure dissipation.

Note that, according to its definition, the characteristic feature of a drained condition is that the pore pressure retains its initial value, i.e. no excess pore pressure develops. Therefore, if the rate of loading is high relative to the soil hydraulic conductivity, so that water cannot escape from the pores during loading, this circumstance is called **undrained condition**.

Formally, undrained condition indicates the circumstance in which a soil element (i.e. locally) cannot exchange water mass with the surrounding ambient. If soil is saturated and both particles and water are assumed to be incompressibile, the above definition means that the undrained condition is a *constant volume condition*. Because of this constraint, an excess pore pressure develops and increments of effective and total stresses do not coincide.

5.3 Soil testing: requirements of laboratory apparatuses

Any laboratory testing apparatus should induce within the tested soil sample a uniform state of stress and strain, because only if this basic requirement is satisfied can the soil sample be considered as a volume element.

An additional basic requirement is to allow a control of drainage conditions. Furthermore, the need to perform arbitrary stress paths would require the use of rather sophisticated devices, allowing an independent control of all three principal stress and strains. These requirements are satisfied by a limited number of soil testing apparatuses and some of them are mainly used in the research field.

References for more sophisticated devices can be found in Jamiolkowski *et al.* (1985), whereas in this section the attention is mainly focused on the apparatuses currently used in practice, with the aim of highlighting advantages and limitations.

5.3.1 The triaxial test

This test (see Figure 5.3) would be more properly defined as a cylindrical compression or extension test, in that a cylindrical soil sample is loaded by applying the axial stress σ_z and the radial stress σ_r. The sample, of height about twice the diameter (a 38 mm diameter sample is commonly used, but a 100 mm is sometimes used in special tests), is encased in a thin rubber membrane to prevent the cell water from penetrating into the soil, being the sample contained in a water-filled cell with a cell pressure σ_r.

In the *hydraulic triaxial cell* shown in Figure 5.3, the axial force is applied by means of a frictionless hydraulic ram incorporated into the base of the cell. Since it is inaccurate to compute the axial load from the pressure in the hydraulic ram,

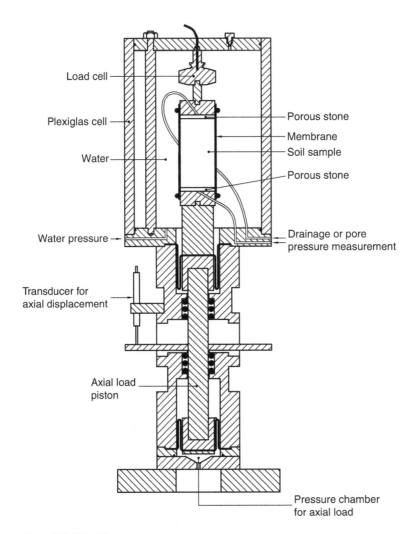

Load cell

Plexiglas cell

Water

Porous stone

Membrane

Soil sample

Porous stone

Water pressure

Drainage or pore
pressure measurement

Transducer for
axial displacement

Axial load
piston

Pressure chamber
for axial load

Figure 5.3 Triaxial apparatus.

a load cell is provided on the top of the sample to allow independent measurements of the applied load.

The ends of the specimen are connected by means of porous stone to a drainage system, so that when the drainage valve is open the sample is drained and the volume of the escaped water can be measured; if the valve is closed the sample experiences an undrained condition and the pore pressure can be measured. These possibilities of controlling drainage conditions, as well as measuring the pore pressure, represent the major capability of the triaxial apparatus and justify why, at present, it is by far the most common test in practice.

There is a recurrent terminology used to describe triaxial tests, as briefly here recalled.

When performing the so-called *consolidated-drained tests*, the soil specimen is reconsolidated during the first stage to a stress state that can be isotropic (*CID* tests) or anisotropic (*CAD* tests and in particular CK_oD tests if the sample is one-dimensionally reconsolidated). It is then sheared slowly so as to prevent the development of any pore pressure excess.

If the second stage is performed in *undrained conditions*, the drainage valve is closed and pore pressure measurements allow us to compute at any step the effective stresses. These tests are called *consolidated-undrained* tests and are referred to with the abbreviation CIU or CAU, depending on the initial stress conditions (isotropic or anisotropic).

The following quantities are objects of direct measurements during the test: the cell water pressure, the ram load, the vertical displacement, the pore pressure (in undrained tests) and the volume of escaped water (in drained tests).

To compute stress and strain components from these quantities, we need to introduce some assumptions. First, the assumption of zero friction between the top cap and the specimen implies $\tau_{zr} = 0$.

The axial stress is obtained by summing up the stress produced by the ram load to the radial stress:

$$\sigma_z = \sigma_r + \frac{P}{A},$$

where A is the current cross area of the sample, allowing for changes of axial ε_z and volumetric ε_v strains (see Bishop and Henkel, 1962):

$$A = A_o \frac{1 - \varepsilon_v}{1 - \varepsilon_z},$$

being A_o the initial value.

If a uniform state of stress develops within the sample, then $\frac{\partial \sigma_r}{\partial r} = 0$ and the indefinite equilibrium equation $\frac{\partial \sigma_r}{\partial r} + \frac{\sigma_r - \sigma_\theta}{r} = 0$ allows us to write $\sigma_\theta = \sigma_r$.

The radial strain ε_r is not usually measured, but calculated from measurements of the axial strain $\varepsilon_z = \frac{-\delta h}{h_o}$ and volumetric strain:

$$\varepsilon_v = \varepsilon_z + 2\varepsilon_r.$$

As already discussed in Section 4.14, the behaviour of a soil sample in triaxial tests is represented by using the mean effective stress and the deviator stress, respectively defined as:

$$p = \frac{\sigma_z + 2\sigma_r}{3}.$$
$$q = \sigma_z - \sigma_r$$

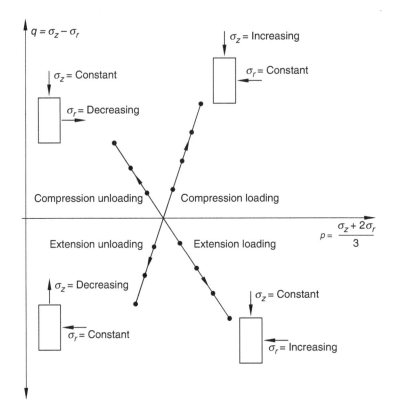

Figure 5.4 Typical stress paths performed with triaxial tests.

The conjugate strain parameters are the volumetric strain and the deviator strain:

$$\varepsilon_v = \frac{-\delta v}{v} = \varepsilon_z + 2\varepsilon_r$$

$$\varepsilon_s = \frac{2}{3}\left(\varepsilon_z - \varepsilon_r\right)$$

Hydraulic triaxial cells allow us to control axial and radial stresses independently and Figure 5.4 illustrates the most common total stress paths, with the aim of clarifying terms such compression, extension, loading and unloading, depending on whether σ_z or σ_r is increasing or decreasing while the other is held constant. Note that by using the above definitions it can be proved that these stress paths have slope $\delta q/\delta p$ equal to 3 or −3/2.

5.3.2 The direct shear test

This kind of test is apparently simple in principle. The soil specimen is encased in a shear box (Figure 5.5), which is divided into two halves in order to allow relative movements. Soil samples (of square or circular cross-section) have a side (or diameter) dimension of 60 or 100 mm and height equal 20 or 40 mm. A normal load is first

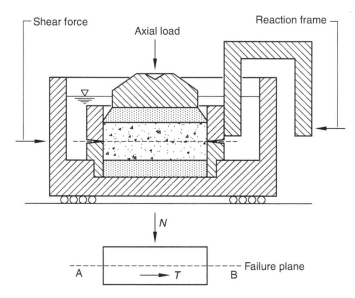

Figure 5.5 Direct shear test.

applied to the soil specimen that is reconsolidated in one-dimensional conditions; then one half of the shear box is pushed horizontally, at constant rate of displacement, while the other half remains fixed. Measurements of the shear load are taken by means of a proving ring, and the vertical and the horizontal displacements are measured through displacements transducers.

One of the major disadvantages of this test is that it is impossible to control the drainage. As a consequence, the test should be run under completely drained conditions. To accomplish this requirement, it has been suggested (Lambe, 1951) to use a rate of relative displacement of the order of 10^{-4} mm/s for clays, and 0.02 mm/s for sands.

In addition, during the test the specimen is forced to fail in the horizontal plane, but the state of stress cannot be determined because only the normal and the shear stress on the horizontal plane are known.

We need to assume that the state of stress is uniform and that the horizontal plane is the plane of maximum obliquity in order to draw the Mohr circle and compute the angle of shear strength.

However, because stress and strain within the sample are not uniform and because it is extremely difficult to measure the pore pressure, the direct shear test is of limited use. It is usually used when testing coarse-grained soils, because the high hydraulic conductivity ensures that drained conditions are attained, and to determine the residual strength of fine grained soils, provided that some aspects are properly taken into account, as discussed in Section 5.9.

5.3.3 The resonant column test

In the triaxial cell, vertical deformations of the sample are usually deduced by observing the movement of the ram through a displacement transducer or a dial gauge.

This procedure causes some errors, because it includes spurious movements of the systems, so that the triaxial test requires special techniques and the use of *local strain gauges* in order to investigate soil behaviour at small strains.

At very low strains, soil stiffness can be obtained from measurements of shear wave velocity using piezoceramic crystals (Viggiani and Atkinson, 1995; Arroyo *et al.*, 2006) or in the resonant column test. In the *resonant column test* (see Richart *et al.*, 1970; Drnevich, 1978; Drnevich *et al.*, 1978; Hardin, 1970; Saada and Townsend, 1981), a solid or hollow cylindrical soil sample (Figure 5.6) is subjected to axial or torsional vibrations at a varying frequency, until the resonance condition is attained.

The most common testing configuration is that with one end of the specimen fixed, because of simplicity of testing equipment. The mass on the free end of the specimen consists of a top platen, the equipment used to apply the torque to the specimen and the transducers to measure the resulting motion.

The use of torsional vibrations that generate shear waves is the most used procedure, and for this reason this way of testing will be described in the sequel.

By using a function generator and a power amplifier a signal is produced, which can be modified both in frequency and amplitude. The signal is converted in a torque through an electromagnetic motor, which consists of eight drive coils interacting with four magnets, these latter being connected to the top sample. The frequency of excitation is gradually increased (usually in a range from 10 to 100 Hz) until a condition of resonance is attained. If the torque and velocity signals are viewed on an oscilloscope as an *x-y* function, the attainment of resonance gives a straight line, whereas frequencies slightly off resonance give an ellipse.

If I is the mass polar moment of inertia of the specimen, I_o is the mass polar moment of inertia of the driving system at the top of the sample, and h is the height of the sample fixed at the base, it can be proved (see Richart *et al.*, 1970) that the circular frequency ω_n of a natural mode of vibration is linked to the shear wave velocity v_s through the relationship:

$$\frac{I}{I_o} = \frac{\omega_n h}{v_s} \tan\left(\frac{\omega_n h}{v_s}\right). \tag{5.5}$$

The resonance frequency is obtained experimentally, then equation (5.5) allows us to obtain the velocity of the shear wave and the shear modulus G of the sample is computed by means of the relation $G = \rho v_S^2$. A piezoelectric accelerometer detects a signal that is proportional to the angular acceleration of the top of the sample, so that by integrating twice in time domain gives the rotation, which is linked to the engineering shear strain γ (see Example 3.3 in Chapter 3). Moreover, testing the soil sample at a progressively larger amplitude gives the possibility to obtain the soil stiffness as function of strain amplitude.

The energy is dissipated in soils by many mechanisms, mainly friction, but a convenient approach is to introduce a viscous damping to take into account for dissipation. Within the context of this assumption, damping is determined from the frequency response curve using the so-called half-power bandwidth method. This method requires us to measure the amplitude at different frequencies in order to obtain the

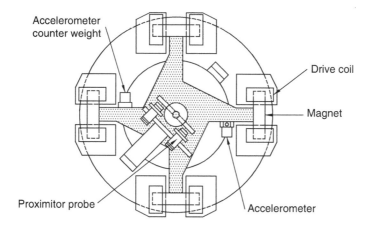

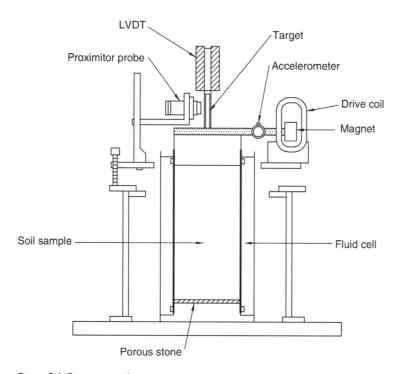

Figure 5.6 Resonant column test.

maximum value, then, by considering the frequencies f_1 and f_2 corresponding to $1/\sqrt{2}$ the maximum value, it can be proved that the damping ratio D is determined by:

$$D = \frac{f_2 - f_1}{2f_n} .$$

(5.6)

where f_n is the resonance frequency.

5.4 Dilatancy, peak and critical state strength

Figure 5.7 summarizes the results of two triaxial tests, performed on Hokksund sand at the same level of confining stress $\sigma_r' = 50\,\text{kPa}$.

If we consider the loosely packed sand specimen, with an initial void ratio $e_o = 0.77$, it can be observed that the soil progressively reduces its volume and tends towards an

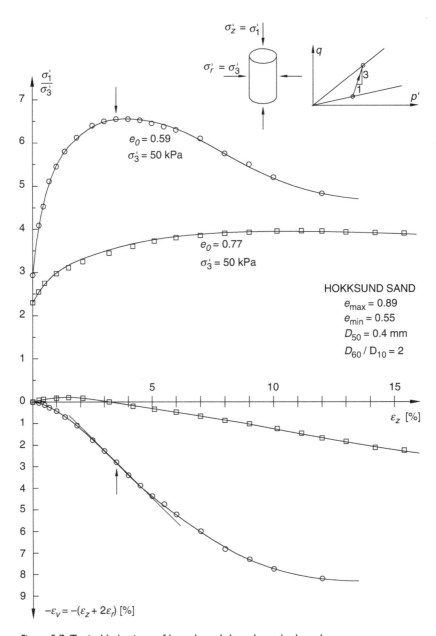

Figure 5.7 Typical behaviour of loosely and densely packed sand.

ultimate condition, reached at large strains, where distortion can occur without any further change of volume or effective stress ratio.

On the contrary, a more densely packed sand (initial void ratio $e_o = 0.59$) is characterized by shearing with an increase of volume and a peak condition is attained at the maximum rate of dilation.

Once the peak has been attained, further deformations prove that peak condition is unstable, because the stress ratio progressively reduces until an ultimate condition

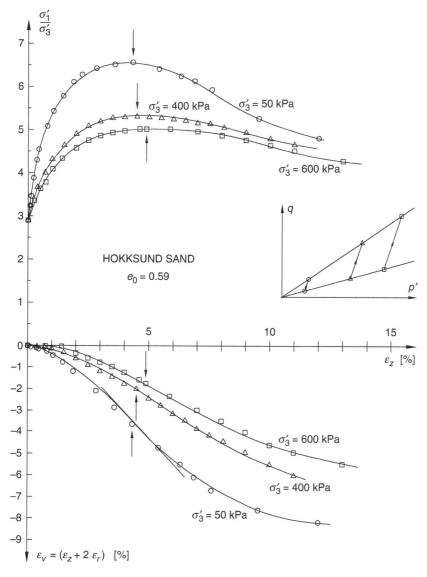

Figure 5.8 Influence of rate of dilation on the peak strength.

is reached where distortions again occur without change of volume or effective stress ratio.

In both cases there is evidence that soil, when sheared, tends towards an *ultimate* (asymptotic) or *stationary condition*, in which shearing can continue without changes of volume or effective stresses. This ultimate condition is referred to as **critical state**, and the corresponding critical state strength is characterized by an angle φ'_{cs} of shearing resistance called *critical state angle* or *angle at constant volume* (for this reason the symbol φ'_{cv} is also used).

If we consider further triaxial tests (Figure 5.8), in which sand specimens have been tested starting from the same initial void ratio $e_o = 0.59$, but subjected to different confining stresses (σ'_r ranges from 50 kPa to 600 kPa), the rate of **dilation** (change of volume referred to as shear distortion) progressively reduces with an increase of the confining stress and, consistently, a similar reduction of peak strength can be observed.

For this reason, if the Mohr circles at peak strength are plotted on the (σ', τ) plane, a curved failure envelope is usually obtained (Figure 5.9), since the peak strength reduces if the stress level increases. Understanding the curvature of failure envelope as well as the difference between peak and critical state strength requires the appreciation of what can be considered the main feature of soils, if compared to other materials: soils change in volume when sheared, and, as it will be proved in the subsequent section, steady friction and the rate of dilation must be summed up to give peak strength. This rate of dilation is currently referred as **dilatancy**, a term that Osborne Reynolds (1885) used to demonstrate change in the volumetric packing of granular materials when subjected to distortion.

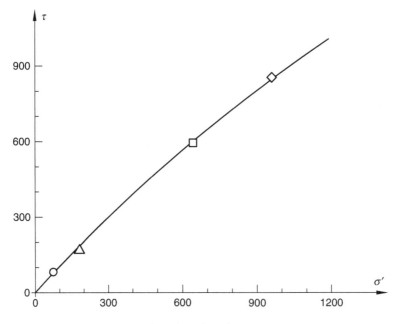

Figure 5.9 Failure envelope of Hokksund sand.

5.5 Taylor expression of dissipation of work

(a) In his book *Fundamentals of Soil Mechanics*, Taylor (1948) used the word *interlocking* to indicate dilatancy and proved that rate of dilation and steady friction must be summed up to give peak strength. To reach this goal, Taylor introduced an expression for dissipation of work, which was later used to generate the original Cam Clay Model (Schofield and Wroth, 1968), as discussed in Section 5.11. Taylor made reference to results of direct shear tests, in which a specimen of Ottawa sand was experiencing a deformation pattern, as shown in Figure 5.10.

Taking into account the external work done by dilation, the loading power supplied to the specimen assumes the expression:

$$\delta W = \tau_{yx} A \delta x - \sigma'_y A \delta y \tag{5.7}$$

and if the total loading power is equated to the frictional work (internal dissipation):

$$\tau_{yx} A \delta x - \sigma'_y A \delta y = \mu \sigma'_y A \delta x, \tag{5.8}$$

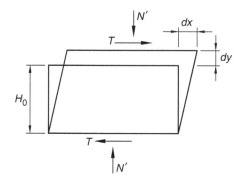

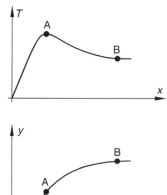

Figure 5.10 Taylor model of dissipation.

the following relation is obtained:

$$\frac{\tau_{xy}}{\sigma'_y} = \mu + \frac{\delta y}{\delta x}. \tag{5.9}$$

This expression states that the mobilized strength is the sum of the stationary friction and the contribution of dilatancy, expressed by the term $\partial y / \partial x$. This latter term is usually also referred in terms of **angle of dilation**, defined as:

$$\psi = \tan^{-1}\left(\frac{-\delta\varepsilon_v}{\delta\gamma}\right) = \tan^{-1}\left(\frac{\delta y}{\delta x}\right). \tag{5.10}$$

In particular, the **peak strength** corresponds to the *maximum rate of dilation* and in terms of angle of shear resistance (5.9) writes:

$$\tan\varphi'_{peak} = \tan\varphi'_{cv} + \left(\frac{\delta y}{\delta x}\right)_{max}. \tag{5.11}$$

Critical state is an asymptotic stationary condition, where distortion occurs without changes of volume or effective stresses, so that critical state strength is defined by the absence of the dilatancy term:

$$\left(\frac{\tau_{xy}}{\sigma'_y}\right)_{cs} = \mu = \tan\varphi'_{cv}. \tag{5.12}$$

The critical state angle φ'_{cv} depends on mineralogy, roughness and grading of particles, but it is independent of the initial conditions. In silica sands it can range from 30° to 38° (higher values apply to angular particles), and in feldspar or carbonatic sands can reach values up to 40°.

On the contrary, since the peak value of the angle of shear resistance is linked to the rate of dilation (Figure 5.11), it depends on soil state and cannot be regarded as a soil property.

The rate of dilation depends on the specific volume and on the effective stress level (Figure 5.12), so that we reach the conclusion that the peak value of the angle of shear resistance increases if the relative density increases, but decreases with higher level of effective stress.

(b) Casagrande (1936) introduced the concept of *critical void ratio* to indicate the ultimate condition, in terms of volumetric packing, towards which granular materials tend irrespective of the initial packing. Data in Figure 5.13, obtained by Wroth (1958) on steel balls tested in simple shear apparatus, support this assumption. However, the critical void ratio e_{cs} is not unique, since it reduces with increasing effective stresses and when the critical void ratio is plotted as a function of the logarithm of the mean effective stress p' the relationship (e_{cs}, p') is a straight line, called **critical state line**.

It is common to plot this relation as a double line, labelled CSL, as shown in Figure 5.14; its position is defined by Γ, the specific volume at a reference pressure p'_a, and the slope λ.

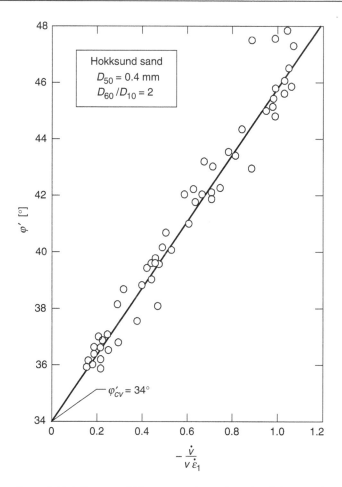

Figure 5.11 Influence of dilatancy on the peak angle of shear resistance (Lancellotta, 1993).

Note that determining the CSL in the (p', e) plane is not a simple matter, because of strain localization (see Section 5.6) and grain crushing at high stress level, and the expected linear relationship is commonly accepted for values of p' up to 1 MPa (see Been *et al.*, 1991; Finno *et al.*, 1997; Harris *et al.*, 1995; Mooney *et al.*, 1998; and Desrue *et al.*, 1996).

Since points lying on this line must have zero rate of dilation, Wroth and Bassett (1965) postulated that the rate of dilation depends on the distance of current state from this line, which means that it is the relation between the actual state and the critical state that has the major influence on soil behaviour.

In Figure 5.14 this distance is shown to be equal to $(\Gamma - v_\lambda)$, if v_λ is the value of the specific volume at the reference pressure p'_a on a line parallel to the CSL passing through the current state (p', v). The parameter v_λ accounts for both variables (p', v), being expressed as:

$$v_\lambda = v + \lambda \ln p',$$

(5.13)

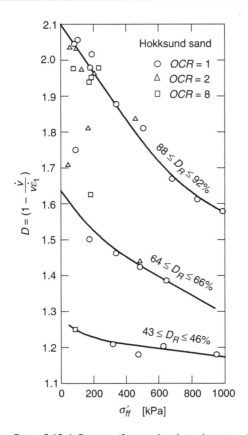

Figure 5.12 Influence of stress level on the rate of dilation (Lancellotta, 1993).

and for this reason it is indicated as **state parameter** (see also Been and Jefferies, 1985).

Therefore, since the peak strength depends on the state of the soil, it can be expected that the peak value of the angle of shear resistance will be a unique function of v_λ, as proved by data reported in Figure 5.15.

Bolton (1986) supplied further experimental evidences supporting all previous arguments and suggested the following relation, which expresses the influence of relative density and stress on the angle of shear resistance:

$$\varphi' - \varphi'_{cv} = mDI < 12°$$
$$DI = D_R \left(Q - \ln p'_f \right) - 1 \tag{5.14}$$

where:

p'_f is mean effective stress at failure (in kPa, in equation 5.14);

m depends on deformation constraints (it is equal to 3 for triaxial strain conditions and equal to 5 for plane strain conditions);

Q depends on grain crushing strength and ranges from 10, for quartz and feldspar sands, to 8 for limestone and 5.5 for chalk.

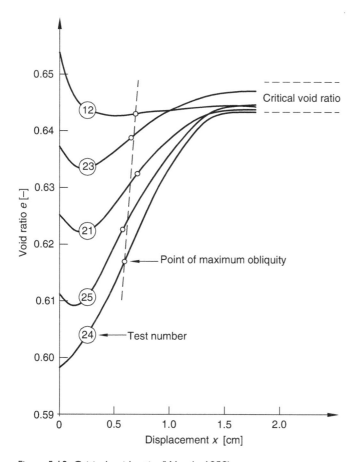

Figure 5.13 Critical void ratio (Wroth, 1958).

(c) Once the difference between the peak and critical state strength has been clarified, a recurrent dilemma is whether ultimate limit state design should be based on peak strength or critical state strength (see Atkinson, 1993; Powrie, 1997; Padfield and Mair, 1984). Figure 5.16 shows some common design situations, which differ from each other because of different stress levels involved and because of different kinematic constraints, as it can be deduced by comparing for example the gravity wall case (a) with the shallow footing case (b), or the deep foundation (c). As an example, when dealing with shallow footings, it is a matter of tradition to express the margin of safety through the introduction of a *load factor*, which is also intended to be a safeguarding measure to avoid excessive settlements under working conditions. In this case the use of the peak strength should have the advantage of avoiding load factors, which changes with the relative density or stiffness.

However, the peak strength should be selected for the maximum stress level that the soil experiences in the field (see Section 8.12). This is particularly relevant for piles, where the stress level below the base is so high as to suppress dilatancy, so that the expected operational strength may be practically coincident with the critical state strength.

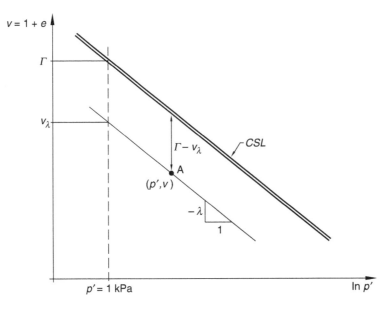

Figure 5.14 Definition of state parameter v_λ.

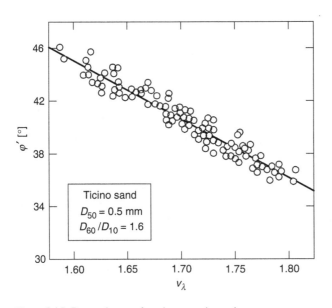

Figure 5.15 Dependence of peak strength on the state parameter v_λ (Lancellotta, 1993).

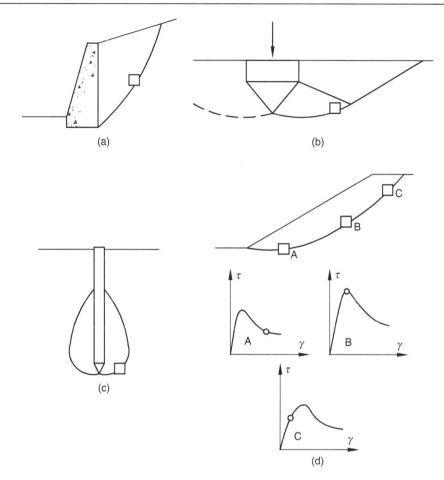

Figure 5.16 Choice of shear strength.

A further aspect that deserves special attention is represented by the **progres-sive failure**. When the failure propagates along a sliding surface (see case d in Figure 5.16), the soil element at 'B' may exhibit its peak strength, but the soil element 'A' may be well past its peak strength, because of large strains and may approach the critical state condition. In this case, it is unrealistic to base the design on peak strength and the choice of the critical state strength will be a safegarding measure against the possibility of a progressive failure.

5.6 Strain localization

When interpreting data from triaxial tests, the main assumptions introduced in Section 5.3 concern the homogeneity of stress and strain, i.e. every material point of the specimen exhibits the same stress and strain. In practice, loose soils tend to barrel (Bishop and Green, 1965), as shown in Figure 5.17a, due to friction constraints on the end platens, and the use of lubricated ends do not completely solve the problem (it may also contribute to bedding errors). Dense soils (densely packed sands and stiff clays) tend to show a localized pattern of deformation (Figure 5.17.b), which further develops in a typical shear band (Arthur *et al.*, 1977; Scarpelli, 1981; Scarpelli and Wood, 1982). The term *shear band* is used to highlight that shear strains are localized within a narrow zone, of thickness of the order of 10 or 20 times the particle dimension, which divides the soil sample in two parts, each one moving relatively to the other as a rigid body.

Once the localization phenomenon develops, the soil sample cannot further be considered as a continuum body, and the experimental results also depend on how much the sample is slender and on the kinematic constraints imposed by the testing apparatus. For this reason, the observed strain softening behaviour is only partially attributed to material behaviour and part of it is attributed to testing constraints (geometric softening, as defined by Drescher and Vardoulakis, 1982).

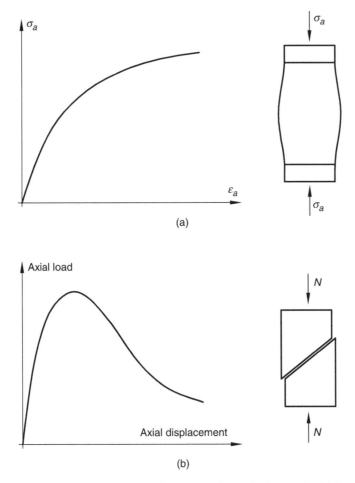

(a)

(b)

Figure 5.17 (a) Deformation of a loose sand or soft clay sample; (b) Strain localization in a sample of dense sand or stiff clay.

5.7 State paths: drained and undrained tests on reconstituted samples

5.7.1 Drained triaxial tests

As already discussed in Section 5.3, when performing *consolidated-drained tests* on clay samples, the soil specimen is reconsolidated during the first stage to a stress state that can be isotropic (*CID* tests) or anisotropic (*CAD* tests and in particular CK_oD tests if the sample is one-dimensionally reconsolidated). It is then sheared slowly to prevent the development of any pore pressure excess (Bishop and Henkel, 1962, provide suggestions about the rate of loading).

As an example, Figure 5.18 shows the experimental data from a *CID* compression test on a reconstituted sample of Pisa clay, and we can describe the soil response as characterized by these features:

- the deviator stress progressively increases until large deformations of the order of 10% to 20% are attained;
- during shearing the specimen continuously reduces its volume and only at large deformations do shear strains occur without any further volume change;
- finally, because there is no excess pore pressure, total and effective stress paths are coincident, and by considering the test conditions (constant radial stress) one finds $\delta q / \delta p' = 3$.

If we analyse the results of other tests on the same clay, considering specimens that have been reconsolidated under different values of the mean effective pressure, it can be observed that the ultimate stationary points (Figure 5.19) all belong to a single locus, called **critical state line**, which in the plane (p', q) is given by the equation:

$$q = Mp', \tag{5.15}$$

where

$$M = \frac{6 \sin \varphi'_{cv}}{3 - \sin \varphi'_{cv}}. \tag{5.16}$$

In the plane (p', v) the critical state condition is represented by the curve shown in the bottom part of Figure 5.19.

If, as is common in soil mechanics, the specific volume is reported as a function of the logarithm of the mean effective stress, the curve of normal consolidation (NCL) and the curve of critical state (CSL) plot as two straight lines, parallel to each other (Figure 5.20), and the critical state line is described by the equation:

$$v = \Gamma - \lambda \ln p'. \tag{5.17}$$

where Γ is the value of the specific volume at the reference mean effective pressure $p' = 1$ kPa.

Note that it can also be proved that loading paths at constant stress ratio $\eta = q/p'$ correspond to a family of parallel lines in the plane $(v, \ln p')$.

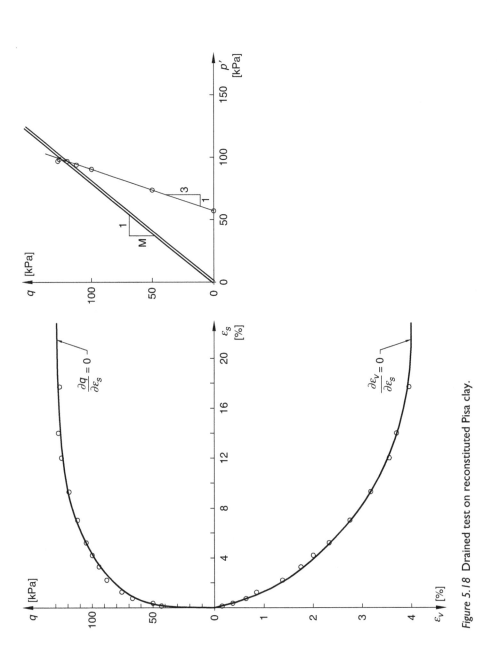

Figure 5.18 Drained test on reconstituted Pisa clay.

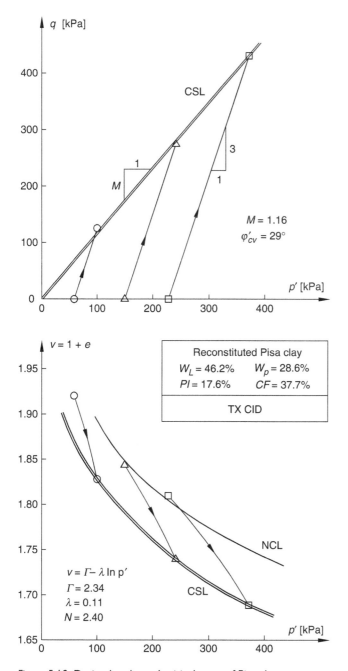

Figure 5.19 Drained paths and critical state of Pisa clay.

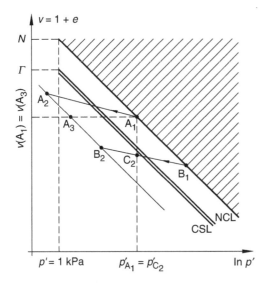

Figure 5.20 Critical state and normally consolidation lines.

The result expressed by (5.15) plays an important role, because it represents a validation of the Mohr-Coulomb criterion. For this reason it is of interest to explore if the critical state locus is independent of the test conditions by also examining the results of consolidated-undrained tests.

5.7.2 Consolidated-undrained tests

When considering the results of a CIU test (see Figure 5.21), total and effective stress paths are no longer coincident, because of the pore pressure excess. We impose the specimen the kinematic constraint that volume cannot change, so that an excess pore pressure develops as reaction to this constraint. One therefore needs to take care of measuring the pore pressure during the shearing stage and, according to the principle of effective stress, only the effective stress path is relevant for the determination of strength parameters.

Referring to the effective stress path, the conclusion is reached that the soil sample tends again to attain a critical state condition, represented by a unique locus, coincident with that obtained with the previously examined drained tests (see Figure 5.22).

5.7.3 Predicting critical state

At this stage it is instructive to show how the introduced critical state concepts are used to predict the ultimate condition reached by following a specified path (drained or undrained), once we have specified the initial condition, as expressed by the state variables (v_o, p'_o, q_o), and the relevant soil parameters (M, Γ, λ). This is done in the following worked examples 5.1 and 5.2.

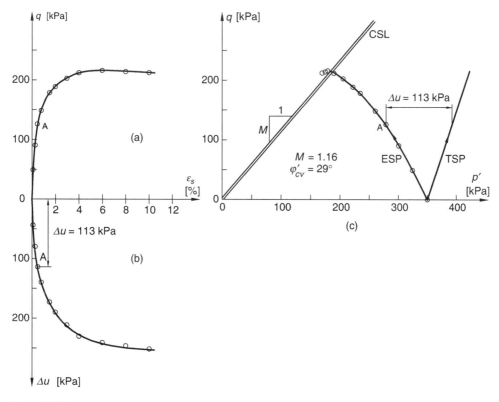

Figure 5.21 Undrained behaviour of Pisa clay.

Example 5.1

Consider the drained test in Figure 5.19, corresponding to the following initial data: $p'_o = 150$ kPa; $v_o = 1.83$. Given the soil parameters $M = 1.16, \Gamma = 2.34, \lambda = 0.11$, it is required to define the effective stress and specific volume at the critical state.

Solution. The loading condition is expressed as $\delta q/\delta p' = 3$, so that at critical state it must be $q_{cs} = Mp'_{cs} = 3(p'_{cs} - p'_o)$.
It follows $p'_{cs} = 3p'_o/(3 - M) = 245$ kPa and $q_{cs} = Mp'_{cs} = 284$ kPa.
When the critical state is reached, both conditions (5.15) and (5.17) must be simultaneously satisfied, so that (5.17) allows to obtain the specific volume $v_{cs} = \Gamma - \lambda \ln p'_{cs} = 2.34 - 0.11(\ln 245) = 1.74$.

Example 5.2

Consider now the undrained test in Figure 5.21. Compute the effective stresses and the specific volume at the critical state, given the initial conditions $p'_o = 350$ kPa, $v_o = 1.76$ and the relevant soil parameters $M = 1.16, \Gamma = 2.34, \lambda = 0.11$.

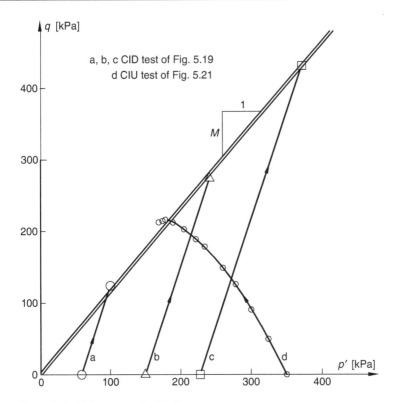

Figure 5.22 Uniqueness of critical state.

Solution. The test conditions impose the kinematic constraint that volume changes are not allowed, so that the specific volume must be constant, i.e. $v_{CS} = v_o$ Then, the mean effective pressure at critical state can be obtained from equation (5.17):

$$p'_{cs} = \exp\left(\frac{\Gamma - v_{cs}}{\lambda}\right) = 195\,\text{kPa}.$$

Finally, by using $q_{cs} = Mp'_{cs}$ we obtain $q_{cs} = 226\,\text{kPa}$.

5.8 State boundary surface

The previously considered loading paths can be thought of as a succession of states, defined by the state variables (p', q, v). If we assume that these variables are the coordinates of an abstract space, we can argue whether or not this space is completely accessible to any arbitrary loading path.

Roscoe *et al.* (1958), on the basis of experimental evidences provided by Rendulic (1936) and Henkel (1960), postulated the existence of a bounding surface, defined **state boundary surface**, which encloses all possible physical states. We now investigate whether drained and undrained paths of *normally consolidated reconstituted clays* belong to the same surface. To do this, we introduce the following working hypothesis: if in the plane (p', q) a drained path intersects an undrained path (as is the case in

Figure 5.22), we can say that these paths belong to the same surface if at the intersection point they have the same specific volume. Rendulic performed drained and undrained tests in Vienna in 1936 and plotted on the $(\sigma'_1, \sqrt{2}\sigma'_3)$ plane contours of water content and found that data from undrained tests were consistent with those of drained tests. In this respect Rendulic established a *generalized principle of effective stress*: for a given clay in equilibrium under given effective stresses at a given initial specific volume, the specific volume after any principal stress increments is uniquely determined by those increments (see Schofield and Wroth, 1968). Later, Henkel (1960) followed the approach of Rendulic, plotting contours of constant water content in the $(\sigma'_z, \sqrt{2}\sigma'_r)$ plane for both drained and undrained paths (Figure 5.23), and again he found that these contours are consistent with each other and have the same form. On this basis Roscoe *et al.* (1958) reached the conclusion that in the space (p', q, v) a unique curved surface is traced by families of drained and undrained paths (Figure 5.24).

The intersection of this surface with the plane $q = 0$ gives the normal compression line *NCL*: NC soil samples, isotropically compressed, move along this line, whereas OC samples move along the unloading curve.

When a NC sample, first reconsolidated under isotropic stress, is successively sheared in **undrained conditions**, its path in the (p', q, v) space can be defined as follows:

- the *stress-path* must lie on the state boundary surface;
- the stress path must also lie on the **undrained plane** of equation $v = v_i = \text{const}$;

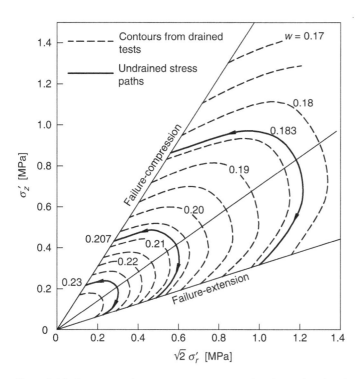

Figure 5.23 Contours of constant water content for drained and undrained tests (after Henkel, 1960).

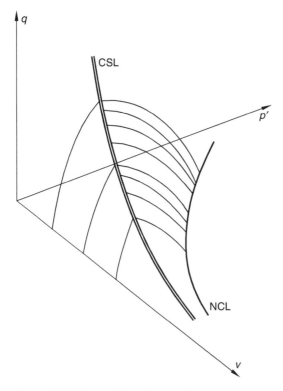

Figure 5.24 The state boundary surface.

- it follows that the stress path is obtained as the intersection of the state boundary surface with the undrained plane, as is shown in Figure 5.25.

When considering many undrained planes (Figure 5.26), the stress paths of NC samples correspond to different sections of the state surface, having different dimensions but all having the same form.

Therefore the experimental results can be normalized with respect to the effective consolidation pressure and experimental evidences show that normalized results merge into a single path.

By using similar arguments, a drained stress path is obtained as the intersection of the state surface with the **drained plane,** defined by a slope 3:1 on the (p', q) plane (see Figure 5.27).

The critical state condition is represented in Figure 5.24 by the **critical state curve.** Its projection into the (p', q) plane is given by the straight line of equation (5.15), and the projection into the (v, p') plane is given by the equation (5.17). Both these equations must be satisfied when the critical state condition is attained.

The isotropic compression line (NCL) is represented by:

$$v = N - \lambda \ln p' \tag{5.18}$$

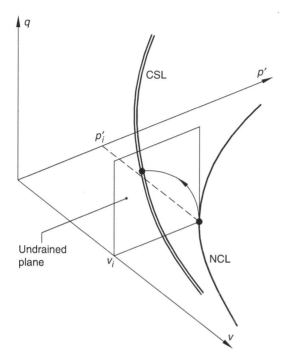

Figure 5.25 Undrained path of a NC soil sample.

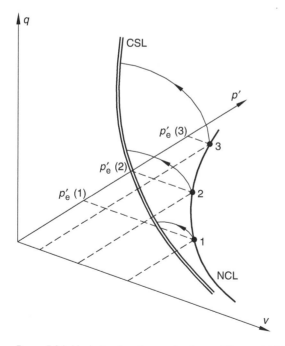

Figure 5.26 Undrained paths moving from different initial states.

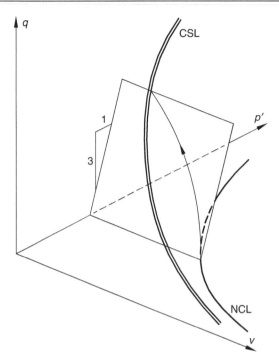

Figure 5.27 Drained path of a NC soil sample.

and the unloading curve is given by:

$$v = v_k - k \ln p',$$
(5.19)

N and v_k being the specific volume at a reference effective pressure of 1 kPa.

5.9 Stiff clays: structural features, peak, post-peak and residual strength

5.9.1 Structural features of stiff clays

The definition of stiff clay generally applies to clays with a liquidity index lower than 0.5 in their natural state (Terzaghi, 1936), but it is also common to describe a clay as stiff if its undrained shear strength exceeds 75 kPa (see Section 5.10).

The first definition identifies, in reality, just a limit for the water content (see Figure 5.28), below which the clay can be defined stiff. Moreover, the same figure also shows that a stiff clay can have a different degree of overconsolidation ratio, and in this respect a normally consolidated clay can also be stiff at greath depths.

However, as most of the time we are concerned with problems at relatively shallow depths, the definition of stiff clay indicates heavily overconsolidated clays, even if in principle the two terms cannot be considered to be synonymous.

It is still necessary to stress that in general a given behaviour cannot be associated with the overconsolidation ratio alone. As it will be shown in the sequel, a more

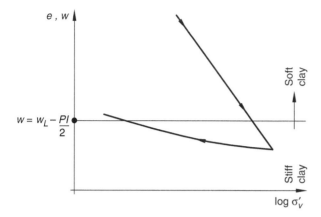

e, w

Soft
clay

$w = w_L - \dfrac{PI}{2}$

Stiff
clay

$\log \sigma'_v$

Figure 5.28 Definition of a stiff clay (Calabresi *et al.*, 1990).

general criterion is given by the difference between the initial state conditions and those corresponding to the critical state: strain hardening or strain softening behaviour can be observed according to these differences.

In the following, reference will be made to soils having a strain softening behaviour, i.e. those clays that in the Critical State framework reach the Horslev surface, before moving towards the ultimate condition.

In addition to the above considerations, it must be observed that stiff clays exhibit discontinuities, in the form of bedding surfaces, small non-systematic fissures and joints. In many cases, stiff clays have also been sheared by tectonic events and landsliding, so that shear zones containing one or more principal surfaces are possible.

These discontinuities play a major role when dealing with stability analyses (see Section 8.20 in Chapter 8), because they give the soil mass surfaces of weakness and the strength of the clay mass can result in being much lower than the strength of the intact clay. For this reason it is important to distinguish between the behaviour observed for intact laboratory soil specimens and the behaviour of the *in situ* soil mass.

5.9.2 Peak strength

Figure 5.29 shows the Mohr-Coulomb failure envelope for Vallericca clay (Burland *et al.*, 1996), which can be considered as a representative behaviour of stiff clays. Vallericca Clay (a few kilometres north of Rome) was deposited in the neritic environment in the depression formed during the late stage of the alpine orogenetic chain, which coincides with the valley of the Tiber river. Average index properties are: $w_L = 60.2\%$; $PI = 33.4\%$; $w_N = 28.6\%$, and the maximum overburden experienced by the clay was about 220 m (Burland *et al.*, 1996).

When analysing these data, a first feature to be considered is the significant curvature near the origin, an aspect relevant when selecting strength parameters. In fact, it is common practice to replace the actual failure envelope with a straight line representing the Mohr-Coulomb failure criterion:

$$\tau_f = c' + \sigma'_f \tan \varphi',$$

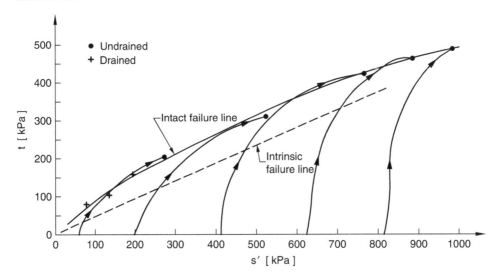

Figure 5.29 Intact Vallericca clay: peak strength and undrained stress paths (Burland *et al.*, 1996).

which in the plane

$$\left(s' = \frac{\sigma_1' + \sigma_3'}{2}, t = \frac{\sigma_1' - \sigma_3'}{2} \right)$$

maps into:

$$t_f = a' + s_f' \tan\alpha',$$

being

$$c' = \frac{a'}{\cos\varphi'}, \quad \sin\varphi' = \tan\alpha'.$$

This produces a cohesion intercept, even if the actual relation curves towards the origin. Therefore, extreme caution is required when selecting a value of cohesion intercept (presumed over the stress range of interest) to be used for stability analyses, or when giving to this intercept any physical meaning (as will be shown later on).

Secondly, it can be observed that samples show a negative excess pore pressure when sheared in undrained conditions, suggesting a pronounced dilatant behaviour, which is peculiar to heavily overconsolidated soils.

Finally, if the failure envelope of the natural clay is compared with the intrinsic failure envelope of the reconstituted clay, the observed difference can be attributed to two factors: the void ratio at failure and the soil microstructure. Recall that in Section 4.4 we defined *intrinsic* as those basic or inherent properties of a remoulded clay, which are independent of its natural state (Burland, 1990).

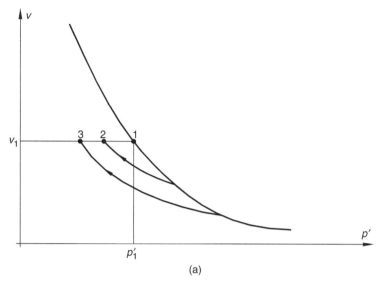

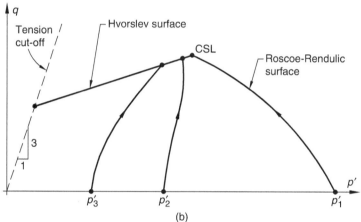

Figure 5.30 Undrained tests on lightly overconsolidated samples: (a) initial conditions; (b) failure states (adapted from Atkinson and Bransby, 1978).

Since peak strength data cannot be properly understood unless they are corrected for the value of state variable at the moment of failure, and the influence of the microstrucure cannot be isolated without an appropriate normalization, the discussion will proceed by considering these aspects.

5.9.3 Normalizing peak states

The state of a normally consolidated clay is uniquely defined by the effective stress. On the contrary, when dealing with overconsolidated clay, Figure 5.20 claims for the need to specify both the effective stress and the void ratio. In fact, if we consider samples at points A_1 and C_2, they have the same effective stress, but very different stiffness; the same conclusion applies to samples A_3 and A_1, they have the same specific volume, but again very different stiffness. It follows that the current state needs to be specified by both variables, the specific volume and the effective stress, or, alternatively by only two of the parameters v, p', OCR.

Accordingly, if two samples at failure have two different values of the water content, their state is represented by points which lie on two different sections of the state boundary surface, so that it is not completely consistent to simply plot the mean effective pressure and the deviator stress on the (p', q) plane, in order to describe the failure envelope, without a normalization criterion.

In order to have the same water content at failure, the experimental programme could use specimens having the same specific volume, but different overconsolidation ratios, sheared in undrained conditions (see Figure 5.30a). The locus of peak strengths, sketched in Figure 5.30b, can be interpreted in this case as the intersection of the undrained plane with the portion of the state boundary surface on the left of critical state, called **Hvorslev surface**. The portion of the state boundary surface on the right of the critical state, previously explored when we considered the behaviour of soft clays, is called **Roscoe-Rendulic surface**. These definitions originated from contributions given by soil scientists to this problem. In Section 5.5 we introduced the state parameter v_λ (Figure 5.14):

$$v_\lambda = v + \lambda \ln p',$$

which accounts for both variables (p', v), and we showed that peak value of the angle of shear resistance is a unique function of v_λ, as supported by data reported in Figure 5.15.

This interpretation of peak strength was based on the understanding that peak states, associated with dense or overconsolidated soils on the dry side of critical state, depend on friction and dilation and the maximum rate of dilation depends on the initial state or on the overconsolidation ratio. Note that lines of constant overconsolidation ratio correspond to equation (5.13).

An alternative procedure is based on the normalization criterion suggested by Hvorslev (1937). If we define as **equivalent stress** p'_e the effective mean stress which on the isotropic compression curve NCL corresponds to the water content at failure (Figure 5.31), then the equation of the NCL gives:

$$p'_e = \exp\frac{N-v}{\lambda} \tag{5.20}$$

and when the peak data are normalized $\left(\dfrac{q}{p'_e}, \dfrac{p'}{p'_e}\right)$ they define a unique failure envelope:

$$\frac{q}{p'_e} = h\left(a + \frac{p'}{p'_e}\right), \tag{5.21}$$

where:

$$h = \frac{6\sin\varphi'_e}{3 - \sin\varphi'_e} \quad a = \frac{c'_e}{p'_e}\cot\varphi'_e.$$

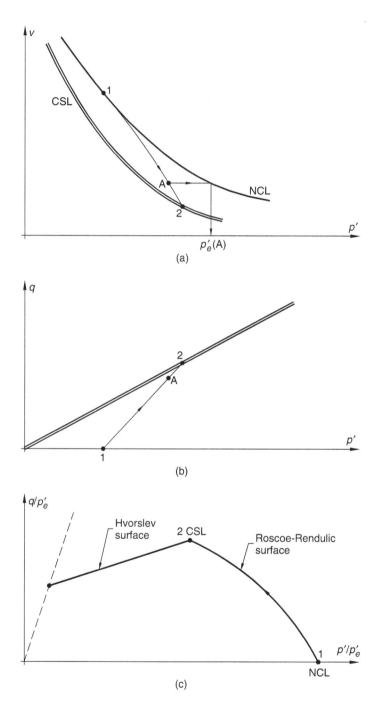

Figure 5.31 Normalization of peak strength: (a) equivalent mean stress (Hvorslev, 1937); (b) drained paths; (c) normalized representation of the State Boundary Surface.

The key aspect of this analysis is represented by the fact that cohesion depends linearly on the equivalent stress p'_e, and, since a linear relation exists between the water content w and $\ln p'$, the apparent cohesion increases exponentially with the decrease of water content.

An even more general conclusion follows. The conventional interpretation of peak data with the use of a Mohr-Coulomb failure envelope gives an *apparent cohesion* intercept. This is not the shear stress which the soil can sustain at zero effective stress, but is merely a parameter to describe the Mohr-Coulomb envelope. Within the framework for *unbonded soils*, the strength is a purely frictional phenomenon.

In this context, it can also be argued that there is a boundary to the Hvorslev surface on the left. Indeed, if it is assumed that soil cannot withstand tensile effective stress, then the condition of $\sigma'_r = 0$ defines the zero tension line:

$$q = 3p',$$

so that the Hvorslev surface is bounded by the zero tension line and the critical state line.

When dealing with **natural stiff clays**, as suggested by Burland *et al.* (1996), the normalization through the equivalent effective pressure gives the possibility to properly take into account the influence of void ratio, so that any difference between the failure envelope of the natural clay and that of the reconstituted clay can be ascribed to soil microstructure. This difference is apparent in Figure 5.32 (Burland, 1990; Rampello, 1989): the dotted line represents the state boundary surface for reconstituted Todi clay, and the normalized peak states for both the intact and freely swelled natural clay lie above the intrinsic Hvorslev surface.

There are further remarks related to the definition of the equivalent effective mean stress and the peak strength interpretation. The equivalent effective mean stress can also be defined assuming it lies on the critical state line, rather than on the NC line,

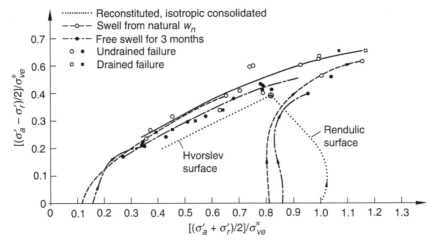

Figure 5.32 Results of triaxial tests on Todi clay, normalized by the equivalent mean stress σ^*_{ve} at failure (reproduced with permission after Burland, 1990).

as suggested by Atkinson (1993, 2007), who argues that the critical state line is unique, whereas the position of the NC line of natural soils can be influenced by structural effects (also see Callisto and Rampello, 2004).

Regarding the second aspect, as an alternative of considering a linearized Mohr-Coulomb failure envelope with an apparent cohesion, a failure envelope strongly curved towards the origin can also be described (Baligh, 1976; De Mello, 1977; Atkinson, 1993, 2007).

In this latter case, Atkinson (2007) suggests the use of the equation:

$$\frac{q}{Mp'_c} = \left(\frac{p'}{p'_c}\right)^{\beta},$$
(5.22)

where p'_c is the equivalent pressure on the critical state line, which accounts for differences in specific volume, and β is a material parameter that describes the degree of curvature of the failure envelope (published data give values of β ranging from 0.17 to 0.73).

5.9.4 Post-peak behaviour

An additional factor complicating the determination of appropriate strength parameters is the **brittle behaviour** exhibited by stiff clays (see Figure 5.33). When the peak strength is reached, the specimen suffers a sudden collapse, displacements are the results of **localized** movements on a newly formed failure surface, and a very stiff loading system is required if the post-peak behaviour is to be detected with accuracy. It is apparent that this behaviour cannot be described any more in terms of continuum mechanics. As reported by Burland *et al.* (1996), slip surface started at the maximum stress ratio (coincident in this case with the peak strength) and was completed when the post-failure strength was reached.

Immediately after the peak stress (at a relative displacement of a few millimetres), the strength drops to values defined by an intercept tending to zero and a friction angle, which, at low to moderate stress level, is similar to that of the reconstituted material (Figure 5.34), also suggesting that the rupture creates a local fabric similar to that of the reconstituted material.

This latter conclusion is of great practical relevance, since back analyses of first sliding failure in fissured clays (Skempton, 1977) suggest that the **operational effective strength** envelope lies between the lower bound envelope for the strength on fissures and the post-peak strength for initially intact samples.

5.9.5 Residual strength

Further relative displacement causes a continuous reduction of shear strength, until, at very large displacements the **residual strength** is reached (Figure 5.35).

Residual conditions are characterized by zero cohesion and by a friction angle φ'_r dependent on the percentage of clay particles that can be reoriented during shearing.

In soils mainly composed of bulky particles ($CF < 20\%$), shearing does not induce any preferred orientation and φ'_r is equal or slightly smaller than φ'_{cv}, whereas in soils

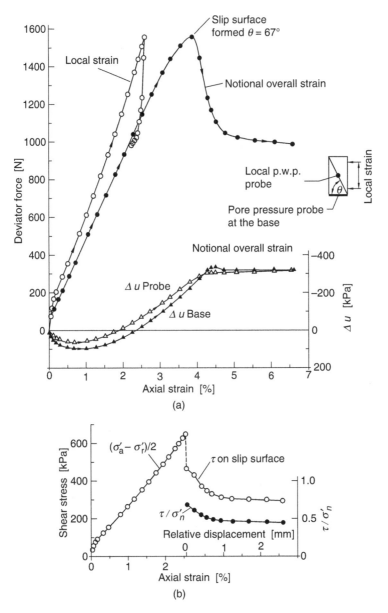

Figure 5.33 Unconsolidated undrained test on Todi clay, with pore pressure measurements (reproduced with permission after Burland, 1990).

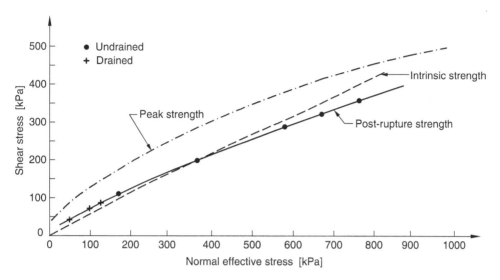

Figure 5.34 Vallericca clay: post-rupture Coulomb strength envelope compared with peak intact and intrinsic Mohr-Coulomb failure lines (Burland *et al.*, 1996).

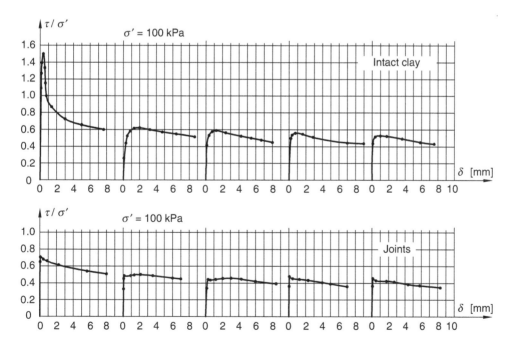

Figure 5.35 Direct shear test on Santa Barbara clay (Calabresi and Manfredini, 1973).

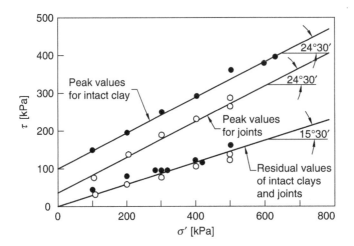

Figure 5.36 Shear strength of intact clay and along joints (Calabresi and Manfredini, 1973).

composed of platy particles ($CF > 50\%$), shearing induces particle orientation and φ'_r is significantly smaller than the critical state value (Figure 5.36) (Lupini *et al.*, 1981; Skempton, 1985).

The direct shear box test has been widely used to this purpose, by repeatedly shearing the specimen and reversing the shear box to its original position (see Figure 5.35), until a steady residual value is attained. An alternative procedure is offered by the ring shear apparatus (Bishop *et al.*, 1971), in which large displacements may be obtained in the same direction.

Since many studies have proved that fully remoulded samples gave the same residual strength as undisturbed samples, it has been assumed that φ'_r can be correlated to the clay fraction of soils (Figure 5.37). However, it must be observed that the development of a smooth shear surface can be inhibited in clay soils containing lithorelicts, as clay shales, and in this case the residual strength of the natural soil may result higher than the residual envelope of the reconstituted material (Leroueil *et al.*, 1997).

Even when the clay fraction controls the residual strength, Kenney (1967, 1977) showed that it is the clay mineral that is the governing factor (φ'_r can be as low as 4° for sodium montmorillonite and up to 15° for kaolinite), and it has been also found that φ'_r increases as salt concentration increases (Ramiah *et al.*, 1970; Moore, 1991; Di Maio, 1996; Di Maio and Fenelli, 1994). In this respect, rather than regarded as a property, the residual strength appears to be a behaviour dependent on the environment.

5.10 Undrained shear strength

In previous sections it has been shown that, when direct measurements of pore pressure are carried out, it is easy to investigate soil behaviour in undrained conditions in terms of effective stress.

But this is not the case in the field, because the excess pore pressure largely depends on local soil features, developments of plastic zones and anisotropy. To avoid these difficulties it is the usual practice to introduce the **undrained shear strength**, expressed in terms of **total stress**. This is equivalent to considering any soil element as a single phase closed system, which does not exchange water with the surrounding ambient.

The usual way of measuring the undrained strength has been, for a long time, to refer to **unconsolidated-undrained tests** (*UU* tests). In these tests a cell pressure is applied to the specimen during the first stage, but *consolidation is not allowed as the drainage is closed*. During shearing, the drainage is still held closed and the deviator stress at failure is measured.

If the *soil is saturated* and the tested soil samples are all from the same depth, they will reach the same value of the deviator stress at failure, independent of the applied total stress (Figure 5.38).

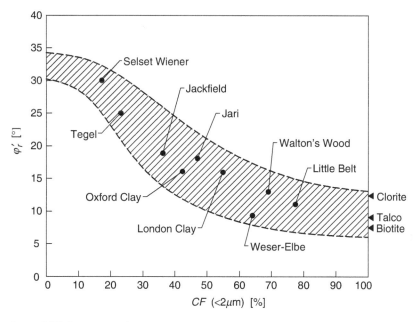

Figure 5.37 Decrease of residual strength with increasing clay fraction (reproduced with permission after Skempton, 1964).

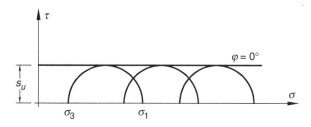

Figure 5.38 Failure criterion in terms of total stress.

The obtained failure envelope, in terms of total stress, is a straight horizontal line, defined by the value of the **undrained strength**:

$$\tau_f = s_u. \tag{5.23}$$

The result expressed by (5.23) can be justified by observing that, since the soil samples have all the same water content (and the same specific volume v_o), which is not changed during the test, at failure it must be:

$$s_u = \frac{1}{2}q_f = \frac{1}{2}Mp'_f = \frac{1}{2}M\exp\left(\frac{\Gamma - v_o}{\lambda}\right), \tag{5.24}$$

where the undrained strength s_u is expressed in the unit of the reference pressure corresponding to Γ.

According to (5.24), Critical State Theory predicts a unique relation between the undrained strength and the specific volume, and because of this relation the conclusion is also reached that the undrained strength cannot be considered as a soil property, but it is rather a soil behaviour.

However, UU tests do not allow to investigate all aspects of this behaviour, and a better choice is to perform **consolidated-undrained tests** (CU tests).

When soil samples of soft clay are reconsolidated under different values of mean effective pressure and then sheared in undrained conditions, based on the previous equation (5.24) we can predict a unique relation between the undrained strength and the consolidation effective pressure, i.e.:

$$\frac{s_u(1)}{p'_o(1)} = \frac{s_u(2)}{p'_o(2)}, \tag{5.25}$$

because the undrained stress paths correspond to parallel sections of the state surface (Figure 5.26).

The result expressed by (5.25) can be justified by observing that:

$$v_o = N - \lambda \ln p'_o, \tag{5.26}$$

so that, by substituting this expression into (5.24), we obtain:

$$\frac{s_u}{p'_o} = \frac{M}{2}\exp\left(\frac{\Gamma - N}{\lambda}\right) = \text{const.} \tag{5.27}$$

As an example, Figure 5.39 shows the experimental results related to CK_oU tests performed on Panigaglia (La Spezia) clay, supporting the relation (5.27). Moreover, it is of interest to observe that

$$p'_o = \frac{1 + 2K_o}{3}\sigma'_{vo}, \tag{5.28}$$

so that equation (5.27) has the implication that in an NC clay deposit the undrained strength increases linearly with depth.

When dealing with **overconsolidated clays**, it is of interest to observe (see Figure 5.40) that, as already discussed in Chapter 4, we can also define an overconsolidation ratio R (equivalent to OCR) in terms of mean effective pressure:

$$R = \frac{p'_o}{p'_{oc}}.$$

(5.29)

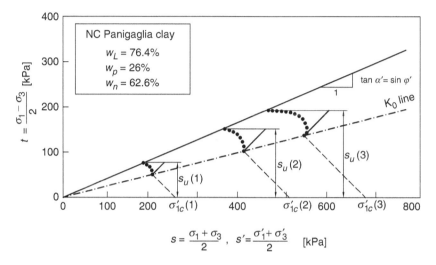

$$s = \frac{\sigma_1 + \sigma_3}{2}, \quad s' = \frac{\sigma'_1 + \sigma'_3}{2} \quad [kPa]$$

Figure 5.39 Undrained strength of Panigaglia clay.

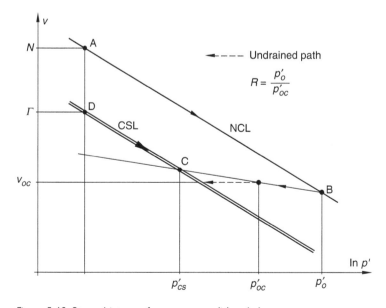

Figure 5.40 Stress history of an overconsolidated clay.

The undrained strength of an overconsolidated sample must depend on its specific volume:

$$s_u = \frac{1}{2} M \exp\left(\frac{\Gamma - v_{OC}}{\lambda}\right) \tag{5.30}$$

and since

$$v_{OC} = N - \lambda \ln p'_o + k \ln R, \tag{5.31}$$

one gets

$$s_u = \frac{1}{2} M \exp\left(\frac{\Gamma - N}{\lambda} + \ln p'_o - \frac{k}{\lambda} \ln R\right). \tag{5.32}$$

If in the r.h. side we add and subtract the quantity $\ln p'_{OC}$, also taking into account (5.27), it follows:

$$\left(\frac{s_u}{p'}\right)_{OC} = \left(\frac{s_u}{p'}\right)_{NC} R^{(\lambda-k)/\lambda}. \tag{5.33}$$

The expected value of the exponent term in (5.33) is of the order:

$$\frac{\lambda - k}{\lambda} \cong 0.8. \tag{5.34}$$

It is interesting to observe how well the trend of the normalized undrained strength with OCR, predicted on the basis of Critical State Theory, may explain the experimental evidences, summarized in Figure 5.41. However, when selecting the operational value of the undrained strength one needs to realize that this parameter depends on the imposed loading path (see Figure 5.42). For example, the soil element A will eventually reach failure in compression loading, but elements B and C will follow paths represented by simple shearing and extension loading respectively. As a consequence, prediction based on triaxial compression tests can be unsatisfactory, because they tend to overpredict the operational strength, mainly when dealing with low plasticity clays (Figure 5.43).

Further, the development of strains along a failure surface forces a large number of soil elements well past the peak strength, leading to a *progressive failure* characterized by an operational strength lower than the peak.

Having considered all these factors, Koutsoftas and Ladd (1985) suggest to evaluate the operational strength by means of the expression:

$$\frac{s_u}{\sigma'_{vo}} = (0.22 \pm 0.03) \cdot OCR^{0.8}. \tag{5.35}$$

The above relation is supported by experimental evidences related to values of OCR up to 10, and a lot of care must be taken when analysing the behaviour of *stiff fissured clays*.

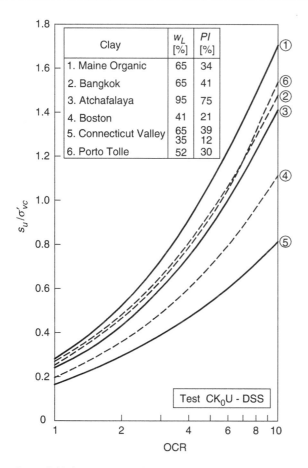

Clay	w_L [%]	PI [%]
1. Maine Organic	65	34
2. Bangkok	65	41
3. Atchafalaya	95	75
4. Boston	41	21
5. Connecticut Valley	65 35	39 12
6. Porto Tolle	52	30

Test CK_0U - DSS

Figure 5.41 Increase in undrained strength with OCR (Ladd and Edgers, 1972).

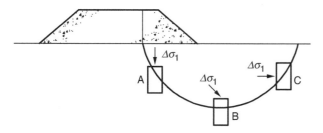

Figure 5.42 Stress states of soil elements along a failure surface.

Researches performed mostly on London clay (Marsland, 1974) indicate that laboratory tests are poorly representative of *in situ* soil behaviour, mainly due to the fact that the influence of fissures and macrofabric features increases with the dimensions of the tested volume. In addition, it must be observed that it is at least doubtful if the behaviour of fissured clays can be defined completely undrained.

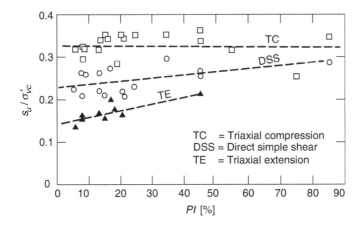

Figure 5.43 Dependence of undrained strength on loading path (Jamiolkowski *et al.*, 1985).

5.11 Predicting soil behaviour: the original Cam Clay Model

In Chapter 3 we stated that in order to identify the plastic deformations it is necessary to introduce a *yield criterion*, a *plastic flow rule* and a *hardening rule*, which allow one to specify when plastic deformations occur, their direction and their magnitude. We further explain these concepts in this section by deriving the mathematical formulation of the original **Cam Clay Model** (Schofield and Wroth, 1968). The intention is to show how a plastic-hardening model can succeed in capturing the essential features of soil behaviour, and in this didactic perspective quantitative aspects will not be addressed. For more insights the reader can refer to Muir Wood (2004) and Yu (2006).

5.11.1 Yield surface

The first requirement of plasticity is to specify when plastic deformations occur. This, for a uniaxial loading condition, corresponds to specifying the yield stress beyond which plastic deformations occur. Certainly, in Soil Mechanics the most familiar example of yield stress is provided by the preconsolidation stress in one-dimensional compression test, but for a more generalized stress path the requirement is to specify the yield surface in the stress space.

Making reference to Figure 5.44, a stress path producing only elastic deformations must lie on an elastic domain, represented in the space (p', q, v) by a vertical wall, the intersection of which with the compression plane (p', v) is given by the unloading curve.

Since the stress path must also be within the yield surface, this latter is obtained as the intersection of the elastic domain with the state boundary surface.

Taylor (1948) introduced an expression for dissipation of work, which was later used to generate the original Cam Clay Model. Taylor made reference to results of direct shear tests, and was able to show that the mobilized strength is the sum of the

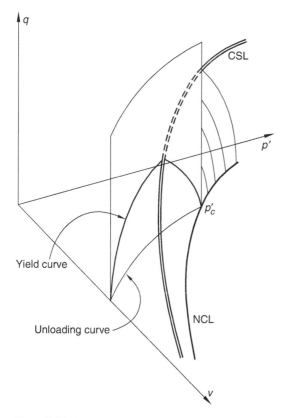

Figure 5.44 Elastic domain.

stationary friction and the contribution of dilatancy:

$$\frac{\tau}{\sigma'} = \mu + \frac{\delta y}{\delta x}.$$

This relation can be rewritten by using stress and strain variables introduced when describing and interpreting triaxial tests, so that it assumes the form:

$$\frac{d\varepsilon_v^p}{d\varepsilon_s^p} = M - \frac{q}{p'}, \qquad (5.36)$$

which, in the context of plasticity theory represents a flow rule, obtained under the assumption that the external work $(q\delta\varepsilon_s + p'\delta\varepsilon_v)$ is dissipated as friction $(Mp'\delta\varepsilon_s)$.

In the original Cam Clay Model (Schofield and Wroth, 1968) dissipation is due to deviator plastic strains; later, in the so-called Modified Cam Clay Model, Roscoe and Burland (1968) assumed that plastic volume deformations also contribute to dissipation.

In Chapter 3 we stated that the Drucker (1959) stability postulate has two important implications: the yield surface must be convex and the plastic strain increment vector must be orthogonal to the yield surface. In this case, the definition of *associated flow rule* is introduced, due to the fact that the yield function plays both the role of specifying when plastic flow occurs and identifying the direction of the plastic strain increment vector. This latter condition implies:

$$\frac{d\varepsilon_s^p}{d\varepsilon_v^p}\frac{dq}{dp'} = -1,$$ (5.37)

and by substituting (5.37) into (5.36) one gets:

$$\frac{dq}{dp'} - \frac{q}{p'} + M = 0.$$ (5.38)

Integration techniques suggest to assume:

$$A(p') = \int -\frac{1}{p'}dp',$$

so that the general integral of (5.38) is given by:

$$q = Ce^{-A(p')} + e^{-A(p')}\int -Me^{A(p')}dp',$$

and:

$$\frac{q}{p'} + M\ln p' = C.$$ (5.39)

The integration constant C can be determined by imposing that $p' = p'_o$ if $q = 0$, so that the yield surface assumes the expression:

$$\frac{q}{p'} + M\ln\frac{p'}{p'_o} = 0.$$ (5.40)

Alternatively, one can also impose the condition $q = q_{cs}$ if $p' = p'_{cs}$, so that the yield surface can also be represented by the equation:

$$\frac{q}{Mp'} + \ln\frac{p'}{p'_{cs}} = 1.$$ (5.41)

5.11.2 Hardening rule

The mean effective pressure p'_o in equation (5.40) plays the role of a *hardening parameter*, because the yield surface changes when p'_o changes, in particular the yield surface modifies its size without changing its shape (isotropic hardening). The hardening rule can be obtained by observing that, when a soil sample is loaded from point O to point A in Figure 5.45, the total volume strain is equal to:

$$\delta\varepsilon_v = \frac{\lambda}{v}\frac{\delta p'_o}{p'_o}. \tag{5.42}$$

The elastic component can be detected by unloading from A to B:

$$\delta\varepsilon_v^e = \frac{k}{v}\frac{\delta p'_o}{p'_o}, \tag{5.43}$$

so that the plastic component is expressed by:

$$\delta\varepsilon_v^p = \frac{\lambda - k}{v}\frac{\delta p'_o}{p'_o} \tag{5.44}$$

which allows to obtain the hardening rule, i.e. how changes of p'_o are related to plastic volume strains:

$$\delta p'_o = \frac{vp'_o}{\lambda - k}\delta\varepsilon_v^p. \tag{5.45}$$

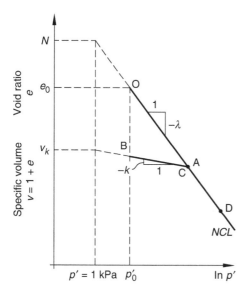

Figure 5.45 Elastic and plastic volume strain.

5.11.3 Computing plastic deformations

By using the normality rule, the plastic deformation components are expressed in the form:

$$d\varepsilon_v^p = \Lambda \frac{\partial g}{\partial p'}$$
$$d\varepsilon_s^p = \Lambda \frac{\partial g}{\partial q}$$
(5.46)

The scalar quantity Λ is determined through the *consistency condition*, which requires that the yield surface must follow the evolution of the stress state:

$$df = \frac{\partial f}{\partial p'}dp' + \frac{\partial f}{\partial q}dq + \frac{\partial f}{\partial p'_o}dp'_o = 0,$$
(5.47)

so that the point representing the current stress state must lie on the yield surface. Since the hardening parameter is linked to the evolution of the plastic deformations through the relation:

$$dp'_o = \frac{\partial p'_o}{\partial \varepsilon_v^p}d\varepsilon_v^p + \frac{\partial p'_o}{\partial \varepsilon_s^p}d\varepsilon_s^p,$$
(5.48)

by combining (5.46), (5.48) and (5.47) gives (also note that $f = g$):

$$\Lambda = -\frac{\dfrac{\partial f}{\partial p'}\delta p' + \dfrac{\partial f}{\partial q}\delta q}{\dfrac{\partial f}{\partial p'_o}\left(\dfrac{\partial p'_o}{\partial \varepsilon_v^p}\dfrac{\partial f}{\partial p'} + \dfrac{\partial p'_o}{\partial \varepsilon_s^p}\dfrac{\partial f}{\partial q}\right)}.$$
(5.49)

At this stage, it is matter of using (5.40) and (5.45) to obtain the expression of the plastic deformations, which in matrix form write:

$$\left\{\begin{matrix} \delta\varepsilon_v^p \\ \delta\varepsilon_s^p \end{matrix}\right\} = \frac{\lambda - k}{Mp'v}\begin{bmatrix} M - \dfrac{q}{p'} & 1 \\ 1 & \dfrac{1}{M - \dfrac{q}{p'}} \end{bmatrix}\left\{\begin{matrix} \delta p' \\ \delta q \end{matrix}\right\}.$$
(5.50)

5.11.4 Elastic soil behaviour

The elastic components of deformation, expressed through a similar matrix form, are given by:

$$\left\{\begin{matrix} \delta\varepsilon_v^e \\ \delta\varepsilon_s^e \end{matrix}\right\} = \begin{bmatrix} 1/K' & 0 \\ 0 & 1/3g' \end{bmatrix}\left\{\begin{matrix} \delta p' \\ \delta q \end{matrix}\right\},$$
(5.51)

so that (5.50) and (5.51) allow one to completely compute the deformation for a given stress increment, paying attention to the fact that (5.50) are only active if the stress point lies on the yield surface. It is also matter of care to consider that the expression of elastic volume strain:

$$\delta\varepsilon_v = k\frac{\delta p'}{vp'}$$ (5.52)

states that the soil bulk modulus:

$$K' = \frac{vp'}{k}$$ (5.53)

depends on the current stress level, so that soil behaviour is non-linear. In addition, if the Poisson ratio is assumed to be constant, then the shear modulus, linked to the bulk modulus through the relation:

$$G = \frac{3(1-2v)}{2(1+v)}K',$$ (5.54)

also depends on the effective mean pressure. This originates a conceptual difficulty, as the assumption of pressure dependent stiffness leads, in non-monotonic loading conditions, to a non-conservative model (see Zytinski *et al.*, 1978). However, when dealing with monotonic loading conditions, it is current practice to account for the soil non-linearity through the introduction of a variable soil modulus, i.e. an equivalent elastic stiffness properly adjusted to account for the current stress and strain level.

5.12 Soil stiffness

As already shown in Chapter 3, modelling soil response requires to specify the link between strain increments and stress increment through a **stiffness matrix C**, i.e.:

$$\sigma_{ij} = C_{ijhk}\varepsilon_{hk},$$ (5.55)

and this incremental link can be considered the most general definition of soil stiffness.

Since soil response is typically non-linear, in order to solve engineering problems, we need to characterize and to model this non-linearity, so that in the sequel we examine the following aspects:

- strain level experienced during engineering work, characterization of soil non-linearity and measurements of stiffness at low strains;
- mathematical modelling of soil response.

5.12.1 Characterization of soil non-linearity and strain measurements

The selection of an appropriate design value of soil stiffness requires the knowledge of an order of magnitude of the strains reached in the ground during engineering work. Even though it is not easy to find a large amount of data, some well-documented cases have been collected (see the General Reports by Atkinson and Sallfors and by Burghignoli *et al.*, the tenth European Conference of SMFE, Florence 1991).

These data prove that, if we disregard the areas of contact with the loading boundaries, where local enclaves of plastic behaviour can take place, most of the subsoil experiences low strains, in the range from 10^{-4} to 10^{-3}.

Investigating soil response at small strains requires special care, since conventional procedures of measuring the vertical deformations of the sample, by observing the movement of the ram through a displacement transducer or a dial gauge, include spurious movements of the system. For this reason, over the past two decades, the technology for measurements of small strains evolved significantly and was able to cover the gap between the soil stiffness, measured in conventional laboratory tests, and the stiffness measured using geophysical techniques in the laboratory or in the field. At present, direct measurements of soil stiffness at low strains is being performed by using local strain gauges in the triaxial tests, or, as an alternative, from measurements of shear wave velocity using piezoceramic crystals or in resonant column tests.

The relevant aspect, which emerges from the available experimental evidence, is the pronounced soil non-linearity even at very low level of strains, which can be described with efficacy by representing how the secant stiffness changes with strain (Figure 5.46).

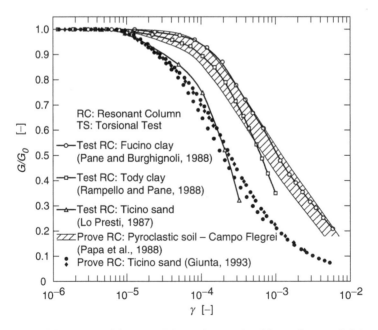

Figure 5.46 Decay of shear modulus with strain level (Lancellotta and Calavera, 1999).

In this respect, we recall that, when dealing with non-linear materials, stiffness can be defined as the **tangent stiffness**:

$$G_t = \frac{d\tau}{d\gamma},$$

which describes the change of stress due to a small perturbation from the current state, or the **secant stiffness**:

$$G = \frac{\tau}{\gamma},$$

which defines an *average stiffness* over a given range of strains from an assumed zero (Example 5.3).

It is usual to represent soil non-linearity by plotting secant stiffness versus strain and, because of the pronounced variation of stiffness at small strains, a logarithmic scale is adopted for the strain (Figure 5.46). Then, for the sake of simplicity and practical convenience, the observed behaviour is described as follows:

1   It is well accepted that there exists a small strain range, bounded by a threshold value γ_t, in which the soil behaviour can be considered linear elastic. In fact, within this range:

  - soil stiffness is constant (and usually denoted as G_o);
  - loading cycles show no hysteresis;
  - volumetric and deviatoric strains are uncoupled;
  - no excess pore pressure is generated during undrained shearing;
  - the soil has no memory of previous deformations.

The observed behaviour also suggests that the very small strain stiffness of soils (G_o) is influenced by the volumetric packing (expressed by the void ratio e), the geometric fabric and the stress state.

An established technique to obtain information about the influence of the mentioned factors is represented by laboratory geophysics, which uses piezoceramics bender elements to generate and detect low amplitude shear waves in soil samples.

A bender element is an electro-mechanical transducer that bends as an applied voltage is changed or generates a voltage as it bends, so that it can play the role of a transmitter or a receiver (see for more details: Viggiani and Atkinson, 1995; Brignoli *et al.*, 1996; Jovicic *et al.*, 1996; Pennington *et al.*, 1997; Arulnathan *et al.*,1998; Arroyo *et al.*, 2003 a,b; Comina, 2005).

The arrangement used for a triaxial test is shown in Figure 5.47: transmitter and receiver elements are placed both in the end platens and through the flexible membrane of the cylindrical specimen, so that shear waves can propagate through the sample from end to end or from side to side. In both cases shear waves can be polarized either vertically or horizontally, so that appropriate interpretation of the test (also see Section 7.12 in Chapter 7) allows us to estimate the shear wave velocity V_{hh} for horizontally propagated wave with horizontal polarization, V_{hv} for horizontally propagated wave with vertical polarization, and V_{vh} for vertically propagated wave with horizontal polarization.

Example 5.3

Figure 5.48 reports the experimental data obtained from a triaxial CK_oD test on Pisa clay. The soil sample, slightly overconsoliated (OCR=2), is characterized by the following initial data:

$$p'_o = 61.59 \, \text{kPa}; \; q_o = 17.56 \, \text{kPa}; \; e_o = 0.93.$$

State and explain how the soil stiffness, corresponding to strain level of 0.1%, can be properly obtained.

Solution. Let us start by introducing, into the deviator strain expression $\varepsilon_s = \dfrac{2}{3}(\varepsilon_z - \varepsilon_r)$, the axial and radial strain components, as given by the Hooke law:

$$\delta\varepsilon_s = \frac{2}{3E'}\left(\delta\sigma'_z - 2\upsilon\delta\sigma'_r - \delta\sigma'_r + \upsilon\delta\sigma'_z + \upsilon\delta\sigma'_r\right) = \frac{2(1+\upsilon)}{E'}\delta q = \frac{\delta q}{3G}$$

The obtained result proves that the slope of the curve (q, ε_s) gives 3G, so that, by using data provided in Figure 5.48, corresponding to a strain level $\varepsilon_s = 0.095\%$, one obtains:

$$G = \frac{30.85 - 17.56}{3 \cdot 0.095}100 = 4663 \, \text{kPa}.$$

By using $\varepsilon_v = \varepsilon_z + 2\varepsilon_r$, the corresponding radial strain is equal to:

$$\varepsilon_r = \frac{0.073 - 0.119}{2} = -0.023$$

and this allows us to compute Poisson ratio $\upsilon = \dfrac{-\varepsilon_r}{\varepsilon_z} = 0.19.$

Since the ratio $\dfrac{\delta\varepsilon_v}{\delta\varepsilon_s}$ can be expressed in the form:

$$\frac{\delta\varepsilon_v}{\delta\varepsilon_s} = \frac{\delta p'}{K'}\frac{3G}{\delta q}$$

and because in a standard triaxial test $\dfrac{\delta q}{\delta p'} = 3$, we also reach the conclusion that the slope of the curve $(\varepsilon_v, \varepsilon_s)$ gives the ratio between the shear and the bulk modulus (Figure 5.48b). In the case under examination one obtains $\dfrac{\delta\varepsilon_v}{\delta\varepsilon_s} = \dfrac{0.073}{0.095} = 0.768$, so that the bulk modulus is equal to 6072 kPa.

If, within the considered very small strain range, the medium is elastic, then the corresponding shear modulus can be estimated:

$$G_{oij} = \rho V_{ij}^2. \tag{5.56}$$

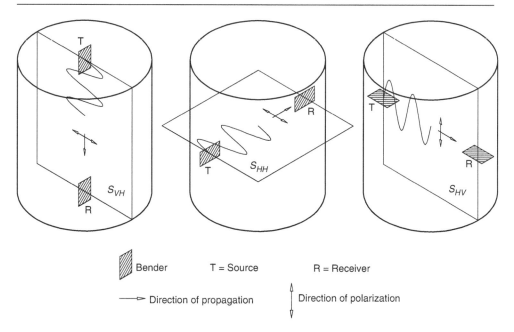

Figure 5.47 Shear wave propagation through a laboratory soil sample.

The need to introduce two indices in the above relationship arises from the anisotropic behaviour of soils: this anisotropy results in an oriented arrangement (**geometric fabric**) of the particles (it may exist before straining, due to the depositional history or it may be induced by strain) and in the forces at contacts between particles (**kinetic fabric**, as defined by Chen *et al.*, 1988). Therefore, in order to account for the volumetric packing, the geometric fabric and the stress state, the following empirical expression is suggested for the shear modulus (Roesler, 1979):

$$\frac{G_{oij}}{p_r} = S_{ij} \cdot F(e) \cdot \left(\frac{\sigma_i' \sigma_j'}{p_r^2}\right)^n , \tag{5.57}$$

where the first two terms S_{ij} and $F(e)$ depend on the geometric fabric and the volumetric packing, and the principal effective stresses σ_i' and σ_j', in the direction of propagation and polarization, take into account the influence of kinetic fabric. The stress p_r is a reference stress and the exponent n is typically of the order of $\cong 0.25$.

2   Beyond the above threshold value γ_t, the irreversible behaviour starts to appear: soil stiffness decreases significantly with strain level and cyclic loading give rise to hysteresis.

3   Finally, beyond a second threshold strain γ_p, the plastic behaviour is predominant: the material exhibits coupling between volumetric and deviatoric strains in drained conditions, excess pore pressure is generated during undrained shearing and memory effects are significant.

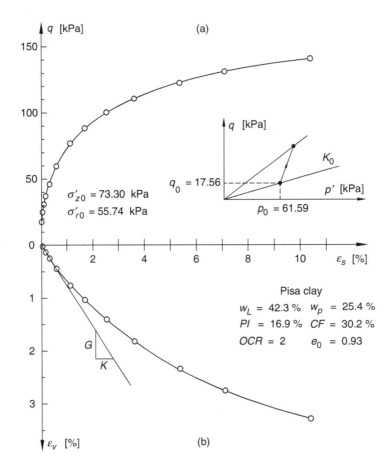

$\Delta\sigma'_z$ [kPa]	q [kPa]	p' [kPa]	ε_v [%]	ε_z [%]	ε_s [%]
7.33	24.89	64.03	0.024	0.044	0.036
11.75	29.32	65.51	0.057	0.105	0.086
13.29	30.85	66.02	0.073	0.119	0.095
15.16	32.73	66.65	0.089	0.142	0.112
19.42	36.98	68.06	0.137	0.208	0.162
28.41	45.97	71.06	0.251	0.407	0.323
42.11	59.68	75.63	0.445	0.742	0.594
59.37	76.94	81.38	0.764	1.385	1.130
70.72	88.29	85.17	1.035	2.038	1.693
82.75	100.32	89.18	1.407	3.015	2.546
93.06	110.63	92.61	1.82	4.22	3.616
105.02	122.58	96.60	2.337	6.15	5.37
113.57	131.14	99.45	2.754	8.019	7.101
123.21	140.78	102.66	3.284	11.486	10.391

Figure 5.48 Drained test on lightly overconsolidated Pisa clay.

5.13 Beyond simple models

In order to include the above discussed non-linearity of soil stiffness in design analysis, simple models are unsuitable and there are basically two strategies which may be followed. The first one assumes empirical expressions to fit the observed non-linear stress strain response (Kondner and Zelasko, 1963; Duncan and Chang, 1970; Jardine *et al.*, 1986, 1991). In this case, parameters included in the empirical expression must take into account variables such as history, current state and stress path, so that it is important to recognize the relationship between these parameters and the appropriate tests. This strategy, often referred to as *variable modulus model*, has the main advantage in its simplicity and its ability to capture the essential features of small strain behaviour: the high stiffness at small strain, the dependence of stiffness on mean effective pressure, the decay of stiffness with strain level.

Disadvantages include the fact that model parameters depend on loading path, so that when applied to boundary problems with complex loading paths the performance of the model cannot be assured; in addition, the model cannot describe the coupling between shearing and volumetric behaviour. For this reason, a more fundamental and alternative strategy is to recognize the inelastic soil response inside the state boundary surface and to develop new mathematical models. In this optic, the simplest approach seems to be the adaptation of the Cam-Clay model, by including some mapping rules or additional yield surfaces.

Models belonging to the class of *bounding surface plasticity* (Dafalias and Herrmann, 1982) introduce plastic strains within the state boundary surface (SBS) through a mapping rule, which associates to each stress state, inside the SBS, an image point on the SBS. Plastic deformations are then computed at a progressive rate, depending on the distance between the current stress point and the image point on the SBS.

Within the class of *kinematic hardening models*, the simplest model involves only two yield surfaces (see for example Al Tabbaa and Wood, 1989). The fully elastic behaviour is bounded by the inner elastic nucleus, which is dragged by the stress point during any loading path. During this path, plastic deformations develop when the inner surface is moving, so that a transitional behaviour can be described. As in the bounding surface model, to each point on the elastic nucleus a unique image point is associated on the SBS, with the rule that outward unit normals be equal. During unloading, the elastic response develops until the nucleus does not move, and plastic deformations start as soon as the nucleus is picked up again.

Kinematic hardening models reproduce a directionality of soil behaviour, since the image points corresponding to different approaching paths are different, so that the incremental response depends on previous recent history (for more details see Al Tabbaa and Wood, 1989; Stallebrass, 1990).

The term *recent history* indicates the influence on soil stiffness of the rotation between the most recent stress path and the current stress path, and that of the rest period (at constant effective stress) between the end of the most recent stress path and the start of the current path (Lambrechts and Leonards, 1978; Atkinson *et al.*, 1990; Atkinson and Sallfors, 1991).

5.14 Summary

(1) When the term *failure* is applied to soils, this term indicates a relative slip occurring on any plane within the soil mass if the ratio of the shear stress and the effective normal stress, acting on this plane, reaches a critical value. The reason is that soil strength is controlled by friction, so that the *failure criterion* (first introduced by Coulomb, 1773) assumes the form:

$$|\tau| = \mu\sigma'.$$

(2) Taylor (1948) recognized the role that volume change plays in defining the peak strength, and, equating the total loading power supplied to a soil element to the internal dissipation, he obtained the expression:

$$\frac{\tau_{xy}}{\sigma'_y} = \mu + \frac{\delta y}{\delta x}.$$

This expression states that the mobilized strength is the sum of the stationary friction and the contribution of dilatancy, expressed by the term $\partial y/\partial x$.

In particular, the **peak strength** corresponds to the *maximum rate of dilation*, while the **critical state** is an asymptotic stationary condition, where distortion occurs without changes of volume or effective stresses. The critical state angle φ'_{cv} depends on mineralogy, roughness and grading of particles, but it is independent of the initial conditions. On the contrary, the peak value of the angle of shear resistance is linked to the rate of dilation; therefore it depends on soil state and cannot be regarded as a soil property.

(3) Considering the peak stress ratio as the sum of the critical state value and the rate of dilation is one possibility of representing peak strength data. It is also common to consider a linearized Mohr-Coulomb failure envelope, with an apparent cohesion, but recognizing that this term represents just an intercept which only gives the position of the linearized envelope. This representation can be unsafe when the stress level in the field is lower than the stress level applied to soil samples in the tests. A safe alternative is to consider a failure envelope strongly curved towards the origin.

(4) When the rate of loading is high relative to the hydraulic conductivity, so that water cannot escape from the pores during loading, this circumstance is called undrained condition. In undrained conditions, the current practice is to refer to total stress analysis due to the difficulties of predicting excess pore pressure in the field. The undrained operational strength must be selected taking into account soil stress history, anisotropy and progressive failure.

(5) Reliable predictions of deformations requires soil non-linearity to be properly taken into account. The use of a variable soil modulus can be satisfactory in monotonic loading, but more sophisticated soil models are required in cyclic loading or to capture the coupling between volume and shear deformations.

5.15 Further reading

There are two major books explaining the conceptual framework referred to in this chapter:

D.W. Taylor (1948) *Fundamentals of Soil Mechanics*. Wiley.
A. Schofield and P. Wroth (1968) *Critical State Soil Mechanics*. McGraw-Hill.

However, further readings on critical state soil mechanics should include:

J.H. Atkinson and P.L. Bransby (1978) *The Mechanics of Soils: An Introduction to Critical State Soil Mechanics*. McGraw-Hill.
D.M. Wood (1990) *Soil Behaviour and Critical State Soil Mechanics*. Cambridge University Press.

Advanced topics on soil modelling are the subject of the following volumes:

D. Muir Wood (2004) *Geotechnical Modelling*. Spon Press.
H.S. Yu (2006) *Plasticity and Geotechnics*. Springer.
R. Nova (2002) *Fondamenti di Meccanica delle Terre*. McGraw-Hill.

Some reference papers, related to the behaviour of natural clays, are the following:

J.B. Burland (1990) On the compressibility and shear strength of natural clays. Rankine Lecture, *Géotechnique*, 40, 3, 329–378.
J.B. Burland, S. Rampello, V.N. Georgiannou and G. Calabresi (1996) A laboratory study of the strength of four stiff clays. *Géotechnique*, 46, 3, 491–514.
G. Calabresi (2004) Terreni argillosi consistenti: esperienze italiane. *Quarta Conferenza* 'Arrigo Croce', RIG, 1, 14–57.
F. Cotecchia and R.J. Chandler (2000) A general framework for the mechanical behaviour of clays. *Géotechnique*, 50, 4, 431–447.
M. Jamiolkowski, C.C. Ladd, J.T. Germaine and R. Lancellotta (1985) New developments in field and laboratory testing of soils. Theme Lecture, XI ICSMFE, San Francisco, 1, 57–153.
S. Leroueil and D.W. Hight (2003) Behaviour and properties of natural soils and soft rocks, in *Characterisation and Engineering Properties of Natural Soils*, Vol. 1, 29–254, Tan *et al.* (eds), Swets & Zeitlinger, Lisse.

There is a topic, usually thought in advanced courses, that deserves special attention: soil liquefaction. This term indicates the phenomenon in which soil rapidly loses much of its strength, due to monotonic or cyclic loading, but not necessarily from earthquakes. A recently published reference which covers in detail this topic is the following book: M. Jefferies and K. Been (2006) *Soil Liquefaction. A Critical State Approach*. Taylor & Francis.

Finally, some reference books related to soil testing:

A.W. Bishop and D.J. Henkel (1962) *The Measurement of Soil Properties in the Triaxial Test*. Edward Arnold.
K.H. Head (1980) *Manual of Soil Laboratory Testing*. Pentech Press, London.
T.W. Lambe (1951) *Soil Testing for Engineers*, Wiley.

Chapter 6

Flow in porous media

The analysis of flow in porous media is aimed at solving many engineering problems. In geotechnical engineering this analysis is currently performed to predict the seepage through the core of an earth dam to evaluate both the water loss and the stability of the dam. It is also relevant to compute the flow into an excavation in order to design the dewatering system, as well as to analyse the stability of the excavation bottom. More generally, the stability of a soil element depends on the seepage forces, so that any kind of stability analysis will require the solution of a flow problem.

When deriving the basic equations of flow in porous media reference is made to the Darcy's law. Darcy proved, in 1856, that the discharge velocity is proportional to the hydraulic gradient and the coefficient of proportionality is termed hydraulic conductivity. No other parameter exhibits so wide a variation when moving from coarse to fine-grained soils. In the case of course-grained soils the hydraulic conductivity is so high that excess pore pressure, due to any external agency, dissipates in a very short time. The consequence is that the transient phase can be disregarded most of the time and soil behaviour can be analysed in drained conditions, by referring to a stationary flow pattern.

On the contrary, fine-grained soils have such low conductivity that the excess pore pressure takes a long time to reach the final equilibrium condition and the transient phase cannot be disregarded. In this case it is important to distinguish between the initial undrained conditions, the transient phase (called *consolidation*) and the final stationary condition. Due to the evolution with time of the excess pore pressure, the effective stresses will also change with time and this will produce an evolution of the stability with time as well as delayed deformations.

This chapter deals with most of the mentioned aspects: Darcy's law (Section 6.2), stability analysis in presence of seepage forces (Section 6.4), stationary flow (Sections 6.6 to 6.9) and the theory of consolidation (Sections 6.10 to 6.13).

6.1 Terminology

A *vector field* in R^n is a map $\mathbf{f}: R^n \to R^m$ that assigns to each $\mathbf{x}$ a vector $\mathbf{f}(\mathbf{x})$. In fluid mechanics, the fluid velocity is an example of vector field:

$$\mathbf{v}(t, \mathbf{x}) : R^4 \to R^3 ,$$

and the vector field $\mathbf{v}(t,\mathbf{x})$ at a fixed time t_o is called the *kinetic field*.

A convenient way to highlight the relevant features of the flow, *at a fixed time* t_o, is to make reference to the *streamlines* or *flow lines* $\Psi(\mathbf{x})$. For a given vector field a flow line is a curve such that, at each of its points, the tangent vector of the curve coincides with the vector field (see Figure 6.1). Thus the flow lines are found by solving the system:

$$\frac{d\mathbf{x}}{ds} = \mathbf{v}(\mathbf{x}, t_o) \,, \tag{6.1}$$

where s is a parameter along the curve.

If a scalar function Φ exists such that $\mathbf{v} = \nabla\Phi$, the line integral of $\mathbf{v}$:

$$C = \int \mathbf{v} \cdot d\mathbf{s} \tag{6.2}$$

depends only on the endpoints of the path:

$$C = \Phi(B) - \Phi(A) \,. \tag{6.3}$$

The scalar function Φ is called a *potential* for the vector field $\mathbf{v}$ and surfaces of equation $\Phi = \text{const}$ are called the *equipotential surfaces*, since the potential is constant on them.

In a simple connected region, the necessary and sufficient condition that the integrand (6.2) is a perfect differential of some function Φ is that:

$$\nabla \times \mathbf{v} = 0 \,, \tag{6.4}$$

i.e. every *irrotational vector field* is the gradient of a scalar field, called the *potential function*.

If $d\mathbf{x}$ is any tangent vector to the potential surface, since:

$$\mathbf{v} \cdot d\mathbf{x} = d\Phi = 0 \,, \tag{6.5}$$

it follows that flow lines must be orthogonal to equipotential lines (Figure 6.1).

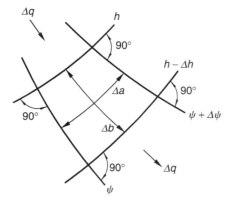

Figure 6.1 Equipotential and flow lines.

If a vector field $\mathbf{v}$ represents the flow of a fluid, then the quantity:

$$\omega = \frac{1}{2} rot \mathbf{v} \qquad (6.6)$$

represents the *angular velocity* of a rigid body, that rotates as the fluid does near that point.

The condition $rot \mathbf{v} = 0$ has the following meaning: if a small paddle wheel is placed in the fluid, it will move with the fluid but will not rotate around its axis.

A *particle path* is the trajectory of an individual particle. Formally, it is expressed by the map $\mathbf{r}$ that assigns to each t the vector $\mathbf{r}(t)$:

$$\begin{aligned} & \mathbf{r}: R \to R^3 \\ & \mathbf{r} = (x, y, z); \ t \in [t_o, t_1] \end{aligned} \qquad (6.7)$$

The determination of the particle paths requires the integration of the following system of first-order differential equations:

$$\frac{d\mathbf{x}}{dt} = \mathbf{v}(\mathbf{x}, t) . \qquad (6.8)$$

It is relevant to note that into (6.8) the time is an independent variable, whereas into (6.1) it is a parameter, the value of which defines a given family of flow lines. In *steady flow*, all flow properties are constant in time at each point, so that each particle path is along one of the unchanging flow lines. In *unsteady flow*, the flow line pattern changes with time and a particle path may not coincide with any flow line.

The quantity of fluid which flows across a surface per unit time, i.e. the rate of fluid flow, is called *flux*. If we consider a surface S in space and we assign a positive side to the surface and we take the normal $\mathbf{n}$ in the positive sense, then the *flux* is given by the scalar quantity:

$$\int_S \mathbf{v} \cdot \mathbf{n} dS . \qquad (6.9)$$

A **perfect fluid** (or *Euler fluid*) is a fluid whose stress tensor is expressed by:

$$\sigma_{ij} = -u\delta_{ij} . \qquad (6.10)$$

where u is a scalar quantity called **pressure**. If the pressure depends on density, that is:

$$u = u(\rho) ,$$

then the fluid is called *barotropic* or *compressible*. On the contrary, if the fluid is incompressible, the pressure does not depend on density and its value is no longer defined by a constitutive equation. It assumes the role of a reaction to an internal constraint and it must be obtained from the momentum balance equation.

6.2 Darcy's law

Water flows from a point, to which a given amount of energy can be associated, to another point at which the energy will be lower. This energy is the kinetic energy plus the potential energy. The first depends on the fluid velocity, the second is linked to the datum and to the fluid pressure.

When we consider flow systems where gravity and elevation provide the driving potential for the flow, the concept of **head** is particularly useful. Accordingly, if reference is made to a *fluid element of unit weight*, then the energy components can be expressed in terms of *heads* (Figure 6.2):

the **elevation head** z depends on the elevation with respect to an arbitrary given datum;

the **pressure head** $\dfrac{u}{\gamma_w}$ is the rise in height due to the fluid pressure;

the **velocity head** $\dfrac{v_w^2}{2g}$ is associated to the kinetic energy.

By summing these terms, the total specific energy can be expressed in terms of **total head**:

$$H = z + \frac{u}{\gamma_w} + \frac{v_w^2}{2g}. \tag{6.11}$$

In steady flow of a perfect incompressible fluid the total head is constant along any seepage line, i.e.:

$$z_A + \frac{u_A}{\gamma_w} + \frac{v_{wA}^2}{2g} = z_B + \frac{u_B}{\gamma_w} + \frac{v_{wB}^2}{2g}, \tag{6.12}$$

and the above equation is widely known as the Bernoulli (1738) equation.

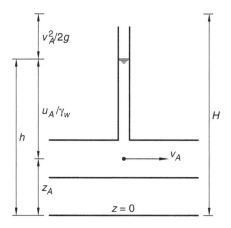

Figure 6.2 Elevation, pressure and velocity head.

It can also be proved that, provided the fluid is incompressible and subjected to gravity, in the case of *irrotational steady flow* the total head remains constant throughout the flow field.

When dealing with flow through soils, a high velocity can be of the order of 0.01 m/s, so that the velocity head will be of the order of $5 \cdot 10^{-6}$m, a negligible term in equation (6.11). Therefore, the total head at any point in the flow domain is assumed to be equal to the **piezometric elevation**:

$$h = z + \frac{u}{\gamma_w} . \tag{6.13}$$

The meaning of the above introduced definitions of pressure head and piezometric elevation can be appreciated if reference is made to **piezometers** (for details see Section 7.4 in Chapter 7). The simplest piezometer is an open standpipe, installed at a given elevation z. If u is the local pore pressure in the surrounding soil, the water will rise within the pipe up to the height u/γ_w, called pressure head; the elevation reached above the arbitrary datum, from which we compute the elevation z, is equal to $(z + u/\gamma_w)$, and this quantity is *the piezometric elevation*, related to the point in which the piezometer is installed.

In groundwater flow, the fundamental equation which relates the velocity to the energy content of the fluid motion is known as **Darcy's law**. Darcy (1856) proved that the rate of flow Q through a soil element (see Figure 6.3) is linearly related to the area of the total section A of the soil element and to the total head loss occurring over the distance L:

$$Q = -KA\frac{h_2 - h_1}{L} , \tag{6.14}$$

where $K(L \cdot T^{-1})$ is the **coefficient of hydraulic conductivity** (see Table 6.1 for typical values).

In physical terms, we can define the coefficient of hydraulic conductivity as a measure of the ease, the water may flow through a porous medium.

If we introduce the *discharge velocity*, defined as the quantity of water which flows through a unit of *total area* of the porous medium in a unit time, i.e. the ratio between Q and A, then the equation (6.14) can be rewritten for the three-dimensional case in the form:

$$\mathbf{v} = K\mathbf{i} , \tag{6.15}$$

and the vector:

$$\mathbf{i} = -\nabla h \tag{6.16}$$

is called the **hydraulic gradient**.

The minus sign introduced into (6.16) means that the water flows in the direction of decreasing head and the hydraulic gradient (a pure number) represents the space rate of energy dissipation per unit weight of fluid.

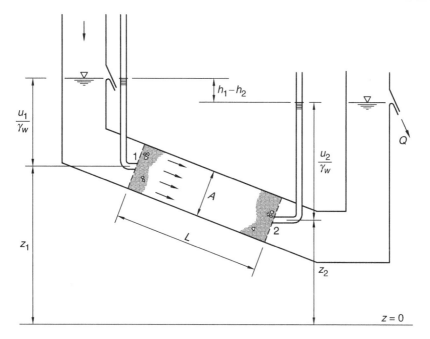

Figure 6.3 Darcy's law.

Table 6.1 Some typical values of the coefficient of hydraulic conductivity

Soil	K (m/s)
Clean gravel	$10^{-2} \rightarrow 1$
Clean coarse sand	$10^{-5} \rightarrow 10^{-2}$
Fine sand	$10^{-6} \rightarrow 10^{-4}$
Silt	$10^{-8} \rightarrow 10^{-6}$
Soft clay	$< 10^{-9}$
Stiff fissured clay	$10^{-8} \rightarrow 10^{-4}$

Remarks

(a) In equations (6.11) and (6.12) the vector $\mathbf{v}_w$ is the mean effective velocity of the water within the pores, so that, if we assume that the ratio of the area of pores to the total area is statistically equal to the volume porosity n, the discharge velocity is linked to the mean effective velocity through the relation:

$$\mathbf{v} = n\mathbf{v}_w .$$ (6.17)

(b) It is relevant to observe that the fluid motion is driven by the difference of total head, not by differences of pressure head. In Figure 6.3, the pore pressure at the location 2 is higher than the pore pressure at location 1 ($u(2) > u(1)$), but, being $h_1 > h_2$, the motion develops from point 1 to 2.

(c) The **coefficient of hydraulic conductivity** K (see typical values in Table 6.1) depends on both the *properties of the solid matrix* (porosity, shape of particles, grading, tortuosity and specific surface), and the *fluid properties* (density ρ_w and viscosity μ).

On the contrary, the **permeability** $k(L^2)$ only depends on the properties of the solid matrix, and is linked to the hydraulic conductivity $K(L \cdot T^{-1})$ through the relation:

$$k = \frac{K\mu}{\rho_w g} . \tag{6.18}$$

(d) The presence in equation (6.15) of the scalar quantity K implies that the soil is isotropic with respect to the flow properties. If we deal with *anisotropic porous media*, the same equation writes:

$$\begin{aligned} \mathbf{v} &= \mathbf{K}\mathbf{i} \\ v_i &= K_{ij}i_j \end{aligned}, \tag{6.19}$$

where $\mathbf{K}$ is now a second-order tensor, called *tensor of hydraulic conductivity*. Expanding the equation (6.19) gives:

$$\begin{aligned} v_x &= K_{xx}i_x + K_{xy}i_y + K_{xz}i_z \\ v_y &= K_{yx}i_x + K_{yy}i_y + K_{yz}i_z \\ v_z &= K_{zx}i_x + K_{zy}i_y + K_{zz}i_z \end{aligned} \tag{6.20}$$

which highlights that in anisotropic soil media the discharge velocity and the hydraulic gradient are no longer collinear.

(e) Equation (6.15) assumes that the fluid motion is observed with respect to solid skeleton, which is fixed in space. If the soil particles are also moving, the equation (6.15) must be rewritten in terms of relative velocity, i.e.:

$$n(\mathbf{v_w} - \mathbf{v_s}) = -\frac{\mathbf{K}}{\rho_w g}(\nabla u + \rho_w g \nabla z) , \tag{6.21}$$

where $\mathbf{v}_w$ and $\mathbf{v}_s$ are the mean effective velocity of the fluid and the soil particles respectively. We shall make use of the equation (6.21) in Section 6.10, when dealing with the theory of consolidation.

(f) A further relevant remark is that equation (6.21) can be regarded as statistically equivalent to the Navier-Stokes equations, because it relates the driving forces, represented by the term $-(\nabla u + \rho_w g \nabla z)$, to the resistance term $n\rho_w g \mathbf{K}^{-1}(\mathbf{v}_w - \mathbf{v}_s)$, which accounts for dissipation developed at the fluid soil interface. In this respect, Darcy's law must be regarded as a macroscopic model of the flow through a porous medium and it plays an important role in the solution of flow problem (this is why so much attention has been devoted to the verification of its validity).

Example 6.1

With reference to Figure 6.4, find the total, elevation and pressure head at location A, B and C.

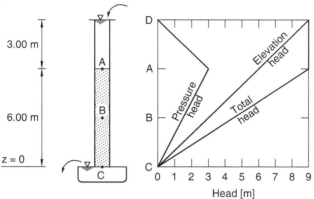

Figure 6.4 Illustrative example of determining the pressure head (downward flow).

Solution. Assuming as a datum the horizontal plane through point C, the total head at A is equal to 9.00 m, because $h_D = h_A$ (there is no energy dissipation along AD). At the location C the total head is zero, and, supposing that the soil sample is homogeneous, the total head will change linearly along the path AC.

The pressure head is obtained by subtracting from the total head the elevation head.

6.3 Coefficient of hydraulic conductivity

We stated in Chapter 1 that engineers are interested in measurements of soil strength, soil stiffness and hydraulic conductivity. The hydraulic conductivity can be thought of in physical terms as the ease with which fluids may flow through porous media, and Darcy's law:

$$\frac{Q}{A} = K\frac{\Delta h}{L},$$
(6.22)

tells us how the hydraulic conductivity K can be determined if the discharge Q is measured in a test in which the hydraulic gradient $\Delta h/L$ is a known quantity.

An example of the so-called *constant head permeability test*, suitable for *coarse-grained soils*, is reproduced in the upper portion of Figure 6.5. The soil sample is placed in a perspex tube and the two ends are connected to reservoirs of water, the height of which can be properly fixed. If the difference Δh is maintained constant with time, the discharge Q, i.e. the quantity of water collected in a given time interval, can be measured and equation (6.22) allows us to compute the hydraulic conductivity.

Usually, the specific discharge Q/A is determined for different values of the hydraulic gradient $\Delta h/L$, and the initial slope of the graph in Figure 6.5a gives the value of the initial hydraulic conductivity. Note that the test can be performed with downward or upward flow. In the first case we may observe a tendency of soil to compact. On the contrary, in presence of upward flow the soil sample may experience a decrease of its initial density, and if the hydraulic gradient reaches a critical value (see Section 6.4) a condition of fluidization may occur with sand sample. This test is unsuitable for *fine-grained soils*, because their low hydraulic conductivity would require a very long time to collect a given amount of water, and evaporation from the measuring cylinder

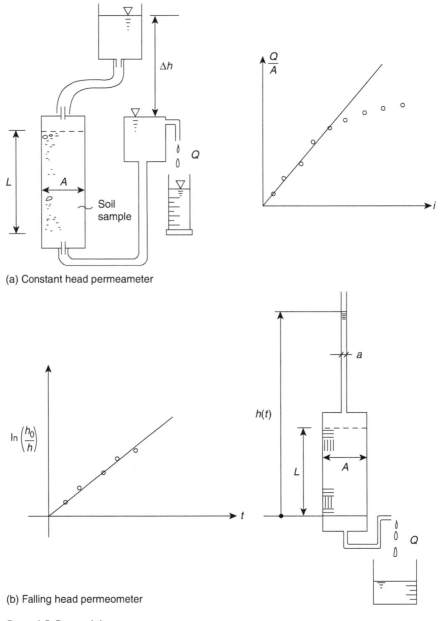

(a) Constant head permeameter

(b) Falling head permeometer

Figure 6.5 Permeability tests.

could lead to errors. In this case the more suitable test is the so-called *falling head test* (Figure 6.5b). In this test, water is driven to flow from a small tube, of cross-section *a*, through a clay sample contained in a larger tube of cross-section *A*.

If at a given time *t* the hydraulic head is equal to *h* above the outlet level, applying Darcy's law to the soil sample will given:

$$Q = K\frac{h}{L}A .$$
(6.23)

Considering the small tube, of cross-section equal to *a*, it also holds:

$$Q = -a\frac{dh}{dt} ,$$
(6.24)

and elimination of *Q* from these two equations give the following differential equation for the head *h*

$$\frac{dh}{dt} = -K\frac{A}{a}\frac{h}{L} .$$
(6.25)

This can be solved by considering the initial condition (i.e. $h = h_o$ if $t = 0$) and the solution can be expressed in the form:

$$\ln\left(\frac{h_o}{h}\right) = K\frac{A}{a}Lt .$$
(6.26)

Therefore, if the experimental data are plotted as a graph of $\ln(h_o/h)$ against *t*, the slope of this graph allows us to compute the hydraulic conductivity.

These examples are instructive to show that the measument of the hydraulic conductivity of a soil sample in a laboratory is relatively straightforward. However, it must be observed that the obtained values are representative of small soil samples, that do not take into account structural features and inhomogeneities of the *in situ* soil mass. For this reason, it is usually required to measure the permeability in the field (see Section 7.14 in Chapter 7).

6.4 Seepage forces

Throughout this book we assume, as is usual in soil mechanics, that the normal stress components are positive if compressive. Then, if the *z* axis is positive downward, the indefinite equilibrium equations write:

$$\frac{\partial\sigma_x}{\partial x} + \frac{\partial\tau_{yx}}{\partial y} + \frac{\partial\tau_{zx}}{\partial z} = 0$$

$$\frac{\partial\tau_{xy}}{\partial x} + \frac{\partial\sigma_y}{\partial y} + \frac{\partial\tau_{zy}}{\partial z} = 0$$
(6.27)

$$\frac{\partial\tau_{xz}}{\partial x} + \frac{\partial\tau_{yz}}{\partial y} + \frac{\partial\sigma_z}{\partial z} - \gamma = 0$$

If we substitute into (6.27) the definition of effective stresses ($\sigma'_{ij} = \sigma_{ij} - u\delta_{ij}$), also considering that:

$$u = \gamma_w\left[h - (H_o - z)\right],$$

where H_o is the distance of the ground surface from an arbitrary datum from which we compute the total head h, equations (6.27) can be rewritten in the form:

$$\frac{\partial \sigma'_x}{\partial x} + \frac{\partial \tau_{yx}}{\partial y} + \frac{\partial \tau_{zx}}{\partial z} + \gamma_w \frac{\partial h}{\partial x} = 0$$

$$\frac{\partial \tau_{xy}}{\partial x} + \frac{\partial \sigma'_y}{\partial y} + \frac{\partial \tau_{zy}}{\partial z} + \gamma_w \frac{\partial h}{\partial y} = 0 \qquad . \tag{6.28}$$

$$\frac{\partial \tau_{xz}}{\partial x} + \frac{\partial \tau_{yz}}{\partial y} + \frac{\partial \sigma'_z}{\partial z} + \gamma_w \frac{\partial h}{\partial z} - \gamma' = 0$$

These equations prove that the soil skeleton can be thought to be in equilibrium under the action of the effective stresses, the volume forces deriving from the gravity field, represented by the unit boyant weight (γ') and the forces arising from the interaction with the flowing water. These latter, of components $\gamma_w \partial h/\partial x_i$, are called **seepage forces** and they represent a second field superimposed to the gravity field. Note that it is common to use the term seepage forces, even though they are not actually forces but forces per unit volume.

The relevance of seepage forces can be appreciated if, as a matter of simplicity, we make reference to the one-dimensional case, in which the flow is upward. The equilibrium equation in the vertical direction now simplifies in the following:

$$\frac{\partial \sigma'_z}{\partial z} = \left(\gamma' - \gamma_w \frac{\partial h}{\partial z}\right), \tag{6.29}$$

which, once integrated with the boundary condition $\sigma'_z = 0 \; if \; z = 0$, gives:

$$\sigma'_z = \left(\gamma' \mp \gamma_w |i|\right) z, \tag{6.30}$$

where, for convenience we have introduced as $\gamma_w|i|$ the absolute value of the seepage force. Then, the negative sign applies when the flow is directed against gravity, and the positive sign applies if the flow is directed according to the gravity field (recall that water flows in the direction of decreasing head).

These results prove that in the presence of an upward flow a critical situation can be generated if anywhere the vertical effective stress is equal to zero. When this occurs, the soil is *fluidized*, i.e. the velocity of the water is so large that the drag forces on the soil particles overcome the submerged weight of the particles that are no longer resting in contact with each other. This condition is called **piping**, if it occurs in localized channels (the transport of particles usually occurs at the ground surface and erosion regresses, forming a pipe-shaped channel in the soil mass), or **quicksand** if it occurs over a large area.

Remarks

1. According to equation (6.30) the effective stress is zero when the hydraulic
    gradient reaches a critical value, given by:

$$i_c = \frac{\gamma'}{\gamma_w} = \frac{G_s - 1}{1 + e} \, ,\tag{6.31}$$

    so that many authors suggest the concept of *critical gradient* as a basis on
    which a safety factor against a quick condition can be defined, i.e.:

$$FS = \frac{i_c}{i_E} \, ,\tag{6.32}$$

    being i_E the *exit gradient*, i.e. the maximum value of the hydraulic gradient
    at the discharge boundary (see Figure 6.6a). However, complex situations,
    such as that shown in Figure 6.6b, in which an inverted filter is placed on
    the surface of a sand layer, requires a more general criterion than the one
    based on the critical gradient. We can in fact observe that in this case the
    effective stresses are greater than zero, even if the upward gradient near the
    surface is equal or greater than the critical value. This suggests that a more
    general criterion remains to check that the effective stresses are everywhere
    greater than zero. With reference to Figure 6.6b, the requirement is to assess
    the equilibrium of a soil mass and to check that the forces acting at the
    boundary do not overcome its weight, in order to avoid a failure mechanism
    called **heave**. By neglecting any friction force acting on the vertical side, this
    requires:

$$(\zeta + \Delta h)\gamma_w < \gamma\zeta + H\gamma_f$$

    where Δh is the excess head to be dissipated along the distance ζ, and H and
    γ_f refer to the filter placed onto the surface. The safety factor then assumes
    the form:

$$FS = \frac{H\gamma_f b + \gamma' b \zeta}{(\Delta h \gamma_w)b} \, ,$$

    which reduces to (6.32) in absence of the filter and if we assume
    $\Delta h = i\zeta$.
2. A **graded filter** consists of layers of granular material that prevent the
    movement of particles subjected to erosion. Successively more permeable
    and coarse-grained soils are placed, such that the fine constituents of each

layer cannot be washed into the voids of the succeeding layer. This can be translated into the following requirements (Somerville, 1986):

(a)   The filter must be more pervious than the protected soil, allowing free seepage of water without head loss, a requirement fulfilled if the ratio:

$$\frac{D_{15}(filter)}{D_{15}(soil)}$$

is greater than 4 and additionally $D_5(filter) > 75\,\mu m$.

(b)   The openings of the protective filter must be small enough to prevent the protected soil from penetrating the filter, causing clogging of the filter. This requirement is fulfilled if the ratio:

$$\frac{D_{15}(filter)}{D_{85}(soil)}$$

is lower than 4. Additionally it is suggested that the particle size distribution curve of the filter should have the same shape as that of the protected soil.

(c)   The filter must be graded so as to prevent internal erosion and the segregation of the material, a requirement fulfilled if the filter material is well-graded with a maximum particle of size up to 75 mm.

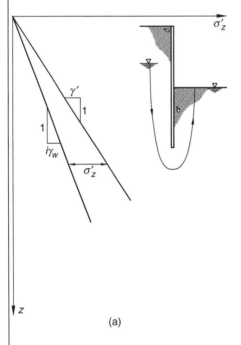
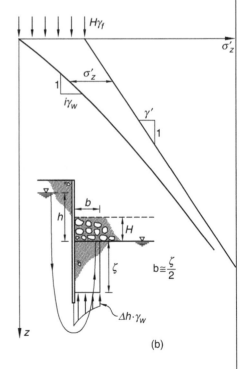

Figure 6.6 Piping and heave.

Example 6.2

With reference to Figure 6.7, it is required to compute the vertical effective stress for the three cases, assuming that the hydraulic conductivity is constant with depth.

Case A. Points A and B have the same total head, so that there is no flow and the vertical effective stress can be computed, taking into account the pore pressure under hydrostatic conditions.

Case B. The total head at point B is greater than that at point A, the difference being Δh. This difference will drive an upward seepage flow and therefore the vertical effective stress can be computed by means of the expression:

$$\sigma'_{vo} = \left(\gamma' - |i| \gamma_w \right) \cdot z .$$

At the depth equal to D we get:

$$\sigma'_{vo} = \left(\gamma' - \frac{\Delta h}{D} \gamma_w \right) \cdot D = \gamma' \cdot D - \gamma_w \cdot \Delta h$$

Case C. The total head at point B is lower than that at point A, so that there is a downward flow. The vertical effective stress is equal to:

$$\sigma'_{vo} = \left(\gamma' + |i| \gamma_w \right) \cdot z ,$$

and at the depth D we obtain:

$$\sigma'_{vo} = \gamma' \cdot D + \gamma_w \cdot \Delta h$$

Example 6.3

Compute the hydraulic gradient in the two cases of steady flow shown in Figure 6.8.

Case A. Let us introduce deliberately the assumption of a *linearized flow pattern*, by considering the flow line *ABCDE*.

The difference of total head at point *A* minus that at point *E* is equal to 3.5 m and, supposing the soil is homogeneous (i.e. the same permeability holds at any location), the approximate value of the hydraulic gradient will be:

$$|i| = \frac{\Delta h}{L} = \frac{3.5}{3.5 + 4 + 4} = 0.304 .$$

The correct solution of this problem will be discussed in Section 6.6, showing that the value of the exit gradient at point *E* is equal to 0.25. Therefore, the linearized flow pattern assumption gives a cautious estimate of the hydraulic gradient.

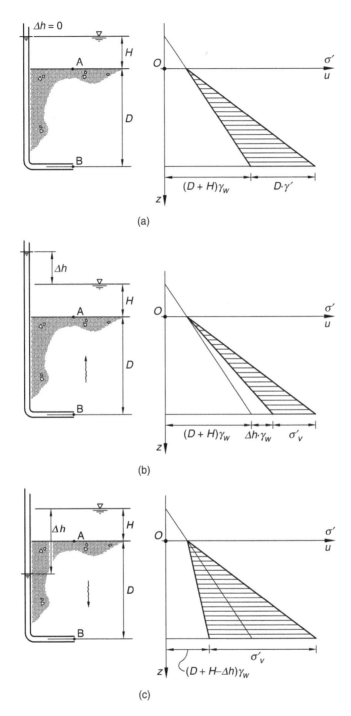

Figure 6.7 Computing the effective stress for the case of hydrostatic condition, upward and downward flow.

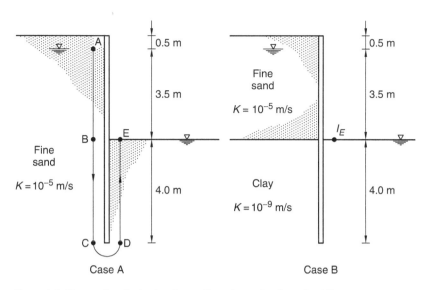

Figure 6.8 Computing the hydraulic gradient through a linearized flow pattern.

Case B. Under the same assumption of linearized flow pattern, the continuity of the flow requires:

$$v = K_s i_s = K_a i_a \,,$$

where $K_s i_s$ represents the discharge velocity within the sand layer and $K_a i_a$ that within the clay layer.

Taking into account the values of the coefficient of hydraulic conductivity we obtain $i_a = 10,000 i_s$, so that we reach the conclusion that the energy is essentially dissipated within the clay stratum. It follows that the approximate cautious estimate of the hydraulic gradient will be $|i| = \dfrac{3.5}{4+4} = 0.438$.

Example 6.4
Figure 6.9 shows a stratigrifical sequence, where failure by uplift of the bottom of the excavation could occur. State and explain how to check the stability against uplift.

The key aspect is the piezometric level within the sand stratum underlying the clay strata. Therefore, the stability of the bottom of the excavation (the clay strata of low permeability) can be checked by comparing the permanent stabilizing action (represented by the weight of the soil cylinder, if we neglect the shear strength along its lateral surface) to the destabilizing actions (represented by the pore pressure acting at the bottom of this cylinder). Using the data shown in Figure 6.9, we obtain a safety factor equal to:

$$FS = \frac{\pi R^2 (4 \cdot 19 + 6 \cdot 18)}{\pi R^2 (9.81 \cdot 13)} = 1.44 \ .$$

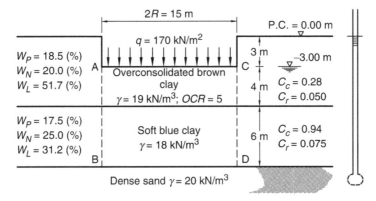

Figure 6.9 Checking against failure by uplift of the bottom of excavation.

6.5 Mathematical modelling of flow in porous media

This section is intended to show how the general principles of continuum mechanics, discussed in Chapter 2, apply to derive the mathematical model which describes the flow in porous media. Subsequent sections deal with applications of this model to both steady state and transitory flow (see in particular Section 6.10 for a detailed treatment of the one-dimensional theory of consolidation).

In what follows we assume isothermal conditions and the porous medium is supposed to be saturated. Under these assumptions, the formulation of the mathematical model requires to take into account the mass conservation and the momentum balance equations of both phases, the state equation of the fluid phase and the constitutive equation of the solid phase. We also introduce the further assumptions that the soil particles are incompressible and that the quantities appearing in the field equations, as well as their gradient, can be regarded as first order quantities, so that we can disregard their power of order $n \geq 2$ and the products among these quantities. Therefore, the obtained model is underpinned by a **linearized theory**.

However, we can state that all the above assumptions can be considered relatively realistic, so that the obtained *storage equation* can be accepted as a reasonable description of the physical phenomenon.

State equation. If β is the water compressibility, then the **state equation**, which in isothermal conditions relates the density to the pore pressure, writes:

$$\rho_w = \rho_{wo} e^{\beta(u-u_o)} , \tag{6.33}$$

where ρ_{wo} *and* u_o are the reference values.

In Soil Mechanics the compressibility of the solid particles is usually neglected, therefore:

$$\rho_s = \text{const}. \tag{6.34}$$

Mass conservation. The **mass conservation principle** postulates that during the motion the mass of any material portion does not change in time. Accordingly, the local formulation of this principle for both the fluid and the solid phase writes:

$$\frac{\partial(n\rho_w)}{\partial t} + \nabla \cdot (n\rho_w \mathbf{v}_w) = 0. \tag{6.35}$$

$$\frac{\partial\left[(1-n)\rho_s\right]}{\partial t} + \nabla \cdot \left[(1-n)\rho_s \mathbf{v}_s\right] = 0. \tag{6.36}$$

If we introduce the state equations into the mass conservation equations, and we neglect the terms $n\mathbf{v}_w \cdot \nabla\rho_w$ and $(1-n)\mathbf{v}_s \cdot \nabla\rho_s$, the previous equations become:

$$\frac{\partial n}{\partial t} + n\beta\frac{\partial u}{\partial t} + \nabla \cdot (n\mathbf{v}_w) = 0 \tag{6.37}$$

$$-\frac{\partial n}{\partial t} + \nabla \cdot \left[(1-n)\mathbf{v}_s\right] = 0. \tag{6.38}$$

Summing up these equations, we eliminate the term $\partial n/\partial t$. Further, taking into account the definition of discharge velocity:

$$\mathbf{v} = n(\mathbf{v}_w - \mathbf{v}_s)$$

and by observing that $\nabla \cdot \mathbf{v}_s = -\partial\varepsilon_v/\partial t$ (the negative sign is needed because the compression is assumed positive, whereas the displacement u_k is positive if oriented as the x_k axis), we obtain the mass conservation of the porous medium:

$$n\beta\frac{\partial u}{\partial t} - \frac{\partial\varepsilon_v}{\partial t} + \nabla \cdot \mathbf{v} = 0. \tag{6.39}$$

This equation is called **storage equation** and can be interpreted as follows: *the rate of compression of a soil element is equal to the rate of compression of the fluid plus the flow of the fluid expelled from the element.*

Incidentally we also note that equation (6.38) can be rewritten as:

$$\frac{\partial n}{\partial t} = -(1-n)\frac{\partial\varepsilon_v}{\partial t}, \tag{6.40}$$

by neglecting second-order terms, so that volume changes can be expressed in terms of porosity changes.

Momentum balance equation. The momentum balance equation states that, for any arbitrary material volume, the rate change of momentum is equal to the resultant of all forces acting on it. Assuming that the inertial terms can be neglected, the momentum balance equations of the overall porous medium reduce to equilibrium equations:

$$\frac{\partial \sigma_x}{\partial x} + \frac{\partial \tau_{yx}}{\partial y} + \frac{\partial \tau_{zx}}{\partial z} + b_x = 0$$

$$\frac{\partial \tau_{xy}}{\partial x} + \frac{\partial \sigma_y}{\partial y} + \frac{\partial \tau_{zy}}{\partial z} + b_y = 0 , \qquad (6.41)$$

$$\frac{\partial \tau_{xz}}{\partial x} + \frac{\partial \tau_{yz}}{\partial y} + \frac{\partial \sigma_z}{\partial z} + b_z = 0$$

where b_i are the component of body forces.

Further, we assign to the Darcy law:

$$\mathbf{v} = -K\nabla h \qquad (6.42)$$

the role of momentum equation for the fluid phase (as already discussed in Section 6.2) and, for convenience, we give to the total h the following form:

$$h = z + \frac{u^{st}}{\gamma_w} + \frac{u}{\gamma_w} , \qquad (6.43)$$

being u^{st} the initial stationary value of the pore pressure (which changes linearly with z), and u is now the **excess pore pressure.**

Constitutive law. Finally, by assuming that the soil behaves as an *isotropic linear elastic body*, the constitutive law writes:

$$\delta\sigma'_{ij} = (\delta\sigma_{ij} - \delta u\delta_{ij}) = \lambda\delta\varepsilon_{kk}\delta_{ij} + 2G\delta\varepsilon_{ij} , \qquad (6.44)$$

the small strain tensor being expressed in terms of displacement field:

$$\varepsilon_{ij} = \frac{1}{2}\left(\frac{\partial u_i}{\partial x_j} + \frac{\partial u_j}{\partial x_i}\right) . \qquad (6.45)$$

The system of **field equations**, composed by equations (6.39), (6.41), (6.44) and (6.45), is now completed, since it represents 16 equations for a total of 16 unknowns (6 independent stress components, 6 strain components, 3 displacement components and the excess pore pressure).

Furthermore, if the equilibrium equations are expressed in terms of displacements, as it can be obtained by substituting (6.45) into (6.44) and the result into (6.41), we can eliminate stresses and strains and the system can be simplified as follows:

$$\lambda u_{k,kj} + G(u_{i,j} + u_{j,i})_{,i} - u_{,j} = 0 \qquad (6.46)$$

$$\frac{K}{\gamma_w}\nabla^2 u + \frac{\partial \varepsilon_v}{\partial t} = 0 . \qquad (6.47)$$

The reader should take care that in equation (6.46) the symbol u, without any index, is the scalar quantity representing the excess pore pressure, whereas the same symbol

Table 6.2 Available solutions of the Biot theory for problems of practical engineering interest

a.   Uniformly loaded strip on half-space (Biot, 1941; McNamee and Gibson, 1960; Schiffman *et al.*, 1969)

b.   Uniformly loaded circle on half-space (De Jong, 1957; McNamee and Gibson, 1960; Schiffman and Fungaroli, 1965)

c.   Uniformly loaded circle on finite layer (Gibson *et al.*, 1967)

d.   Uniformly loaded rectangle on half-space (Gibson and McNamee, 1957)

e.   Circular loading on half-space overlain by a permeable layer (Mandel, 1957, 1961)

with the pedex, u_i , represents the component of the displacement vector. Moreover, in order to obtain the equation (6.47), Darcy's law (6.42) has been substituted into the balance equation (6.39), by also neglecting the fluid compressibility.

Equations (6.46) and (6.47) give four partial differential equations (PDE) in which the four unknowns are represented by the excess pore pressure u and the three components of displacement u_i. They represent the mathematical model of the **three-dimensional theory of consolidation**, as initially formulated by Biot (1941). Since this model allows us to analyse the simultaneous dissipation of excess pore pressure and the time evolution of the soil deformation, it is known in literature as the **coupled model of consolidation**. However, because of its complexity only a few solutions of practical interest are available in literature, among which we recall those listed in Table 6.2.

6.5.1 Undrained conditions and pore pressure parameters

In Section 5.2, we defined as **undrained condition** the circumstance in which a soil element does not exchange any amount of water with the surrounding ambient.

If, according to this assumption, in equation (6.39) the flux term vanishes, i.e.:

$$\nabla \cdot \mathbf{v} = 0 \qquad (6.48)$$

we obtain the relation:

$$n\beta \frac{\partial u}{\partial t} = \frac{\partial \varepsilon_v}{\partial t}, \qquad (6.49)$$

which states that the volume change of the soil element must be equal to the volume change of the fluid phase, due to its compressibility (recall that soil particles are supposed to be incompressible).

As an example, if reference is made to the 1D case (oedometric compression), the previous result allows us to write:

$$n\beta \delta u = (\delta \sigma_z - \delta u)m_v \qquad (6.50)$$

and we can deduce for the excess pore pressure the expression:

$$\delta u = \frac{\delta \sigma_z}{1 + \dfrac{n\beta}{m_v}} \qquad (6.51)$$

Since the water compressibility β is negligible when compared with that of the soil skeleton, we reach the important conclusion that, in an undrained loading process, the excess pore pressure is equal to the applied total vertical stress. It can be proved that the same conclusion applies if the soil element is subjected to the spherical total pressure δp. In axisymmetric conditions, the above requirement that the volume change of the soil skeleton must be equal to the volume change of the pore fluid gives the following expression for the excess pore pressure, first suggested by Skempton (1954):

$$\delta u = B\left[\delta\sigma_3 + A\left(\delta\sigma_1 - \delta\sigma_3\right)\right] \tag{6.52}$$

where the parameters A and B are known in literature as **pore pressure parameters**.

The form given to equation (6.52) is such that it allows to distinguish the effect due to the application of the spherical component (cell pressure in a conventional triaxial test) $\delta\sigma_3$, from that produced by a deviator stress ($\delta\sigma_1 - \delta\sigma_3$). The pore pressure parameter B depends on the degree of saturation and, provided the soil element is fully saturated, it is equal to unit. The pore pressure parameter A would be equal to 1/3 in the case of an isotropic elastic behaviour. In practice, soft clays exhibit values from 0 to 1, and stiff clays exhibit values ranging from -0.5 to 0 (as expected, a negative value indicates a dilatant behaviour).

6.6 Steady state flow

When the flow characteristics do not change with time, the following two basic relationships apply (see equation 6.39):

$$\nabla \cdot \mathbf{v} = 0, \tag{6.53}$$

$$\mathbf{v} = -K\nabla h . \tag{6.54}$$

These conditions state that the velocity field is a solenoidal and a potential field at the same time, and for this reason it is said to be an *harmonic field*. Furthermore, in a simple connected region, the necessary and sufficient condition for the existence of a potential function for the vector field $\mathbf{v}$ is given by:

$$\nabla \times \mathbf{v} = \mathbf{0} . \tag{6.55}$$

If for sake of simplicity a plane problem is considered, then the condition (6.55) of irrotational flow and the condition (6.53) are expressed as:

$$\left. \begin{array}{l} \dfrac{\partial v_x}{\partial y} - \dfrac{\partial v_y}{\partial x} = 0 \\[2mm] \dfrac{\partial v_x}{\partial x} + \dfrac{\partial v_y}{\partial y} = 0 \end{array} \right. \tag{6.56}$$

If we introduce the scalar potential, represented by a function such:

$$\Phi(x, y) = -Kh(x, y) + C,$$ (6.57)

substituting (6.57) into the equation (6.53) gives the **Laplace's equation:**

$$\frac{\partial^2 \Phi}{\partial x^2} + \frac{\partial^2 \Phi}{\partial y^2} = 0.$$ (6.58)

But we can also observe that, if we introduce the function $\Psi(x, y)$, called *flow function*, such that:

$$v_x = \frac{\partial \Psi}{\partial y}$$
$$v_y = -\frac{\partial \Psi}{\partial x}$$ (6.59)

this function satisfies equation (6.53) and Laplace's equation.

Can we give the function $\Psi(x, y)$ any physical meaning? If a curve $\Psi(x, y) = \mathbf{const}$ is considered, then:

$$d\Psi = \frac{\partial \Psi}{\partial x} dx + \frac{\partial \Psi}{\partial y} dy = 0,$$ (6.60)

and taking into account (6.59) we obtain:

$$\frac{v_x}{dx} = \frac{v_y}{dy}.$$ (6.61)

This equation proves that the tangent at any point on the curve $\Psi(x, y) = \text{const}$ represents the direction of the discharge velocity, so that the curves $\Psi(x, y) = \text{const}$ are **flow lines.**

Moreover, since:

$$\nabla \Phi \cdot \nabla \Psi = 0$$ (6.62)

it follows that equipotential and flow lines are othogonal, as already proved in Section 6.2.

Finally, when referring to Figure 6.1, the flow through a channel between two flow lines is given by:

$$q = \int_{\Psi_1}^{\Psi_2} (v_x dy - v_y dx) = \int_{\Psi_1}^{\Psi_2} d\Psi = \Psi_2 - \Psi_1,$$ (6.63)

and this result states that along a channel the flux is constant, being the difference $(\Psi_2 - \Psi_1)$.

6.6.1 Boundary conditions

An important question related to differential equations refers to the types of *auxiliary conditions* that lead to a well posed problem. Recall that a problem is **well posed** if there exists a solution, the solution is unique and depends continuously on the data. In this respect, the choice of appropriate boundary conditions is of paramount importance in determining the well posedness of the problem, and the following two common types of boundary conditions give well posed problems. When at the boundary we prescribe the value of the hydraulic head, this is a *Dirichlet boundary condition*. An alternative could be to prescribe a condition on the normal derivative of h along the boundary, and this is called a *Neumann boundary condition*.

The case of seepage under a diaphragm wall, sketched in Figure 6.10, is now considered as an example to highlight the boundary conditions. Along the boundaries CD and FG the total head is constant, so that these boundaries are equipotential lines and we can assign a Dirichlet boundary condition:

$$h(\mathbf{x}) = f(\mathbf{x}) \tag{6.64}$$

The impervious base AB requires no flow component in its orthogonal direction, so that the line AB must be a flow line (and the equipotentials must intersect this line at right angles). The same condition applies to the boundaries along the diaphragm wall: the line DEF must therefore be a flow line.

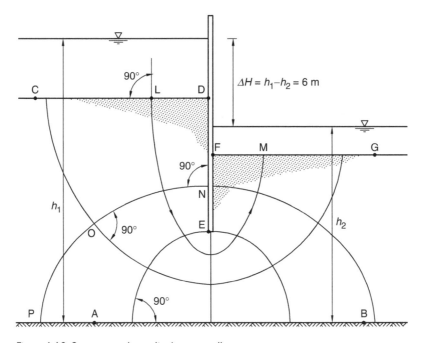

Figure 6.10 Seepage under a diaphragm wall.

If we try to express this boundary condition in a more general context, let the boundary surface be represented by the equation:

$$F(\mathbf{x}, t) = 0 . \tag{6.65}$$

The unit normal to this surface is given by:

$$\mathbf{n} = \frac{\nabla F}{|\nabla F|} , \tag{6.66}$$

and the condition of no flow through the surface writes:

$$\mathbf{v} \cdot \mathbf{n} = 0 , \tag{6.67}$$

or in terms of total head:

$$\nabla h \cdot \frac{\nabla F}{|\nabla F|} = 0 . \tag{6.68}$$

It follows that prescribing a condition on the flow gives a *Neumann* boundary condition.

The graphical representation of the family of *flow lines* with their corresponding *equipotential lines* is called a **flow net**. Figure 6.10 provides an example of such a flow net. It is relevant to observe that, once the total head h is known at a given point, it is possible to derive the pore pressure and the effective stress at that point.

In order to derive the quantity of seepage, by referring to Figure 6.1, we can deduce that:

$$\Delta q = -K \frac{h_j - h_i}{\Delta b} \cdot \Delta a . \tag{6.69}$$

Moreover, if we construct the flow net in such a way that flow lines and equipotential lines, in addition to having right-angle intersections and satisfying the boundary conditions, originates **curvilinear squares**, that means:

$$\Delta a = \Delta b , \tag{6.70}$$

then equation (6.69) tells us that $q = K \Delta h$. Because q is constant along a flow channel, it follows that $\Delta h = \text{const}$, so that the loss in head between two equipotential lines can be computed as the total loss in head divided by the number of equipotential drops, i.e. $\Delta H / N_e$. Therefore, if N_f is the number of flow channels, the total seepage is given by:

$$Q = K \frac{\Delta H}{N_e} \cdot N_f . \tag{6.71}$$

6.6.2 Solutions of practical interest

The case of flow around a sheet pile wall (sketched in Figure 6.11a) is discussed in detail by Schofield and Wroth (1968) (see pages 59–64 of their book *Critical State Soil Mechanics*). An exact mathematical solution is obtained by assuming that the

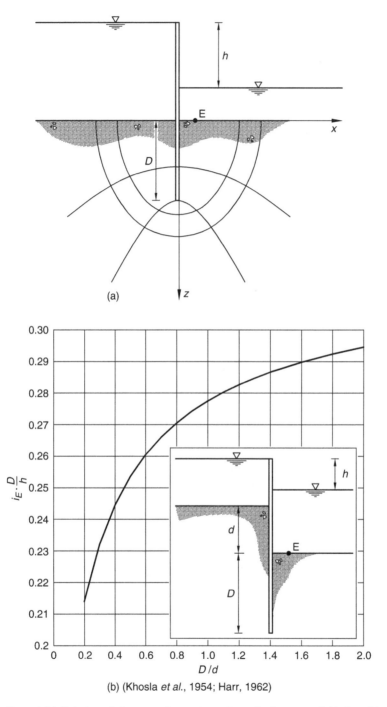

(a)

(b) (Khosla *et al.*, 1954; Harr, 1962)

Figure 6.11 Solution of the case of seepage under a diaphragm wall (deduced from Harr [1962] and Schofield and Wroth [1968]).

soilbed is of infinite extent both laterally and in depth. The obtained solution shows that *flow lines* are *confocal ellipses* and *equipotential lines* are *confocal hyperbolae*. The hydraulic gradient at any point in the soilbed is given by the following expression:

$$|i| = \frac{h}{\pi\left[\left(D^2 + x^2 - z^2\right)^2 + 4x^2z^2\right]^{0.25}}, \tag{6.72}$$

and, in particular, its value at the surface of the soilbed immediately behind the sheet-pile (the exit point E, which is of concern for the stability against piping), assumes the value:

$$|i_E| = \frac{h}{\pi D}. \tag{6.73}$$

A more general case, depicted in Figure 6.11b, is discussed by Harr (1962) and the solution has been obtained by means of *conformal mapping techniques* (in this case the Schwarz-Christoffel transformation). The hydraulic gradient at the exit point E assumes the expression:

$$|i_E| = \frac{h}{D}\frac{D}{d}\frac{m}{(1-m)}, \tag{6.74}$$

where m has to be deduced from:

$$\pi\frac{D}{d} = \frac{\sqrt{1-m^2}}{m} - \cos^{-1}m. \tag{6.75}$$

Figure 6.12a sketches a second case of engineering interest: an impervious structure resting on the surface of a soilbed of infinite depth. Polubarinova-Kochina (1952) obtained the solution for this case, showing that *flow lines* are *confocal ellipses* (with foci at $x = \pm b$) and *equipotential lines* are *confocal hyperbolas*. The solution obtained for the exit gradient:

$$|i_E| = \frac{h}{\pi\sqrt{x^2 - b^2}} \tag{6.76}$$

shows that at $x = b$ there is in this area the danger of piping. To avoid this danger, a sheet pile wall has been introduced at the toe of the structure, as shown in Figure 6.12b, and the solution obtained by Khosla *et al.* (1954) gives the following expression for the exit gradient (see Harr, 1962):

$$|i_E| = \frac{h}{D}\frac{1}{\pi\sqrt{\lambda}}$$

$$\lambda = \frac{1 + \sqrt{1 + (2b/D)^2}}{2}. \tag{6.77}$$

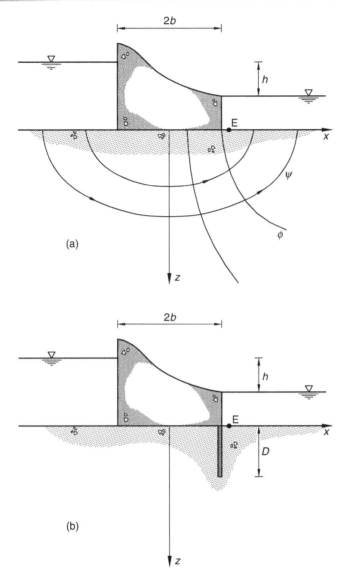

Figure 6.12 Seepage under an impervious structure.

Remarks. When using the above referred theoretical solutions, it is relevant to recall that they are based on the assumption that the soil is *homogeneous*. If this is not the case, and, as an example, the soil at the downstream side is less permeable, the exit gradient can be locally much higher than the value computed for the homogeneous case. This conclusion can be reached by observing that, since on the basis of continuity $K_1 i_1 = K_2 i_2$, the local gradient will be inversely proportional to the hydraulic conductivity, so that the soil may be locally eroded, attracting more water and leading to the formation of a *pipe* below the structure (this is the origin of the term *piping* introduced to describe this kind of hydraulic failure). For this reason it is not a simple matter to prescribe in a code of practice a fixed value for the exit gradient (a safety factor of 2 or larger is usually recommended) and some safeguarding measures should always be taken into consideration. These include a blanket of clay on the upstream side, in order to increase the resistance to flow and a gravel pack at the downstream side, acting as a drainage blanket.

6.7 Numerical solution of Laplace's equation

The *method of finite differences* is a widely used numerical method for finding approximate solutions of partial differential equations. The essence of the method consists of approximating the partial derivative of a function by finite difference quotients.

Let us assume the domain is covered by a mesh of rectangles of sides $\Delta x, \Delta z$ (see Figure 6.13). By using the Taylor's series expansion for $h(x)$, the hydraulic head at points 1 and 3 can be expressed as:

$$h_1 = h_o + \Delta x \left(\frac{\partial h}{\partial x}\right)_o + \frac{(\Delta x)^2}{2!}\left(\frac{\partial^2 h}{\partial x^2}\right)_o + \frac{(\Delta x)^3}{3!}\left(\frac{\partial^3 h}{\partial x^3}\right)_o + \dots \qquad (6.78)$$

$$h_3 = h_o - \Delta x \left(\frac{\partial h}{\partial x}\right)_o + \frac{(\Delta x)^2}{2!}\left(\frac{\partial^2 h}{\partial x^2}\right)_o - \frac{(\Delta x)^3}{3!}\left(\frac{\partial^3 h}{\partial x^3}\right)_o + \dots . \qquad (6.79)$$

Adding these two expressions and by neglecting terms of higher order, the second order partial derivative assumes the form:

$$\left(\frac{\partial^2 h}{\partial x^2}\right)_o \cong \frac{h_1 + h_3 - 2h_o}{(\Delta x)^2} \qquad (6.80)$$

and, in like manner, in the z direction we obtain:

$$\left(\frac{\partial^2 h}{\partial z^2}\right)_o \cong \frac{h_2 + h_4 - 2h_o}{(\Delta z)^2} . \qquad (6.81)$$

Substitution of (6.80) and (6.81) into Laplace's equation gives:

$$\frac{h_1 + h_3 - 2h_o}{(\Delta x)^2} + \frac{h_2 + h_4 - 2h_o}{(\Delta z)^2} \cong 0 \qquad (6.82)$$

and, suppose the domain is covered by a mesh of squares of side $\Delta x = \Delta z$, we get:

$$h_1 + h_2 + h_3 + h_4 - 4h_o \cong 0 , \qquad (6.83)$$

where the equality is approached as the size of the mesh decreases.

Equation (6.83) holds for all internal nodes and states that the hydraulic head at any node is the average of the four surrounding nodes. Note that this equation can also

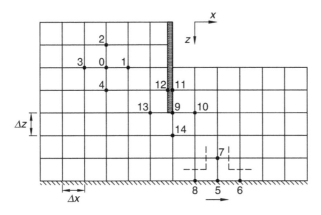

Figure 6.13 Numerical solution of Laplace's equation.

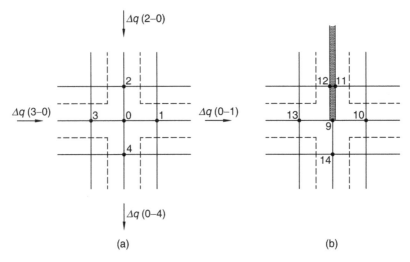

(a)   (b)

Figure 6.14 The condition of continuity.

be found directly from Darcy's law. Referring to Figure 6.14a the following relations hold:

$$\Delta q(2-0) = K\frac{h_2 - h_o}{\Delta z}\Delta x$$

$$\Delta q(0-4) = -K\frac{h_o - h_4}{\Delta z}\Delta x$$

$$\Delta q(3-0) = K\frac{h_3 - h_o}{\Delta x}\Delta z$$ (6.84)

$$\Delta q(0-1) = -K\frac{h_o - h_1}{\Delta x}\Delta z$$

and, since the continuity condition requires that $\sum \Delta q = 0$, we obtain (6.83).

This observation is of value when imposing the boundary conditions. As an example, considering point 5 at the impervious boundary (Figure 6.13), the continuity requires:

$$+\left(\frac{h_8 - h_5}{\Delta x}\frac{\Delta z}{2}\right) - \left(\frac{h_5 - h_6}{\Delta x}\frac{\Delta z}{2}\right) - \left(\frac{h_5 - h_7}{\Delta z}\Delta x\right) = 0. \tag{6.85}$$

Similarly, considering point 9 at the bottom of the sheet pile wall (Figure 6.14b), we obtain (assuming a square mesh):

$$(h_{13} - h_9) - (h_9 - h_{10}) + (h_{14} - h_9) + 0.5(h_{12} - h_9) - 0.5(h_9 - h_{11}) = 0. \tag{6.86}$$

The above outlined procedure shows that it is possible to write a finite difference equation for each nodal point and to produce a system of n linear algebraic equations in n unknown, represented by values of h at the n nodal points. Note that, even if it is more convenient to assemble the nodal equation into a matrix form and to solve the problem using a computer, the values at each node could be obtained iteratively by hand.

The finite difference method is convenient when the aim is to solve the hydraulic aspects of the problem. If the interest is to combine the hydraulic solution with deformations resulting from changes in effective stress, then the finite element method proves to be more convenient (for these aspects the reader can refer to Zienkiewicz and Taylor, 1994).

6.8 Flow through anisotropic and inhomogeneous media

In the previous section the assumption has been introduced that the hydraulic conductivity is the same in all directions. Most natural soil deposits are anisotropic and inhomogeneous due to depositional processes, therefore there is a need to analyse seepage problems where the hydraulic conductivity in the horizontal direction (k_x) is different (usually greater) from that in the vertical direction (k_z), or where there are zones of different hydraulic conductivity.

Anisotropic medium. The continuity equation:

$$K_x\frac{\partial^2 h}{\partial x^2} + K_z\frac{\partial^2 h}{\partial z^2} = 0 \tag{6.87}$$

can be reduced to Laplace's equation:

$$\frac{\partial^2 h}{\partial z^2} + \frac{\partial^2 h}{\partial y^2} = 0, \tag{6.88}$$

if any x dimension is mapped into:

$$y = x\sqrt{\frac{K_z}{K_x}}. \tag{6.89}$$

Equation (6.89) defines a scale factor to be applied to the x dimensions to transform the anisotropic flow domain into a fictitious isotropic flow region, where Laplace's equation applies. However, when using this transformation technique, some aspects deserve special attention.

(i) In the transformed region the seepage is:

$$\Delta q = K_e i \Delta y = K_e \Delta y \left(\frac{\Delta h}{\Delta z} \right) ,$$

whereas in the real domain one must have:

$$\Delta q = K_z \left(\frac{\Delta h}{\Delta z} \right) \Delta x = K_z \left(\frac{\Delta h}{\Delta z} \right) \Delta y \sqrt{\frac{K_x}{K_z}} ,$$

from which it follows that the hydraulic conductivity that must be applied in the transformed region is given by:

$$K_e = \sqrt{K_x K_z} . \tag{6.90}$$

(ii) In the transformed region, equipotential and flow lines have right-angle intersections, whereas this is not the case in the real anisotropic domain.

Zones of different hydraulic conductivity. Figure 6.15 refers to a boundary between two isotropic zones of hydraulic conductivity, K_1 and K_2. The flow and equipotential lines are continuous across the boundary, but their slope changes. In order to find the change of direction, let us impose the condition that the rate of flow is constant along the channel:

$$\Delta q = K_1 \left(\frac{\Delta h}{l_1} \right) b_1 = K_2 \left(\frac{\Delta h}{l_2} \right) b_2 ,$$

from which we obtain the relation:

$$\frac{b_2}{l_2} = \frac{K_1}{K_2} \frac{b_1}{l_1} , \tag{6.91}$$

that specifies the dimensions of the flow net in region 2, given the flow net in region 1. Moreover, since $l_1/b_1 = \tan\alpha_1$ and $l_2/b_2 = \tan\alpha_2$, one gets:

$$\frac{\tan\alpha_1}{\tan\alpha_2} = \frac{K_1}{K_2} , \tag{6.92}$$

which proves that the change of flow direction through the boundary is similar to the law of refraction in optics.

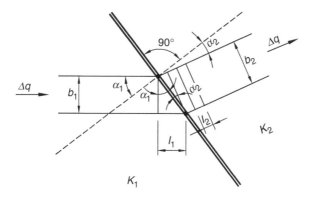

Figure 6.15 Boundary between two isotropic zones of different hydraulic conductivity.

6.9 Unconfined steady state flow

Consider the seepage through an earth dam resting on an impervious base, as sketched in Figure 6.16, and examine in detail the boundary conditions. At the impervious boundary AD the velocity component normal to the boundary must vanish, so that this boundary is the locus of a flow line and the boundary condition is expressed by equation (6.68). The surface AB is an equipotential line, since at any point such as M along this boundary the following condition applies:

$$z + \frac{u}{\gamma_w} = h_1 = \text{const}.$$

The same conclusion also applies to the boundary CD.

The surface EC (called *surface of seepage*) intersects the flow lines, so that it cannot be a flow line; neither it is an equipotential line, since the pressure on this surface is atmospheric, so that the condition applies that $h - z$. Finally, the surface BE is a *free surface*, on which the following two conditions hold: in a cross-section it must be a top flow line; in addition, the pore pressure at any point along this surface is equal to the atmospheric pressure, so that the total head must be equal to the elevation head.

When one boundary of the flow domain is a free surface, the flow domain is said to be **unconfined**, in contrast to flow domains, examined in the previous sections, where all the boundary conditions were known initially and the flow was said to be *confined*. Note that the location of the free surface depends on the flow regime, and the flow regime depends in its turn on the position of the free surface. It follows that determining the position of the free surface requires an iterative procedure, as shown in the sequel.

Back to the requirements to be satisfied on this surface, since the pressure at any point along this surface is equal to the atmospheric pressure, this surface will be represented by the equation:

$$S(x, y, z) := h(x, y, z) - z = 0.$$ (6.93)

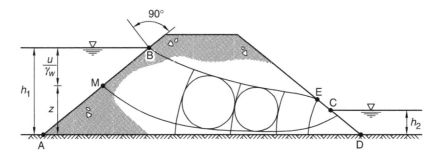

Figure 6.16 Unconfined flow.

Furthermore, the requirement that this surface in a cross-section must be a top flow line corresponds to the Neumann condition:

$$\nabla S \cdot \mathbf{v} = 0 \,. \tag{6.94}$$

Both these conditions allow the determination of its position, through an iterative procedure: initially, select an arbitrary location of the free surface and impose the requirement that this surface is a flow line, so that all boundary conditions are known. Solve the flow problem and compute the total head and the pore pressure at any point of the domain. Check the pore pressure along the free surface: a positive value will indicate that the assumed location was wrong and the flow was excessively confined, so that the location must be adjusted upward; the adjustment will be downward in case of negative pore pressure.

When drawing the top flow line, it is of help to observe that, according to (6.93), the total head along this line varies linearly with elevation head, so that the intersection points of the flow line with successive equipotential lines of equal drops have constant vertical distance from each other.

In practice, this class of problem is often analysed by referring to the **Dupuit theory** (1863), based on the assumption that, if the free surface is characterized by a small gentle slope, the equipotential lines can be assumed to be vertical, so that the total head does not depend on z, i.e.:

$$h(x, y, z) \cong h(x, y) \,. \tag{6.95}$$

Taking into account the requirement expressed by (6.93), one gets:

$$h = h(x, y) = z \,, \tag{6.96}$$

which states that the total head (constant along a given vertical) is equal to the elevation of the free surface and the flow lines are horizontal.

If a soil element bounded by the free surface is considered, the mass conservation equation in the x direction writes:

$$(q_x - q_{x+dx}) = K \frac{\partial}{\partial x} \left(h \frac{\partial h}{\partial x} \right) dx dy,$$

and in the y direction:

$$\frac{\partial}{\partial x}\left(h\frac{\partial h}{\partial x}\right)+\frac{\partial}{\partial y}\left(h\frac{\partial h}{\partial y}\right)=0\,,$$

from which:

$$\frac{\partial^2\left(h^2\right)}{\partial x^2}+\frac{\partial^2\left(h^2\right)}{\partial y^2}=0\,,\qquad(6.97)$$

which means that in the case of unconfined flow the function h^2 must satisfy Laplace's equation.

The two-dimensional case of Figure 6.17 is now considered as an example. Equation (6.97) reduces to:

$$\frac{d^2\left(h^2\right)}{dx^2}=0\,,\qquad(6.98)$$

and by integrating:

$$h^2=Ax+B\,.\qquad(6.99)$$

The boundary conditions require:

$$h=h_1\quad for\quad x=0$$
$$h=h_2\quad for\quad x=b$$

so that:

$$h^2=h_1^2-\frac{\left(h_1^2-h_2^2\right)x}{b}\,,\qquad(6.100)$$

which proves that the free surface is a parabola, a result that has been used extensively in order to define a priori the position of a free surface when dealing with flow through

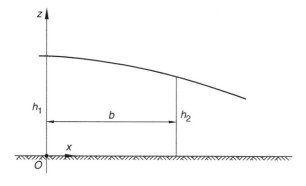

Figure 6.17 Parabolic free surface.

earth dams (Casagrande, 1937). In this respect, Figure 6.18 briefly recalls that the confocal parabolas are an example of conjugate functions, so that if the boundaries AB and FC are equipotential lines, curves such as AC are flow lines. These conditions apply to the case of Figure 6.19, where a horizontal underdrain is provided at the exit. Casagrande (1937) suggested that the focus should be fixed at point F and the segment EB should be assumed to be equal to 0.3BC. With these assumptions it follows that the focal distance is equal to:

$$S = \sqrt{d^2 + b^2} - d, \tag{6.101}$$

and the free surface is given by:

$$x = \frac{z^2 - S^2}{2S}. \tag{6.102}$$

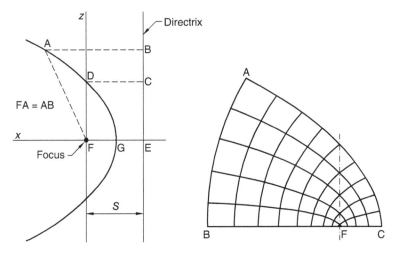

Figure 6.18 Confocal parabolas.

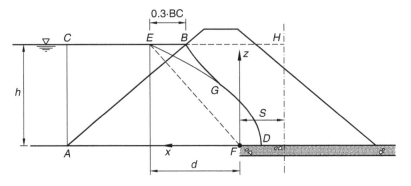

Figure 6.19 Flow through an earth dam.

Once the parabola has been obtained, the reversed curvature at point B can be sketched, in order to satisfy the boundary condition.

6.10 Transient flow: One-dimensional theory of consolidation

We observed in Chapter 4 (Section 4.8) that in many circumstancies the settlement of structures interacting with fine-grained soils can be computed by approximating the problem to a one-dimensional process. However, in addition to computing the magnitude of the settlement, an estimate is also required of the rate at which the settlement occurs. When dealing with coarse-grained soils, characterized by high values of permeability (the ease with which water can flow), it can be assumed that the settlement occurs during the construction time. On the contrary, with fine-grained soils the time required for the final settlement to develop may be of the order of months or even years (recall that the hydraulic conductivity coefficient K is of the order of 10^{-8} m/s).

The **transient coupled phenomenon** of the evolution with time of the deformation of the soil skeleton and of the fluid flow is called **consolidation** and its mathematical formulation is known, in soil mechanics, as **consolidation theory**.

The theory developed by Terzaghi (1923, 1925) is based on the following assumptions:

1   The porous medium is saturated.
2   Both particles and fluid are incompressibles.
3   Deformations are infinitesimal (for a finite deformation theory see Gibson *et al.*, 1967; Lancellotta and Preziosi, 1997; Pane and Schiffman, 1985).
4   The water flow and the soil displacement occur in the vertical direction only.
5   Soil behaviour is expressed through the constitutive relation $\delta\varepsilon_{zz} = m_v \delta\sigma'_{zz}$, the compressibility m_v being constant within the considered stress range.
6   The hydraulic conductivity is also assumed to be constant within the considered stress range.

With such assumptions, the continuity equation for the solid phase requires:

$$\frac{\partial(1-n)}{\partial t} + \frac{\partial[(1-n)v^s]}{\partial z} = 0\,, \tag{6.103}$$

and, similarly, for the pore water:

$$\frac{\partial n}{\partial t} + \frac{\partial(nv^w)}{\partial t} = 0\,, \tag{6.104}$$

and by combining these two equations the continuity condition for the mixture will be:

$$\frac{\partial[(1-n)v^s + nv^w]}{\partial z} = 0\,. \tag{6.105}$$

Taking into account the remarks of Section 6.2, Darcy's law plays the role of the momentum balance equation of the fluid phase:

$$\frac{\partial h}{\partial z} + \frac{n}{K}(v^w - v^s) = 0, \tag{6.106}$$

where, for convenience, we give to the total h the following form:

$$h = z + \frac{u^{st}}{\gamma_w} + \frac{u}{\gamma_w}, \tag{6.107}$$

being u^{st} the initial stationary value of the pore pressure (which changes linearly with z), and u is now the **excess pore pressure**. Substitution of this expression into (6.106) gives:

$$\frac{\partial^2 u}{\partial z^2} + \frac{\gamma_w}{K}\frac{\partial[n(v^w - v^s)]}{\partial z} = 0$$

and combining with (6.105) leads to:

$$\frac{\partial^2 u}{\partial z^2} - \frac{\gamma_w}{K}\frac{\partial v^s}{\partial z} = 0. \tag{6.108}$$

The gradient of the solid velocity $\partial v^s/\partial z$ can be expressed in terms of deformation rate:

$$\frac{\partial v^s}{\partial z} = -\frac{\partial \varepsilon_{zz}}{\partial t}, \tag{6.109}$$

and, by introducing the constitutive relation for the soil skeleton:

$$d\varepsilon_{zz} = m_v d\sigma'_{zz}, \tag{6.110}$$

the previous equation can be written as:

$$\frac{\partial^2 u}{\partial z^2} + \frac{\gamma_w}{K}m_v\frac{\partial \sigma'_{zz}}{\partial t} = 0. \tag{6.111}$$

If the applied load is constant with time, the equilibrium equation requires:

$$\frac{\partial \sigma_{zz}}{\partial t} = 0 \quad \forall z.$$

Since the total stress is the sum of the effective stress and the pore pressure, the previous result tells us that any increases of effective stress must be balanced by a decrease of the excess pore pressure, i.e. $\frac{\partial u}{\partial t} + \frac{\partial \sigma'_{zz}}{\partial t} = 0$, so that, by substituting the effective stress rate with the pore pressure change into (6.111), we obtain the **1D consolidation equation:**

$$\frac{\partial u}{\partial t} = c_v \frac{\partial^2 u}{\partial z^2} \,.$$

(6.112)

The coefficient:

$$c_v = \frac{K}{m_v g \rho_w} \,,$$

(6.113)

condenses the relevant soil and fluid parameters and is called **coefficient of consolidation.**

In order to derive the solution of (6.112) we have to analyse the following **initial and boundary value problem,** known in mathematical physics as the **Cauchy-Dirichlet problem** (Figure 6.20):

$$
\begin{aligned}
\frac{\partial u}{\partial t} &= c_v \frac{\partial^2 u}{\partial z^2} && 0 < z < 2H, \quad t > 0 \\
u(0,t) &= u(2H,t) = 0 && t > 0 \\
u(z,0) &= f(z) && 0 \le z \le 2H \quad t = 0
\end{aligned}
$$

(6.114)

The specified **boundary conditions** refer to a clay layer, having thickness $2H$, between two free draining strata (i.e. sand strata), so that the excess pore pressure at the boundaries is zero at any time.

In order to derive the **initial conditions,** let us recall equation (6.39):

$$n\beta \frac{\partial u}{\partial t} + \nabla \cdot \mathbf{v} = \frac{\partial \varepsilon_v}{\partial t} \,,$$

which states that the rate of volume deformation of the porous medium $(\partial \varepsilon_v / \partial t)$ is due to the rate of fluid compression $(n\beta \, \partial u / \partial t)$ plus the flow $(\nabla \cdot \mathbf{v})$ of fluid expelled

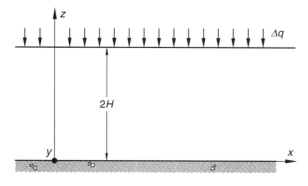

Figure 6.20 One-dimensional consolidation problem.

from the soil element. If the **initial conditions** are supposed to be **undrained**, due to the low hydraulic conductivity of the porous medium, then the flow term vanishes, so that it must be (note that in 1D conditions $\varepsilon_v = \varepsilon_z$):

$$n\beta\delta u = m_v(\delta\sigma_z - \delta u) \ .$$

The initial excess pore pressure is then defined as:

$$\delta u = \frac{1}{1 + \dfrac{n\beta}{m_v}}\delta\sigma_z \ ,$$

and, because the water compressibility is significantly lower than that of the soil skeleton, it follows:

$$\delta u(z, 0) = \delta\sigma_z(z) \ .$$

As already mentioned in Section 6.5, in soil mechanics it is usual to give to the ratio $C = 1/(1 + n\beta/m_v)$ the name of **pore pressure parameter** (a definition introduced by Skempton, 1954).

If the initial excess pore pressure is uniform (as is the case when the load on the surface is of infinite extent):

$$u(z, 0) = \text{const} = u_o \ , \tag{6.115}$$

the solution of the problem (6.114) is expressed in the form:

$$u = \sum_{n=0}^{\infty} \frac{2u_o}{M}\sin\left(\frac{Mz}{H}\right)\exp\left(-M^2T_v\right) \tag{6.116}$$

where the dimensionless **time factor** is defined as:

$$T_v = \frac{c_v t}{H^2} \tag{6.117}$$

and M depends on the summation index n through the relation:

$$M = \frac{2n+1}{2}\pi \ .$$

Note that, since when $z = H$, $\partial u/\partial z = 0$, the thickness H represents a **characteristic length,** from which the excess pore pressure may dissipate by drainage towards one of the boundaries (i.e. the plane $z = H$ is a plane of symmetry). It follows that the same solution also applies to the case of a soil thickness equal to H when the drainage occurs toward a single draining surface (as an example, a stratum resting on an impermeable bottom with a permeable top surface). To highlight the engineering relevance of the consolidation theory, it is convenient to introduce the following indices. The first one, called **degree of pore pressure dissipation,** is a *local* measure of the rate of the process:

$$U_z := 1 - \frac{u}{u_o}. \tag{6.118}$$

The second one, called **average degree of consolidation,** is a *global* measure of the process and is given by the ratio between the consolidation settlement at time t and the value attained at the end of the consolidation process, i.e.:

$$U_s = \frac{w(t)}{w_c}. \tag{6.119}$$

In Terzaghi's theory, the following relations hold true:

$$\frac{\partial \sigma'}{\partial t} = -\frac{\partial u}{\partial t} \quad ,$$

$$\delta \varepsilon_v = m_v(-\delta u)$$

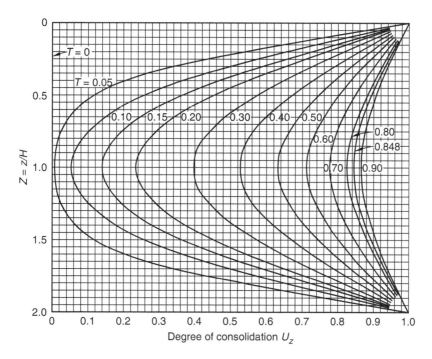

Figure 6.21 Solution of one-dimensional consolidation.

so that the average degree of consolidation, expressed in terms of settlements, equals the average degree of pore pressure dissipation:

$$U_s = U_p := \frac{\int\limits_0^{2H} \left(u_o - u(t)\right) dz}{\int\limits_0^{2H} u_o dz},$$ (6.120)

and for the case of initial uniform excess pore pressure:

$$U_s = 1 - \sum_{n=0}^{\infty} \frac{2}{M^2} \exp\left(-M^2 T_v\right).$$ (6.121)

Values of U_z, as defined by equation (6.118), are given in Figure 6.21, for the analysed case of uniform initial excess pore pressure. The shown curves represent the spatial distribution of the excess pore pressure, at a fixed instant of time, and are called **pore pressure isochrones**. For ease of computation, the relation between the average degree of consolidation and the dimensionless time factor is sketched in Figure 6.22,

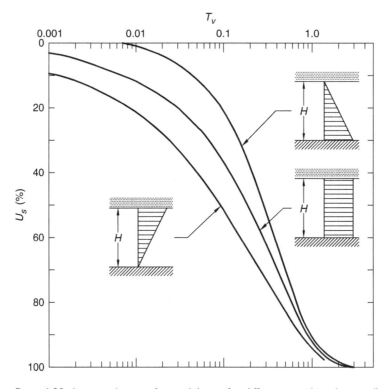

Figure 6.22 Average degree of consolidation for different initial isochrones (Janbu et al., 1956).

where the dependence of this relation on the initial distribution of excess pore pressure is also highlighted.

It must be noticed that, theoretically, the consolidation process ends when $t \to \infty$. But, as can be deduced from Figure 6.22, in practice we can consider the process as finished when $c_v t / H^2 \cong 2$. This latter is an important result, because it shows the influence of the main parameter on the consolidation process. In particular, this result tells us that if the soil hydraulic conductivity is half as large, the consolidation process will be twice as long. More important, if the drainage length is reduced by a factor of 2, the consolidation process will be reduced by a factor of 4. This highlights the importance of reducing the drainage path in order to improve the consolidation process, an argument that will be further addressed in Section 6.13, when dealing with vertical drains.

Example 6.5
Consider the stratigraphical sequence shown in Figure 6.23 and suppose that the estimated consolidation settlement is equal to 1.00 m. *Find* the time required to reach amounts of settlement equal to 0.3 m and 0.5 m. Furthermore, *evaluate* the time required for the excess pore pressure to reach a value equal to 0.75 of its initial value, at the depth of 6.25 m from the top of the clay stratum.

Solution. The definition of **average degree of consolidation** allows to obtain:

$$U_s = \frac{w(t)}{w_c} = \frac{0.3}{1.00} = 0.3$$

and the solution plotted in Figure 6.22 gives a value of the dimensionless time factor T_v equal to 0.0707. Then, since:

$$T_v = \frac{c_v t}{H^2}$$

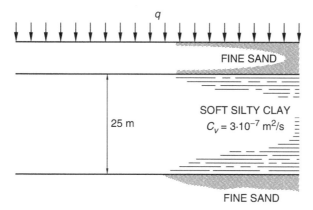

Figure 6.23 Illustrative example of consolidation induced by surface load.

we get:

$$t = \frac{0.0707 \cdot (12.5)^2}{3.10^{-7}} = 3.68 \cdot 10^7 \, s \cong 1.17 \; years.$$

Similarly, when considering the amount of settlement of 0.5 m we obtain $T_v = 0.196$ and the corresponding time:

$$t = \frac{0.196 \cdot (12.5)^2}{3 \cdot 10^{-7}} = 10.21 \cdot 10^7 \, s \cong 3.24 \; years.$$

By computing other couples, the evolution of settlement with time can be described.

At the required dimensionless depth

$$\frac{z}{H} = \frac{6.25}{12.5} = 0.5,$$

the *local* value of the degree of consolidation $U_z = 0.25$ (see Figure 6.21) corresponds to a dimensionless time factor $T_v = 0.10$, so that the required time will be:

$$t = \frac{0.1 \cdot 12.5^2}{3 \cdot 10^{-7}} = 5.21 \cdot 10^7 \, s \cong 1.65 \; years.$$

Example 6.6

Figure 6.24 shows a clay stratum underlaying a sand aquifer. Suppose that the groundwater level in the upper sand horizon is constant due to recharge, whereas the underlaying

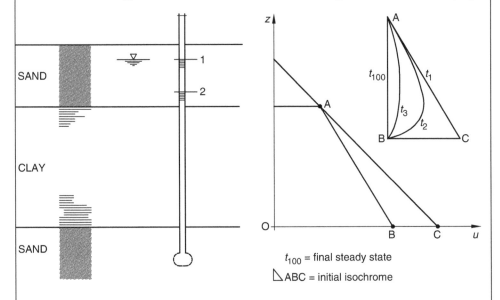

Figure 6.24 Illustrative example of subsidence induced by underdrainage.

acquifer is significantly lowered below that in the upper one, for water supply purposes. Describe and evaluate the consolidation phenomenon associated to the *underdrainage* of the clay stratum.

Solution. When the steady state condition is reached, the pore pressure distribution within the clay stratum will be that shown by the line *AB*, so there will be a steady flow of water downward through the clay stratum. However, before reaching this steady state condition, the excess pore pressure, represented by the triangular distribution, must be dissipated. This will cause an equivalent increase of effective stresses, which will give rise to a subsidence of the underdrained region, as was the case for Mexico City (see Zeevaert, 1983).

   The solution for this case is sketched in Figure 6.24, and we can observe that the curves are no longer symmetrical with respect to the mid-depth. However, it can be proved that, if the top and the bottom are both free draining boundaries, the average degree of consolidation is the same for any linear initial distribution of the excess pore pressure, so that the solution (6.121) also applies to this case (see Lambe and Whitman, 1969).

6.11 Experimental determination of the coefficient of consolidation

The worked Example 6.5 highlights that in order to compute the evolution of the consolidation process, we need to assess an appropriate value of the coefficient of consolidation c_v. The most common procedure is to apply the 1D consolidation theory to the results of oedometer tests, by performing essentially a back analysis of the compression dial readings versus time (see Figure 4.7), obtained for each increment of loading.

   There are two procedures commonly used in engineering practice. The first one, suggested by Casagrande (1938) and known as the **log time method**, uses the compression dial readings plotted against the logarithm of time, as shown in Figure 4.7, and the coefficient of consolidation is determined by matching the experimental results with the theoretical solution when the degree of consolidation reaches 50%. This value corresponds to a dimensionless time factor T_v equal to 0.196, so that the coefficient of consolidation is computed using the relation:

$$c_v = \frac{T_v H^2}{t} = \frac{0.196 H^2}{t_{50}},$$
(6.122)

where H is half the thickness of the specimen.

   In order to find the time corresponding to a degree of consolidation equal to 50%, i.e. t_{50}, an estimate of the final compression reading is required. When dealing with the conventional incremental loading test, each increment is maintained for one day, but the end of consolidation occurs in less than one hour. Therefore, the final asymptote of the consolidation compression is not clearly defined due to creep, and rather arbitrarily the point B in Figure 4.7, obtained by the intersection of the two linear parts of the curve, is assumed as the point corresponding to the end of the primary compression.

Once the final reading has been defined, the reading corresponding to 50% of consolidation can be determined provided we know the initial reading. This can be done by observing that, during the early stage of consolidation, T_v is proportional to U_s^2, so that:

$$\frac{\delta H(t_1)}{\delta H(t_2)} = \sqrt{\frac{t_1}{t_2}}.$$

Therefore, if we choose any two values of times in the ratio of 4, i.e. $t_2 = 4t_1$, since $\delta H(t_2) = 2\delta H(t_1)$, marking off a distance equal to $\delta H(t_2) - \delta H(t_1)$ above $\delta H(t_1)$ defines the zero reading.

Alternatively, we make use of the **square root of time method** suggested by Taylor (1948), based on observation that during the early stage of the process the compression readings are proportional to $\sqrt{t}$ (Figure 6.25). However, a straight line drawn through the experimental points up to about 60% of consolidation would give a value of $\sqrt{T_v} = 0.798$ when $U_s = 0.9$, because the parabolic approximation corresponds to the relation:

$$U_s = \sqrt{\frac{4T_v}{\pi}}. \tag{6.123}$$

The correct solution would give $\sqrt{T_v} = 0.921$, with a ratio of 1.15 with respect to the approximate value. This suggests to draw a second line with abscissas 1.15 times as large as the corresponding values of the first one. The intersection of this

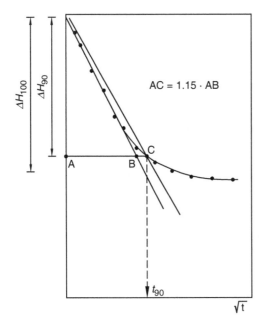

Figure 6.25 Determining the consolidation coefficient of consolidation by means of square root method.

second line with the experimental curve defines the time t_{90} corresponding to 90% of consolidation.

Then, the coefficient of consolidation is determined from the relation:

$$c_v = \frac{0.848H^2}{t_{90}} .$$

(6.124)

There is no reason why the above procedures should give the same answer (even if they usually give close values of the coefficient of consolidation), considering that both are based on approximations. One advantage of the square root of time method is that, by plotting the dial readings during the test, as soon as the t_{90} is reached it is possible to add the next load increment. This will reduce the time for testing and will minimize the contribution of secondary compression (Leonards, 1976; Holtz and Kovaks, 1981).

However, the most relevant aspects related to the determination of C_v are highlighted by the experimental data shown in Figure 6.26: the coefficient of consolidation depends on the stress history and significantly decreases as the applied stress approaches the yield stress; in addition, soil disturbance has a dramatic effect because it obliterates the soil memory. But even when using high quality samples, since the permeability of the specimen may be unrepresentative of the field value, due to the influence of many structural features that cannot be captured by the small laboratory specimen, the data obtained from oedometer tests must be used with caution. This aspect is highlighted by the data shown in Figure 6.27 (Burghignoli and Calabresi, 1975): the coefficient of consolidation obtained from tests on large specimens (of thickness equal to 250 mm and diameter of 500 mm) was about twice the values obtained from conventional tests and in a quite good agreement with the values backfigured from pore pressure and settlement records.

6.12 Solution of consolidation PDE

The partial differential equation (*PDE*) of consolidation is known in mathematics as a *parabolic PDE* and models diffusion processes. Therefore, it is of interest to highlight some basic aspects in a rather general way. Assume that the quantity u, which previously was representing the excess pore pressure, is now the *density* of a given substance, i.e. the amount per unit volume. In a 1D case $u = u(z,t)$. If a scalar function $\Psi(z,t)$ represents the *flux* of u, i.e. the quantity of u flowing through the unit area per unit time, in absence of source terms, the conservation law in differential form writes:

$$\frac{\partial u}{\partial t} + \frac{\partial \Psi}{\partial z} = 0 .$$

(6.125)

Equation (6.125) is a partial differential equation relating two unknowns, the density and the flux, so that another equation is required to have a determined system. This additional equation is usually related to the properties of the system and for this reason

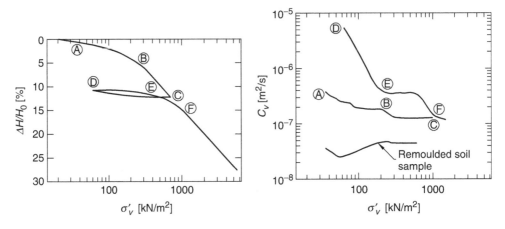

Figure 6.26 Influence of stress history and soil disturbance on C_v.

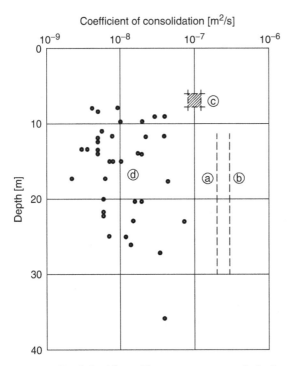

(a) C_v backfigured from pore pressure dissipation
(b) C_v backfigured from settlement evolution
(c) C_v from lab tests on large samples
(d) C_v from lab tests on conventional soil samples

Figure 6.27 The consolidation coefficient as backfigured from field measurements and lab tests (Burghignoli and Calabresi, 1975).

is called *constitutive equation*. In many circumstances it is found that the flux depends on the density gradient, so that we can write:

$$\Psi = -D\frac{\partial u}{\partial z} \, . \tag{6.126}$$

As an example, when dealing with chemical concentration, equation (6.126) is known as **Fick's law**, and the constant D is called **diffusion coefficient**. In the context of heat conduction equation (6.126) is known as **Fourier's law**, and the coefficient is called **thermal conductivity**.

Substituting (6.126) into (6.125) gives the **diffusion equation**:

$$\frac{\partial u}{\partial t} = D\frac{\partial^2 u}{\partial z^2} \, , \tag{6.127}$$

where the coefficient $D(L^2 T^{-1})$ can be used to define a **characteristic time** of the evolution of the process:

$$T_v = \frac{H^2}{D} \, . \tag{6.128}$$

If H is a characteristic length for the analysed problem, then T_v gives the time needed in order to appreciate any significative change.

An important question related to differential equations refers to the types of *auxiliary conditions* that lead to a well-posed problem. Recall that a problem is **well posed** if there exists a solution, the solution is unique and depends continuously on the data.

An auxiliary condition on the excess pore pressure at time $t = 0$ represents the *initial condition*, whereas conditions prescribed at $z = 0$ and $z = 2H$ are called *boundary conditions*. When at the boundary we prescribe the value of the excess pore pressure, this is a *Dirichlet boundary condition*. An alternative could be to prescribe a condition on the normal derivative of u along the boundary, and this is called a *Neumann boundary condition*.

The problem of consolidation is a **Cauchy-Dirichlet problem**:

$$
\begin{aligned}
\frac{\partial u}{\partial t} &= c_v \frac{\partial^2 u}{\partial z^2} & 0 < z < 2H, \quad t > 0 \\
u(0,t) &= u(2H,t) = 0 & t > 0 \\
u(z,0) &= f(z) & 0 \le z \le 2H \quad t = 0
\end{aligned}
\tag{6.129}
$$

and is a well-posed problem.

The solution of diffusion equation can be obtained by using the Laplace transform method or the *method of separation of variables*, as is the case in the sequel. It means that we look for solutions of the form:

$$u(z,t) = Z(z) \cdot T(t) \tag{6.130}$$

where the function Z depends on z and the function T on t.

When substituting (6.130) into (6.129) the partial derivatives appear as ordinary derivatives:

$$Z(z) \cdot T'(t) = c_v Z''(z) \cdot T(t)$$

and rearranging:

$$\frac{T'(t)}{c_v T(t)} = \frac{Z''(z)}{Z(z)} . \tag{6.131}$$

Since the l.h.s. of (6.131) does not depend on z, neither can the r.h.s., and the same argument can be used with respect to the variable t. Then the only function, which is independent of both z and t is a constant, so that:

$$\frac{T'(t)}{c_v T(t)} = K; \quad T'(t) - c_v K T(t) = 0$$
$$\frac{Z''(z)}{Z(z)} = K; \quad Z''(z) - K Z(z) = 0 \tag{6.132}$$

Let's now consider the boundary value problem:

$$Z''(z) - K \cdot Z(z) = 0$$
$$Z(0) = Z(2H) = 0 \tag{6.133}$$

The values of K for which the boundary value problem has a non-trivial solution, i.e. $Z(z) \neq 0$, are called *eigenvalues*, and the corresponding functions are called *eigenfunctions*. If we pose $Z(z) = e^{\alpha x}$, we derive the *auxiliary equation*:

$$\alpha^2 - K = 0 , \tag{6.134}$$

which gives a non-trivial solution only if $K < 0$.

Therefore the roots of (6.134) are:

$$\alpha = \pm i \sqrt{-K}$$

and the general solution of (6.133) is:

$$Z(z) = A \cos(\sqrt{-K} z) + B \sin(\sqrt{-K} z) . \tag{6.135}$$

The boundary conditions define the system:

$$A = 0$$
$$B \sin(\sqrt{-K} \cdot 2H) = 0 \, '$$

and a non-trivial solution can be obtained only if $2H\sqrt{-K} = n\pi$, so that:

$$K = -\left(\frac{n\pi}{2H}\right)^2 .$$

The eigenfunctions are:

$$Z(z) = a_n sin\left(\frac{n\pi z}{2H}\right),$$
(6.136)

where a_n are arbitrary non-zero constants.

Considering now the second equation:

$$T'(t) + c_v\left(\frac{n\pi}{2H}\right)^2 T(t) = 0$$

we get:

$$T(t) = b_n exp\left[-c_v\left(\frac{n\pi}{2H}\right)^2 t\right]$$

so that the solution assumes the form:

$$u(z,t) = c_n exp\left[-c_v\left(\frac{n\pi}{2H}\right)^2 t\right] sin\left(\frac{n\pi z}{2H}\right).$$
(6.137)

Because the *PDE* is linear and the boundary conditions are homogeneous, if two functions are solutions of (6.129), any linear combination of these functions is still a solution. Therefore, we can consider an infinite sum of these functions:

$$u(z,t) = \sum_{n=1}^{\infty} c_n exp\left[-c_v\left(\frac{n\pi z}{2H}\right)^2 t\right] sin\left(\frac{n\pi z}{2H}\right).$$
(6.138)

Moreover, if we impose the initial condition:

$$u(z,0) = f(z) = \sum_{n=1}^{\infty} c_n sin\left(\frac{n\pi z}{2H}\right),$$
(6.139)

we obtain a *Fourier sine series*, whose constants are given by:

$$c_n = \frac{2}{2H} \int_0^{2H} u(z,0) sin\left(\frac{n\pi z}{2H}\right) dz.$$
(6.140)

For the case of *uniform initial excess pore pressure*:

$$u(z,0) = u_o,$$

the solution becomes:

$$u(z,t) = \sum_{n=1}^{\infty} \frac{2u_o(1-\cos n\pi)}{n\pi} exp\left[-c_v\left(\frac{n\pi}{2H}\right)^2 t\right] sin\left(\frac{n\pi z}{2H}\right).$$
(6.141)

Finally, by observing that the quantity $(1 - \cos n\pi)$ is zero when n is even, whereas it becomes 2 when n is odd, it is convenient to write the solution in the form:

$$u = \sum_{n=0}^{\infty} \frac{2u_o}{M} \sin\left(\frac{Mz}{H}\right) \exp\left(-M^2 T_v\right). \tag{6.142}$$

where:

$$M = \frac{2n+1}{2}\pi; \qquad T_v = \frac{c_v t}{H^2}.$$

6.13 Vertical drains

We have already discussed in Section 6.10 that, in practice, we can consider the process of consolidation as finished when the dimensionless time factor $c_v t / H^2 \cong 2$. This result shows the influence of the main parameter on the consolidation process and, in particular, it tells us that if the drainage length is reduced by a factor of 2, the consolidation process will be reduced by a factor of 4. This highlights the importance of reducing the drainage path in order to improve the consolidation process, and, moving from this observation, the primary purpose of **vertical drains** is to speed up the consolidation process by *shortening the drainage path*.

In addition, most natural deposits are anisotropic with respect to flow properties, the horizontal hydraulic conductivity being typically higher than the vertical one $(K_{xx} = K_{yy} >> K_{zz})$. Therefore, the horizontal flow in the presence of vertical drains offers this further advantage combined with the geometrical change of the drainage boundary (see Conte, 1998).

The American engineer D.J. Moran first suggested in 1925 the use of sand drains for soil stabilization, and the first practical installation was carried out in California a few years later (Porter, 1936). In the middle of the 1930s, Kjellman in Sweden also began experiments with prefabricated cardboard drains. In 1971 Wager improved the Kjellman wick drain using a polyethylene core in place of the cardboard, and in the following years a large number of prefabricated drains were introduced (see Holtz *et al.*, 1991).

Figure 6.28 summarizes the main features related to the well-established soil improvement technique, based on the combined use of preloading and vertical drains, a technique mainly used to improve soft inorganic clays as well as compressible silts, organic clays and peats, i.e. all those soil deposits which under ordinary loading conditions can exhibit excessive settlements or cannot satisfy safety requirements (Hegg *et al.*, (1983) provide a detailed case history, which supports the advantages of this technique).

It is rather common practice to provide a sand bed, offering both the drainage function of collecting and draining away the water supplied by drains, as well as that of providing a working platform. Regarding the installation methods, it can be observed that *driven displacement sand drains* are still in use, but their installation

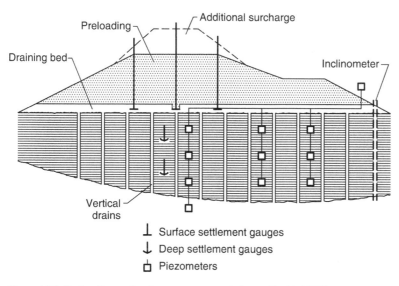

⊥ Surface settlement gauges

↓ Deep settlement gauges

⧄ Piezometers

Figure 6.28 Preloading embankment on vertical drains (Ladd, 1976).

causes severe disturbance to the surrounding soil, which minimizes the performance of the soil-drain system. Sand drains with a continuous flight hollow stem auger cause soil disturbance to a lesser extent and their performance can be considered intermediate to full-displacement and jetted drains. *Jetted sand drains* (of diameter 0.2 to 0.3 m and spacing ranging from 2 to 5 m) minimize the disturbance of the surrounding clay, but their installation is quite complex, requiring an experienced contractor and close supervision. *Prefabricated drains* have much smaller dimensions (the equivalent diameter ranges from 60 to 150 mm with spacing from 1.2 to 4 m) and they are usually installed using a closed-ended mandrel. The amount of soil disturbance is in this case related to the size and the shape of the mandrel, but it is always less than in the case of conventional displacement driven drains.

In order to check the evolution with time of the consolidation process, it is good practice to install piezometers, as well as point surface settlement gauges, deep settlement gauges or multipoint settlement gauges (see Chapter 8 of the book by Leroueil *et al.*, 1990).

By observing the geometrical pattern sketched in Figure 6.29, it is quite satisfactory to assume that the problem of soil consolidation can be addressed by referring to an equivalent cylinder of radius r_e, with an impervious vertical outside surface and an inner cylindrical drain. Furthermore, assuming that the vertical flow can be neglected if compared with the horizontal one, the problem reduces to a pure **radial flow**, so that the consolidation equation writes:

$$\frac{\partial u}{\partial t} = \frac{K_r}{\rho_w g \cdot m_v} \frac{1}{r} \frac{\partial}{\partial r} \left(r \frac{\partial u}{\partial r} \right) . \tag{6.143}$$

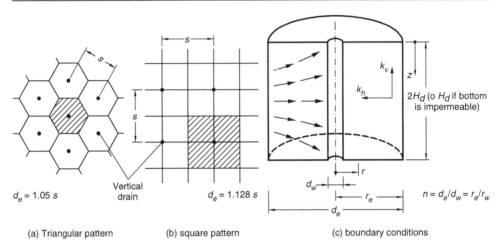

$d_e = 1.05\ s$ Vertical drain $d_e = 1.128\ s$ $n = d_e/d_w = r_e/r_w$

(a) Triangular pattern   (b) square pattern   (c) boundary conditions

Figure 6.29 Consolidation process in presence of vertical drains.

It follows that the initial-boundary value problem can be formulated as follows:

$$\frac{\partial u}{\partial t} = c_r \frac{1}{r}\frac{\partial}{\partial r}\left(r\frac{\partial u}{\partial r}\right) \quad r_w < r < r_e,\ t > 0$$

$$u(r,0) = u_o \qquad\qquad t = t_o$$

$$u(r_w,t) = 0 \qquad\qquad t \geq 0 \tag{6.144}$$

$$\frac{\partial u(r_e,t)}{\partial r} = 0 \qquad\qquad t \geq 0$$

Barron (1948) provided the solution for both cases of free vertical strains and equal vertical strains. The simpler solution refers to the case of *equal strain* and is reported in the sequel, in terms of excess pore pressure and average degree of consolidation:

$$u = \frac{u_o}{r_e^2 F(n)}\left[r_e^2 \ln\left(\frac{r}{r_w}\right) - \frac{r^2 - r_w^2}{2}\right]\cdot e^{\lambda}$$

$$U_p = 1 - e^{\lambda} \tag{6.145}$$

where:

$$\lambda = -\frac{8T_r}{F(n)}$$

$$F(n) = \frac{n^2}{n^2 - 1}\ln(n) - \frac{3n^2 - 1}{4n^2}$$

$$n = \frac{r_e}{r_w} \tag{6.146}$$

$$T_r = \frac{c_r t}{d_e^2} = \left(\frac{K_r}{\rho_w g \cdot m_v}\right)\frac{t}{d_e^2}$$

The above solution assumes that the installation of drains does not change the soil properties. In reality, soil disturbance can be more or less severe, depending on soil sensitivity and macrofabric. Barron (1948) and Hansbo (1979, 1981) have obtained the solution that includes the **smear effect** in the factor $F(n)$:

$$F(n) = \ln\left(\frac{d_w}{d_s}n\right) - 0.75 + \frac{K_r}{K_s}\ln\left(\frac{d_s}{d_w}\right) , \tag{6.147}$$

where d_s is the diameter of the remoulded annulus and K_s is the reduced conductivity within this annulus.

Prefabricated vertical drains generally consist of a flat plastic core wrapped in a filter sleeve made of paper or non-woven geotextile filter fabric. Due to its reduced cross-section, the discharge capacity q_w can be very low and the consolidation process can be delayed. Barron (1948) first considered the case of well resistance and his solution shows that the degree of pore pressure dissipation is no longer constant with depth. Hansbo (1981) presented a similar solution in the form of equation (6.145), where the term $F(n)$ is now given by:

$$F(n) = \ln(n) - 0.75 + \pi z(2l - z)\frac{K_r}{q_w} \tag{6.148}$$

where the characteristic *length l* of the drain is equal to half the drain length for open-ended drains and to the full length for closed-ended drains (for further details see Holtz *et al.*, 1991).

Example 6.7
Consider here again the soil profile shown in Figure 6.23, and compute the time required to obtain an average degree of consolidation equal to 90%, if the vertical drain pattern is such that $d_w = 80\,\text{mm}$ and $d_e = 2.00\,\text{m}$.

Solution. By using (6.145) and (6.146) we obtain:

$$n = \frac{d_e}{d_w} = 25 \quad F(n) = 2.475 \quad \lambda = \ln(1 - U_p) = \ln(0.1) = -2.3 .$$

The dimensionless time factor corresponding to the obtained value of λ is equal to $T_r = -\lambda F(n)/8 = 0.712$, and assuming rather conservatively $c_r = c_v$, the definition of the dimensionless time factor $T_r = c_r t/d_e^2$ allows us to obtain $t \cong 110$ days. This result proves the effectiveness of vertical drains in reducing the time required for consolidation, because without their use the required time would have been equal to about 14 years.

6.14 Three-dimensional consolidation theory

Biot theory. As already discussed in Section 6.5, the following equations:

$$\lambda u_{k,kj} + G(u_{i,j} + u_{j,i})_{,i} - u_{,j} = 0 \tag{6.149}$$

$$\frac{K}{\gamma_w} \nabla^2 u + \frac{\partial \varepsilon_v}{\partial t} = 0 \tag{6.150}$$

are four partial differential equations (PDE) in which the four unknown are represented by the excess pore pressure u and the three components of displacement u_i. The volume strain ε_v is not an independent variable, because:

$$\varepsilon_v = \frac{\partial u_x}{\partial x} + \frac{\partial u_y}{\partial y} + \frac{\partial u_z}{\partial z}.$$

Equations (6.149) and (6.150) represent the mathematical model of the *three-dimensional theory of consolidation*, as initially formulated by Biot (1941) and, since this model analyses the simultaneous dissipation of excess pore pressure and the time evolution of the soil deformation, it is known in literature as the *coupled model of consolidation*. As can be expected, solving this problem is a difficult task, and only a few solutions of practical interests are available in literature, among which we recall those listed in Table 6.2.

Uncoupled theory. Because of the difficulties encountered when trying to solve the Biot problem, many researchers have introduced some simplifying assumptions. The most common is the one we investigate in the sequel. When dealing with an isotropic material, the volume strain is linked to the effective mean stress through the relation:

$$\varepsilon_v = \frac{p'}{K_{SK}} = \frac{p - u}{K_{SK}}. \tag{6.151}$$

If the total stress is assumed to be constant with time, then by substituting equation (6.151) into (6.150) the *storage equation* reduces to the simplest *diffusion equation*:

$$\frac{K}{\gamma_w} K_{SK} \nabla^2 u = \frac{\partial u}{\partial t}, \tag{6.152}$$

with the only unknown represented by the excess pore pressure. This assumption was first made by Rendulic (1936) and the simplified theory is known as **uncoupled theory**. This definition arises from the fact that the pore pressure can be determined from equation (6.152) and, successively, the deformation problem can be solved by using the equilibrium equations, in which the gradient of the excess pore pressure acts as a known body force.

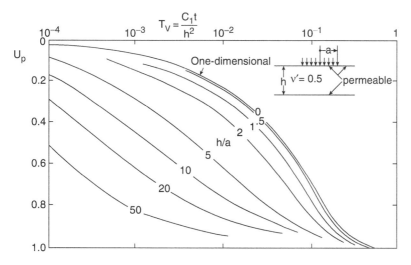

Figure 6.30 Rate of consolidation of a circular footing for drainage boundaries PTPB (reproduced from Davis and Poulos (1972) with permission).

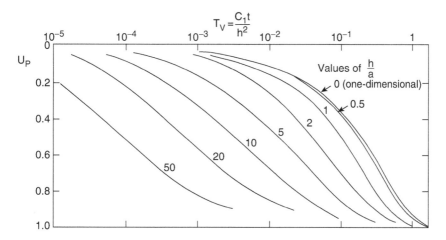

Figure 6.31 Rate of consolidation of a circular footing for drainage boundaries PTIB (reproduced from Davis and Poulos (1972) with permission).

Figures 6.30, 6.31, 6.32 and 6.33 reproduce the solutions provided by Davis and Poulos (1972) for the rate of settlement for the cases of a circular (of radius a) and a strip footing (of width $2b$), with the following boundaries conditions:

1   PTPB: permeable top, permeable bottom.
2   PTIB: permeable top, impermeable bottom.

Note that, when dealing with the diffusion theory, the only way of obtaining the average degree of consolidation in terms of settlement, i.e. U_s, is by a-priori

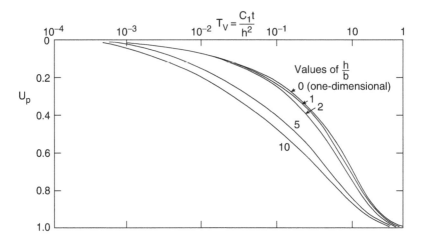

Figure 6.32 Rate of consolidation of strip footing for drainage boundaries PTPB (reproduced with permission from Davis and Poulos, 1972).

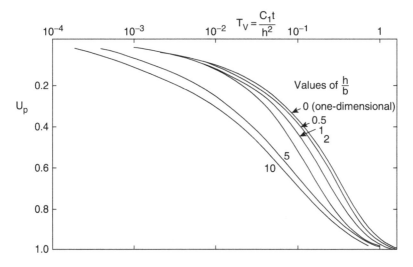

Figure 6.33 Rate of consolidation of strip footing for drainage boundaries PTIB (reproduced with permission from Davis and Poulos, 1972).

assuming $U_s = U_p$, being U_p the average degree of consolidation in terms of excess pore pressure (for a homogeneous soil this would be the case if there were no redistribution of the total stress during the consolidation). However, as proved by Davis and Poulos (1972), this can be considered an acceptable assumption if a suitable adjustment of the time scale is made, by simply defining the dimensionless time factor T_v in terms of $c_1 = c_v$, the one-dimensional coefficient of consolidation, for all cases, one, two and three-dimensional. Further, it must be observed that, although the figures present curves for the centre of the circular and the strip footing, there is little difference for the centre and the edge of the footing, so that these solutions can also be applied to rigid footings with reasonable accuracy.

Solutions provided by the uncoupled theory may be considered to give a reasonable approximation of the overall behaviour in the long term, if local effects in the early stage of the problem are of no concern. In particular, the Rendulic theory cannot capture the so-called Mandel Cryer effect, i.e. the increase of pore pressure before it decays, as predicted by Cryer (1963) and Mandel (1957).

6.15 Summary

(1) The hydraulic conductivity can be thought of in physical terms as the ease with which fluids may flow through porous media, and the fundamental equation which relates the velocity to the energy content of the fluid motion is Darcy's law. Darcy's law states that discharge velocity, defined as the quantity of water which flows through a unit total area of the porous medium in a unit time, is linearly related to the hydraulic gradient:

$$v = Ki \, ,$$

and the coefficient K is defined as hydraulic conductivity K.

The *hydraulic gradient* is defined as the (negative of the) rate of change of total head with distance. The *coefficient of hydraulic conductivity K (LT^1)* depends on both the *properties of the solid matrix* (porosity, shape of particles, grading, tortuosity and specific surface), and the *fluid properties* (density ρ_w and viscosity μ). On the contrary, the *permeability k (L^2)* depends only on the properties of the solid matrix, and is linked to the hydraulic conductivity K through the relation:

$$k = \frac{K\mu}{\rho_w g} \, .$$

(2) In the presence of seepage, the soil skeleton can be thought to be in equilibrium under the action of the effective stresses, the volume forces deriving from the gravity field, represented by the unit boyant weight (γ'), and the forces arising from the interaction with the flowing water. These latter are called *seepage forces* and they represent a second field superimposed to the gravity field. Note that it is common to use the term seepage forces, even they are not actually forces but forces per unit volume.

If we consider an upward flow, a critical situation can be generated if anywhere the vertical effective stress is equal to zero. When this occurs, the soil is *fluidized*, i.e. the velocity of the water is so large that the drag forces on the soil particles overcome the submerged weight of the particles that are no longer resting in contact with each other. This condition is called *piping*, if it occurs in localized channels (the transport of particles usually occurs at the ground surface and erosion regresses forming a pipe-shaped channel in the soil mass), or *quicksand*, if it occurs over a large area.

(3) When the flow characteristics do not change with time, we are in the presence of a steady state (or stationary) flow. Steady state flow is governed by Laplace's

equation and the solution of this problem can be derived by sketching a net of flow lines and equipotential lines.

(4) When dealing with coarse-grained soils, characterized by high values of hydraulic conductivity, it can be assumed that the settlement occurs during the construction time. On the contrary, with fine-grained soils the time required for the final settlement to develop may be of the order of months or even years (recall that the hydraulic conductivity coefficient K is of the order of 10^{-8} m/s).

The *transient coupled phenomenon* of the evolution with time of the deformation of the soil skeleton and of the fluid flow is called *consolidation* and its mathematical formulation is known as *consolidation theory*.

6.16 Further reading

General aspects related to groundwater flow are discussed in all textbooks for undergraduate students.

T.W. Lambe and R.V. Whitman (1969) *Soil Mechanics*, Wiley, offers a presentation with many worked examples, and R.D. Holtz and W.D. Kovaks (1981) *An Introduction to Geotechnical Engineering*, Prentice-Hall, gives a clear presentation of the method of fragments for the solution of confined flow problems.

M.E. Harr (1962) *Groundwater and Seepage*, McGraw-Hill, is a reference that gives the reader an analytical approach to the solution of flow problems. In particular, this book emphasizes the usefulness of advanced mathematics, such as the theory of complex variables, conformal mapping and elliptic functions.

The volume of J. Bear (1972) *Dynamics of Fluids in Porous Media*, Elsevier (Dover ed. 1988), conceived for advanced undergraduate and graduate students, offers a comprehensive treatment of the subject through a macroscopic description.

Chapter 7

In situ investigations

Field investigation is aimed at assessing enough information to select the most appropriate foundation solution, to highlight problems that could arise during construction, and, more in general, to highlight potential geological hazards in the examined area (as an example, slope instability or soil liquefaction during earthquakes).

This requires the definition of a design soil profile, throughout the identification of subsurface soil stratification, the detection of groundwater regime and the description of relevant structural features.

The extension of the investigation programme depends on the importance of the project. In some cases the designer can take advantage of the available information at the site and apply a conservative approach without specific or extensive *in situ* investigation. But most of the time, the nature of the geological formations is the relevant factor which determines the character and the extension of the programme. Therefore, any carefully done investigation programme should be aimed at answering questions linked to the environment of sedimentation and synsedimentary and post-sedimentation events, including tectonic events, uplift, erosion and weathering.

This outlines the multidisciplinary nature of the approach, requiring cooperation between geologists and geotechnical engineers, and there is no need to emphasize that answering the above mentioned questions is the most fascinating aspect of field investigation.

Unfortunately, the importance of field investigation is quite often underestimated, because it may require an elaborate and costly programme, and, surprisingly, we generally rely on limited information.

This is certainly an unsatisfactory and unsafe approach, by considering that most failures have been caused by lack of field investigation or undetected essential features (such as groundwater regime, structural discontinuities, flexural slip and many others).

In addition, it must be recalled that quite often an unsatisfactory site investigation is a matter of litigation; therefore, authorities charged to make official decisions should always be encouraged to consider the importance of site investigation.

7.1 Exploration programme

The investigation programme should be planned in order to collect information to select the most appropriate solution, but should also consider possible changes of the solution. The results of the investigation could in fact prove that the initial planned project is not the most suitable and, if the investigation has been restricted to this specific project, no available information could result for the analysis of other solutions. For this reason, the experienced engineer knows that the best result is obtained if the programme is developed in stages and it is adapted during the investigation, when preliminary data start to be available.

As an example, a starting point is represented by desk studies, i.e. the collecting of all information sources such as geological maps and relative sections, air photographs and reports on previously done investigations. Preliminary investigations are aimed at confirming or reviewing the findings of desk studies, by means of trial pits, some preliminary borings, exploratory geophysical seismic tests or other *in situ* tests, and are also aimed at defining groundwater conditions.

The final step consists of a detailed investigation, including borings, sampling and well-focused *in situ* and laboratory tests. An obvious consequence of planning in staging the investigation is that only general guidelines can here be developed, according to what has been suggested by many authors (Hvorslev, 1949; Tomlinson, 1963; Terzaghi and Peck, 1967; Clayton *et al.*, 1982; Weltman and Head, 1983), and the reader should verify how appropriate they are for each case encountered in practice.

We also observe that the extent of site investigation is always a matter of debate. There is no simple answer, but the designer should consider whether additional investigations produce economical benefits or increases confidence about safety and performance of the structure.

For ordinary buildings it is recommended to plan four borings, one at each corner of the structure. As discussed in more detail in Section 7.2, **borings** are vertical holes, of small diameters, from which soil samples or cuttings are recovered in order to identify the stratigraphical profile. These borings are also usually used for the installation of sensors (piezometers) to detect the groundwater regime (see Section 7.4), to obtain undisturbed soil samples for laboratory tests (Section 7.3) and to execute most common *in situ* tests (Section 7.5) at the selected depth during the rest of the drilling process.

The depth of the borehole usually depends on the depth at which the soil is still affected by the stresses induced by the foundation loads. In general, it can be observed that at a depth of 1.5 times the foundation width the induced vertical stress is of the order of 20% of the unit load applied at the surface, so the depth of the borehole should be at least 1.5 times the foundation width (Figure 7.1a).

This depth will be relatively shallow for widely spaced *spread footings* (Figure 7.1b), but if the same footings are so closely spaced that there is overlap of the induced stresses (Figure 7.1c), the depth of the boring should be compared to the overall area, as in the case of a *raft foundation*.

Alternatively, in order to account for the applied bearing stress, the induced vertical stress at the required minimum depth should not exceed 10% of the overburden effective stress at this depth.

When dealing with *piled foundations* (Figure 7.1d), the depth of the boring should be computed by considering the piled foundation as a whole, i.e. a *block* in which the whole group is considered as a single foundation. If a rock horizon is encountered within this depth, the boring should penetrate the rock by at least 3 m, to ascertain that the bedrock, rather than a large boulder, has been reached.

For earth *retaining structures* (Figure 7.1e), the investigation should be extended to a depth within which surfaces of failure can develop. For this reason a preliminary

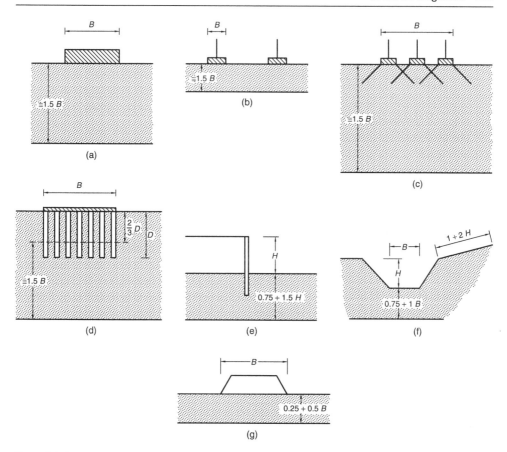

Figure 7.1 Planning for subsoil exploration.

estimate of a depth equal to 0.75 to 1.5 of the height of the wall is suggested, but this depth could increase to two times the height if the wall is founded on compressible strata which may cause excessive settlements. In addition, if the *diaphragm wall* has a hydraulic function, the depth of the boring should be related to the need to explore the seepage problem.

For *deep cuts* (Figure 7.1f), the depth of a sliding surface depends on the width of the cut, whereas the lateral extension is related to the depth of the cut.

For *railroad embankments* (Figure 7.1g), the depth of borings should be higher than the thickness of the weathered strata, and sufficient to allow the analysis of stability, the drainage conditions and possible frost effects.

7.2 Methods of investigation

Boring is intended to be any vertical, inclined or horizontal hole drilled with the purpose of obtaining *representative samples*, that allow to define the stratigraphical profile of the investigated material, or *undisturbed samples* that allow to investigate the stress-strain characteristics of soils.

There are two basic operations to make a boring: the first one is to advance the hole to the depth at which samples are planned to be obtained, and the second one is the sampling of soil.

As will be described in the sequel, there are several methods for advancing the hole and, usually, these methods are grouped depending on the technique with which the material is removed. Accordingly we describe these methods as continuous sampling, augering, wash boring, percussion drilling and rotary drilling.

Samplers used to recover the investigated material include buckets, hollow-stem augers, solid and split-tube drive samplers and the more sophisticated samplers described in Section 7.3.

In general we can observe that cuttings from drill holes cannot give information about thickness and depth of strata, and a proper identification of soil profile requires to recover *representative* or *undisturbed soil samples* by means of a sampling tube. When recovering *representative* (but not undisturbed) samples of clay soils, the samples retain partially the characteristics of the soil in the field, so that they are representative for proper identification and classification, but not suitable for the determination of the stress–strain characteristics. To reach this scope, we need *undisturbed samples*, as explained in Section 7.3. Coarse-grained soils cannot be retained without the addition of a spring core catcher to the sampling tube, so that the samples may be representative but they are sampled in a completely altered state.

Before discussing techniques used to make a borehole, it is important to recall that **trial pits** represent one of the simplest and most reliable methods of soil investigation. It allows one to directly explore soil conditions, so that the sequence of strata as well as macrofabric features can be accurately detected. This procedure is a great advantage when dealing with stiff fissured clays, and in addition it enables one to obtain block samples, which can be cut from the bottom or from the sides of the pit. Obviously, the shortcoming of this procedure is the depth of the investigation, which is confined to a maximum of 5–6 m in stiff clays, and to shallow depth in coarse-grained soils, also depending on groundwater table.

Continuous sampling method. Where no information is available, as well as where bedrock has to be investigated, borings are made by using the continuous sampling method. In this case the advancement of the hole is achieved by continuous sampling, and samplers producing either disturbed or undisturbed samples may be used. Obviously, this method is more expensive than those described in the successive points.

Auger boring. The *short-flight auger*, used to advance the hole, is a helic of limited length attached to a *Kelly bar*. This is a drilling rod which is rotated and pushed into the soil to obtain the advancing of the hole. As expected, the shorter the flight the more time consuming is the procedure, because the auger must be lowered and raised for each reached depth to remove the soil. This disadvantage is overcome by using the *continuous flight auger*, which allows depths of up to 50 m to be reached. In this case, additional sections of spiral flight are successively added, as the hole is advancing. The soil rises to the surface along the helix and is sampled as it emerges. The diameter of the auger is usually from 75 mm to 300 mm and this technique is mainly used in clays, because the hole does not require support. Bentonite slurry is introduced to support unstable holes, whereas casing is generally avoided because it requires the removal of the auger before introducing the casing. Difficulties with this procedure generally arise when cobbles and boulders are encountered and the main limitation comes from the fact that different soil strata are mixed, therefore it is difficult to recognize where stratigrafical changes occur. To avoid this shortcoming, one makes use of the *continuous hollow-stem auger*. This consists of a hollow stem (75 to 150 mm internal diameter) with a spiral flight and a centre rod, running inside the stem, to which a plug is attached at the lower end. The centre rod and the plug can be removed at any time to allow disturbed or undisturbed soil sampling, by means of a sample tube which is lowered down the stem and is pushed into the soil below the auger. Drilling can also be made through the hollow stem, when bedrock is encountered.

Wash boring.   *Wash boring* is a simple procedure to obtain relatively deep holes. In this case, water is pumped through a a hollow rod and is forced through small slots in the drill bit. The soil fragments are carried to the surface through the annular space between the casing and the washing rod. The hole is generally encased, but mud can also be used as an alternative. The soil fragments carried to the surface do not allow an accurate identification of soil type, therefore the method has to be considered only as a borehole making procedure.

Percussion boring.   Percussion boring is the oldest method of making a borehole. This procedure is particularly used in penetrating coarse gravels, boulders and hard strata. In this procedure a heavy drilling bit is raised and dropped by means of a cable. The used bits are the bailer and the clay cutter; the bailer, used in cohesionless soils, is a steel tube with a clack valve at the tip; the clay cutter is an open steel tube with a cutting shoe. Depths of up to 60 m can be reached, with borehole diameters ranging from 150 to 300 mm. The advantage of this method is its simplicity, but it is difficult to identify thin soil layers or macrofabric features.

Rotary boring.   Rotary drilling is a procedure which was developed for drilling deep holes for oil production. The development of smaller drilling rigs also made the method available for geotechnical exploration. The procedure is generally used to advance boreholes and to clean the bottom of the hole before taking samples. With this method, boreholes can be performed in rock and in soils. A rotating bit cuts the material at the bottom of the hole, and soil fragments are raised at the surface by means of a circulating fluid. Casing is not generally required, as the collapse of the hole is prevented by the drilling bentonite mud. The procedure is difficult to apply in the presence of a high percentage of gravel.

Stabilization of boreholes.   The need to prevent the sides and bottom of the hole from collapse is common to all previously mentioned drilling methods. Dry boreholes can be stable above the water table and at relatively shallow depths, but as soon as the depth increases and the water table is encountered a stabilization procedure must be adopted.

(a) When the choice is that of detecting the groundwater level or carrying out permeability tests, the solution requires the use of a *casing*. Even though it is an expensive procedure, casing represents the safest method for stabilizing the sides of a hole. It must be stressed that, when the borehole is below the groundwater table, a casing does not prevent the collapse of the bottom, therefore the stability of the bottom must be assured by introducing water. At the same time, it must be observed that this is not required if the hole is above the water table and that the introduction of water produces softening of the soil to be sampled.

(b) Uncased boreholes are usually stabilized by using drilling *bentonite mud*. **Bentonite** is a highly plastic, expansive clay, with a high content of montmorillonite, which is a product of the alteration of volcanic ash (the *plasticity index* ranges from 50 to 100% and the *liquid limit* from 300 up to 700%). Usually, a suspension of bentonite is obtained by using a bentonite percentage from 3 to 5%, the density of the drilling mud being of 1.09 to 1.15 Mg/m^3. The higher density of the mud with respect to that of water allows removal of cuttings. In addition, the thixotropic character of the mud also prevents the accumulation of cuttings at the bottom during the rest period between drilling and sampling.

Ground investigation reports.   Civil engineers need a knowledge of soil behaviour in order to predict the performance of structures interacting with soils and under what conditions a structure may reach failure. Therefore, engineers are primarily interested in basic mechanical parameters, such as strength, stiffness and hydraulic conductivity. The assessment of these parameters requires specific tests, but an accurate *description* of the soil can help the engineer by providing him with general guidance about the expected soil behaviour.

We discussed in Chapter 1 that *describing a soil profile* means describing what you see and how soil reacts to simple tests, so that any comprehensive description should include the characteristics (size, shape, grain-size distribution, plasticity) of soil material, as well as the structural features (bedding, fissures, joints) of the soil mass, which can be inferred in the field.

It is also important to recall that, in order to benefit from any description, a common language is essential, and for this reason a detailed scheme is provided by BS 5930-(1981), reproduced in Table 1.4, and it is relevant to distinguish between factual and interpretative reports (Atkinson, 1993). *Factual reports* summarize the results of desk studies, field investigations, *in situ* and laboratory soil testing, without comments and interpretations. *Interpretative reports* add to the information contained in the factual reports the interpretation of the results, with the aim of providing a design profile, which includes the parameters to be used to satisfy design requirements for the structure under consideration.

7.3 Sampling techniques

A soil sample can be considered an **undisturbed soil sample** if it has been so carefully sampled to preserve both the structure and the water content of the soil in the field. However, the process of sampling, trimming and mounting the soil sample in the test equipment has some influence on the structure of the soil and the associated changes are called **disturbance**.

The quality of the sample is by far the most important requirement when determining the yield stress and mechanical parameters, since the *disturbance* of a soil sample tends to modify the original soil structure and all information related to this structure, implying that the soil partially loses its memory.

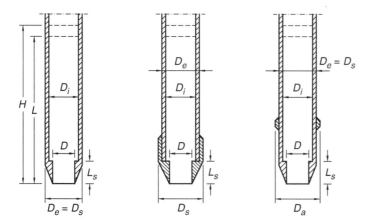

Figure 7.2 Characteristics of samplers.

Where the soil can be exposed by excavation (trial pits) and a *block* of sample can be cut by hand, the process results in high quality samples. But usual situations require the samples of soil to be extracted by pushing into the soil through *borehole* open-ended cylindrical tubes, called **samplers**. The main characteristics of such samplers are summarized in Figure 7.2. The inner diameter D_i depends on the need to obtain samples for laboratory testing apparatuses, so that it usually ranges from 80 to 100 mm. The degree of disturbance depends on the dimension of the sampler, as is indicated by the area ratio:

$$C_p = \frac{D_s^2 - D^2}{D^2} \cdot 100, \tag{7.1}$$

which is less than 15% when using the so called *thin-wall tubes*.

The length of the sample should be higher than a minimum, in order to guarantee that its central part can be considered undisturbed; but as soon as the length tends to increase there is also disturbance due to the internal friction. In order to reduce this friction, the diameter of the *cutting edge* is slightly smaller than the internal diameter of the tube, so that the degree of disturbance also depends on the *inside clearance ratio*:

$$C_i = \frac{D_i - D}{D} \cdot 100. \tag{7.2}$$

This ratio should be higher than 0.5% to reduce friction, but lower than 1.5% to avoid excessive lateral expansion of the sample.

Further, to minimize friction acting on the external lateral surface of the tube, sometimes the external diameter of the cutting edge is higher than that of the tube, but the so-called *outside clearance ratio*:

$$C_a = \frac{D_s - D_e}{D_e} \cdot 100 = \frac{D_a - D_e}{D_e} \cdot 100, \tag{7.3}$$

should be lower than 2% or, at a maximum, 3%.

In clay soils, the *open ended thin-wall tube samplers* are usually pushed at a constant rate of the order of 150 to 300 mm/s (Hvorslev, 1949). Once the sample has been recovered, the ends of the tube are sealed in order to avoid moisture changes. Limitations of the open-ended samplers arise from the fact that remoulded material cannot be prevented from entering the tube from the sides and from the bottom of the hole.

The quality of the samples can be increased by using the *stationary piston sampler* (Figure 7.3); a thin-walled tube is fitted, in this case, with a piston which closes the end of the tube until the bottom of the borehole is reached. In such a way, the remoulded soil

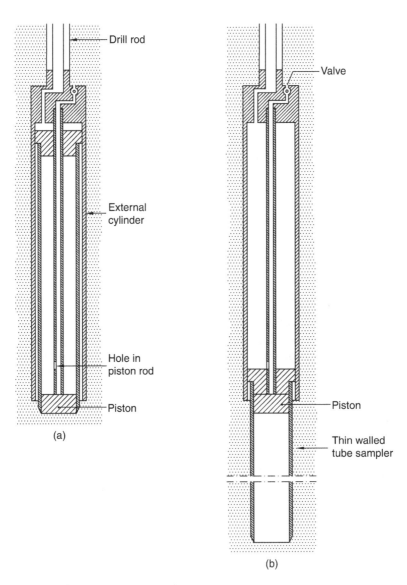

Figure 7.3 Stationary piston sampler.

is prevented from entering the tube; the tube is then pushed past the piston to sample the soil. The available diameters usually range from 35 to 100 mm (but larger diameters have also been successfully used), and the length of the sample is of the order of 1 m.

When the aim is to obtain detailed stratigraphical information, in order to also detect the presence of thin lenses or macrofabric features, it is possible to use the *continuous samplers* (see Kjellman *et al.*, 1950 and Figure 7.4). This kind of sampler is used in soft clays, and enables one to obtain samples of up to 20 m in length. In this case, when the sampler is pushed into the soil, thin strips encase the samples in order to eliminate the internal friction.

Thin-wall tubes cannot be used to obtain undisturbed samples in stiff clays and soft rocks, where *double-core barrels* are generally employed. A core drill (see Figure 7.5)

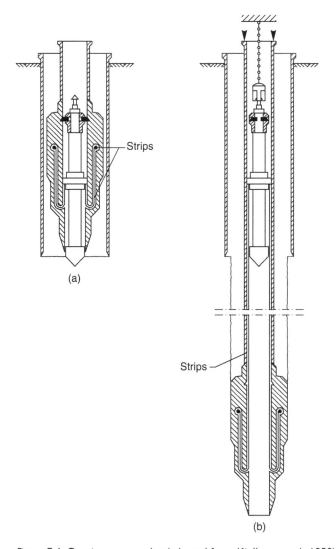

Figure 7.4 Continuous sampler (adapted from Kjellman *et al.*, 1950).

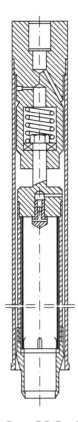

Figure 7.5 Double tube core barrels (from Tornaghi, 1981).

is connected to the lower end of an outer core barrel to cut an annular ring; the soil fragments are washed with the drilling fluid and the soil sample is protected from the circulating fluid by the inner tube.

There is no possibility of obtaining undisturbed samples in coarse-grained soils by using the above described techniques, as the soil structure is completely destroyed when driving or pushing the sampler. At present, the only possibility of obtaining high-quality undisturbed samples of saturated loose and dense sands is offered by an *in situ freezing method* (Yoshimi *et al.*, 1978, 1985). This is a very sophisticated and costly technique, which uses liquid nitrogen to freeze the soil; then, a core barrel with hardened metal teeth is advanced into the frozen soil. The recovered frozen sample is cut into blocks and transported in a refrigerated truck to the laboratory, where the frozen specimens are allowed to thaw after being assembled into the testing apparatus.

7.4 Groundwater conditions

Soil behaviour depends on effective stresses and to compute these stresses a prerequisite is the measurement of pore pressure. Pore pressure needs to be measured during and after the construction, by considering that natural groundwater conditions are likely to be significantly altered during construction, as can be the case when dealing with excavations (see Somerville, 1986). Furthermore, pore pressure measurements are likely to be measured over long periods of time when the scope is to analyse changes occurring with municipal water supply purposes (inducing subsidence phoenomena), or subsequent water level rising, when pumping rates are reduced (see Simpson *et al.*, 1989, for the engineering implications of rising groundwater levels in the deep aquifer below London).

In general, groundwater conditions are described by using some basic definitions, here briefly recalled.

The **water table** is defined as the surface on which the *water pressure is atmospheric*. If the pore pressure is measured by means of a pressure gauge (i.e. a pressure above atmospheric), then on the water table the pressure is zero.

The term **aquifer** (see Figure 7.6) is used to indicate a saturated permeable geological unit which can transmit significant quantities of water under ordinary hydraulic gradients.

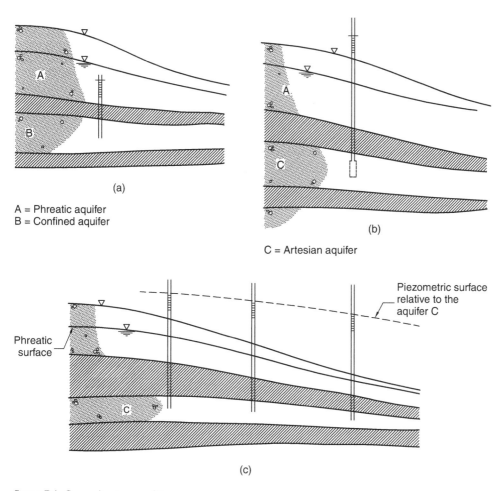

(a)

A = Phreatic aquifer
B = Confined aquifer

(b)

C = Artesian aquifer

Piezometric surface
relative to the
aquifer C

Phreatic
surface

(c)

Figure 7.6 Groundwater conditions.

On the contrary, the term **aquiclude** is used to designate a saturated geological unit which cannot transmit significant water quantities, due to its very low hydraulic conductivity.

The term **aquitard** is also in use to indicate less permeable strata, which can transmit water but not in such quantities as to allow the installation of production wells.

An **unconfined aquifer** is an aquifer which has its upper boundary represented by the water table (Figure 7.6a). On the contrary, a **confined aquifer** is an aquifer confined between two aquitards (Figure 7.6a). In this case, the water table rises above the top of the aquifer, and when this level rises above the ground surface the aquifer is called **artesian** (Figure 7.6b).

Water levels in a number of observation tubes define an imaginary surface, which represent the **piezometric surface** of the investigated aquifer (Figure 7.6c).

Pore water pressure is measured by means of piezometers, and measurements are based on the following two assumptions:

1   The piezometer is a porous element filled with water at the same pressure as the surrounding soil.
2   The presence of the piezometer in the soil does not modify the pore pressure regime.

The validity of these assumptions must be verified in relation to the type of piezometer, the installation procedure and the soil nature, as discussed in the sequel.

7.4.1 Piezometers

The basic instrument for the measurement of hydraulic level is a tube or pipe in which the water level can be measured, and this tube or pipe is called a **piezometer**. The simplest type of piezometer is the open standpipe (Figures 7.7 and 7.8). It consists

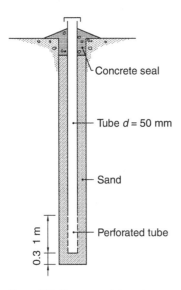

Figure 7.7 Open standpipe piezometer.

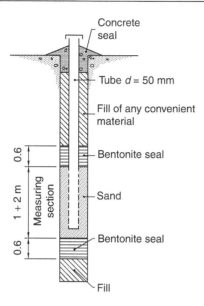

Figure 7.8 Sealed open standpipe piezometer.

of a pipe with a perforated section installed in a surrounding sand filter. If the perforated section is not sealed along its entire length (Figure 7.7), the open standpipe piezometer acts as an observation well it creates a vertical connection between the strata and the obtained measurements may be misleading. Therefore it should be installed only in homogeneous permeable strata, in which it is assumed that the pore pressure is hydrostatic. For reliable measurements of local pore pressure, it is necessary to seal the perforated section with the surrounding filter, as shown in Figure 7.8.

In order to have the same pore pressure value in the pipe and in the surrounding soil, a long period of time is required to allow the water to flow into the piezometer, so that this type of piezometer is usually installed in relatively pervious soils. In order to reduce the response time, Casagrande (1949) designed the standpipe piezometer shown in Figure 7.9, consisting of a cylindrical porous ceramic tube, 300 mm long and 38 mm in diameter, connected to a standpipe of 10 mm in diameter.

In many cases, however, the response time of the Casagrande piezometer is still inadequate. In addition, a shortcoming of standpipe piezometers is that they are subjected to damage during construction of embankment fills. For these reasons, other types of piezometers have been introduced (see the Proceedings of the Conference on Pore Pressure and Suction in Soils (1961), held in London in 1960).

The so-called closed hydraulic piezometer (Figure 7.10) has been developed (Bishop *et al.*, 1961) for installation during the construction of embankments dams. It is connected to two plastic tubes, with a Bourbon tube pressure gauge at each end, which allows one to make readings at remote points. The porous stone is of fine ceramic with pore dimensions of less than 1 micron, so that it can be

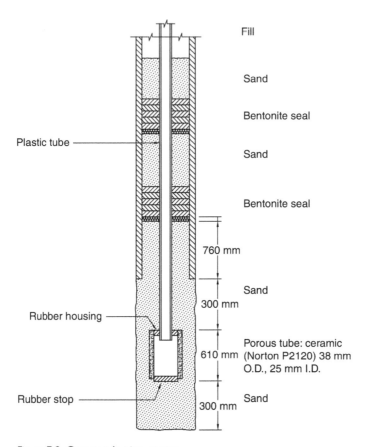

Fill

Sand

Bentonite seal

Sand

Bentonite seal

760 mm

Sand

300 mm

Porous tube: ceramic
(Norton P2120) 38 mm
O.D., 25 mm I.D.

610 mm

Sand

300 mm

Plastic tube

Rubber housing

Rubber stop

Figure 7.9 Casagrande piezometer.

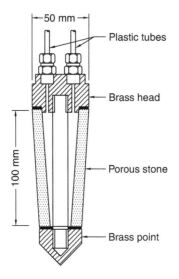

50 mm

Plastic tubes

Brass head

100 mm

Porous stone

Brass point

Figure 7.10 Bishop-type piezometer.

considered a high *entry value* piezometer (see Section 7.4.2) that can measure *suction* up to 75 kPa. Moreover, the twin-tube allows the system to be flushed to remove air.

When the need arises for measuring pore pressure changes in low permeability soils, so that the time response must be extremely low, then *diaphragm type piezometers* can be installed. In this case the probe is extremely rapid, because the volume change required to activate the instruments is only a few mm^3.

Depending on the characteristics of the instrument, diaphragm piezometers are classified as pneumatic piezometers, vibrating wire piezometers and electrical resistance piezometers (see Dunnicliff, 1988). These piezometers require special care when operated, in order to minimize zero drift and errors caused by moisture and temperature. It is therefore recommended that their reliability should always be checked by comparing the results with those from standpipe piezometers installed in parallel. Due to these problems, diaphragm piezometers are well suited for short-term application; twin-tube hydraulic piezometers are usually preferred by most engineers for long-term monitoring.

7.4.2 Unsaturated soils

We discussed in Chapter 4 that pore pressure is usually compressive, but water can also sustain negative pore pressure. In fact, if we consider the water rise in a narrow capillary tube, at the level of the free water surface the pressure is zero, whereas the pressure u, acting at an elevation z above this level, is negative.

Soils are random aggregates of particles and the resulting void distribution hardly bears resemblance to a capillary tube. However, we argued that this analogy still explains capillary phenomena observed in soils, and the approximate values of the height of capillary rise, reported in Table 4.1, are also of practical utility.

One consequence of the values quoted in Table 4.1 is that coarse-grained soils above the phreatic surface (on which the pore pressure is zero) tend to be unsaturated and little water is contained in the pores. On the contrary, fine-grained soils may remain saturated for many metres above the phreatic surface, with negative pore pressure.

Soil physicists usually call the negative pore pressure *suction*, the related head *suction head* and use the term *tensiometer* to indicate the instrument for measuring negative pore pressure.

In order to measure pore pressure in unsaturated soils, we need to take care of the following aspects. If we consider a saturated filter, the amount by which the *air pressure* on one side must exceed the *water pressure* on the other side to force air through the filter is called **air entry value**. If the difference between the air pressure and the subatmospheric pressure in the soil exceeds the air entry value, air enters the saturated filter, water is drawn from the filter into the soil and the measured pore pressure will correspond to the pore air pressure rather than to the pore water pressure. Therefore, in order to prevent the air for entering into the measuring device, the requirement is that porous stones must have sufficiently high air entry values.

7.4.3 Time-lag analysis

The water pressure in the piezometer can differ from the value in the surrounding soil because of soil disturbance induced by the installation of the piezometer, or because of changes induced by perturbations, such as applied loads at the surface, excavation, compaction or pile driving. Therefore it is necessary that some water flows towards or from the piezometer in order to reach an equilibrium condition. During this transient flow, the water pressure in the piezometer differs from the pore water pressure in the soil, and any measurement during the transient phase will not be representative of the soil pore pressure.

The time required to reach the equilibrium condition depends on the soil properties and on the characteristics of the piezometer and represents the time response of the instrument defined as **time-lag**.

A simplified analysis of the phenomenon can be made referring to Figure 7.11 and neglecting soil compressibility, as suggested by Hvorslev (1951). If at a time t the piezometric level is represented by point B, the flux into the piezometer will depend on the difference of height h:

$$q = FKh = FK(z_o - y) . \tag{7.4}$$

In equation (7.4) K is the *hydraulic conductivity* of the soil, z_0 is the difference of height at time $t = 0$ and the constant F is an *intake coefficient*, which depends on the shape and dimensions of the intake section of the piezometer, and can be evaluated from Figure 7.12 and Table 7.1.

If the area of the cross-section of the pipe is equal to A, during the time interval dt the following also holds true:

$$q \cdot dt = A \cdot dy, \tag{7.5}$$

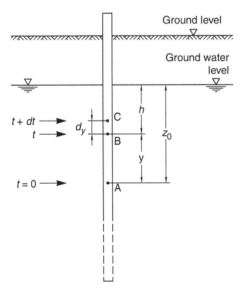

Figure 7.11 Time-lag analysis.

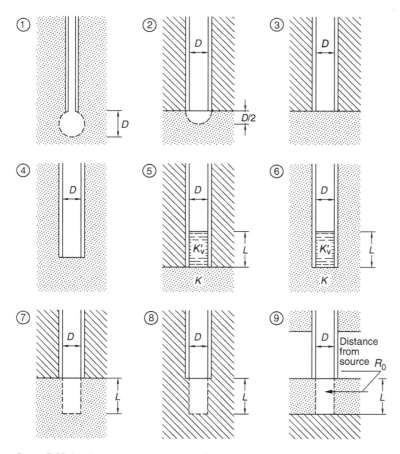

Figure 7.12 Intake sections (Hvorslev, 1951).

so that:

$$\frac{dy}{z_o - y} = \frac{FKdt}{A}.$$

(7.6)

The integration of equation (7.6), with the initial condition $y = 0$ *if* $t = 0$, gives:

$$1 - \frac{y}{z_o} = \exp\left(-\frac{FKt}{A}\right).$$

(7.7)

The equilibrium condition:

$$\frac{y}{z_o} = 1$$

(7.8)

Table 7.1 Intake coefficients F suggested by Hvorslev (1951) and Mathias and Butler (2006)

Case of Figure 7.13	F
1 Spherical soil bottom in homogeneous soil	$2\pi D$
2 Hemispherical soil bottom at impervious boundary	πD
3 Flush bottom at impervious boundary	$2D$
4 Flush bottom in uniform soil	$2.75D$
5 Soil in casing at impervious boundary	$\dfrac{2D}{1+\dfrac{8}{\pi}\dfrac{L}{D}\dfrac{K_h}{K_v}}$
6 Soil in casing in uniform soil	$\dfrac{2.75D}{1+\dfrac{11}{\pi}\dfrac{L}{D}\dfrac{K_h}{K_v}}$
7 Well point or hole extended at impervious boundary	$\dfrac{2\pi R}{\sinh[\tanh^{-1}(R/L)]\ln\{\coth[0.5\tanh^{-1}(R/L)]\}}$
8 Well point or hole extended in uniform soil	$\dfrac{2\pi D}{\sinh[\tanh^{-1}(D/L)]\ln\{\coth[0.5\tanh^{-1}(D/L)]\}}$
9 Hole fully penetrating a confined aquifer	$\dfrac{2\pi L}{\ln\left(\dfrac{R_0}{R}\right)}$

would require an infinite time interval to be reached, but for practical purposes a 90% time response is considered to be adequate, therefore:

$$t_{90} = -\frac{A}{FK}\ln(0.10).\tag{7.9}$$

An estimate of the time response of the various types of piezometers can be made by using this relation and it can be proved that the time lag can vary from a few minutes, if the surrounding soil is sand, to some months if it is a clay. When using the intake coefficients, it must be observed that the shape factors originally listed by Hvorslev for cases 7 and 8 have been proved to be incorrect (see Mathias and Butler, 2006; Ratnam *et al.*, 2001), because they are incompatible with cases 1, 2 and 3. For these reasons, the intake coefficients listed in Table 7.1 for cases 7 and 8 are those suggested by Mathias and Butler (2006).

7.5 *In situ* testing devices

In situ tests have recently received a great deal of attention, due to an improvement of the capability of the existing techniques, to the introduction of new devices and to the development of more appropriate theoretical approaches.

In particular, examples of new testing methods are related to techniques introduced in the area of environmental geotechnics, mainly aimed at identifying contamination of polluted sites. In addition, recent advances have also been made in the area of geophysical testing, as represented by shallow geophysical testing (SASW method is one of the examples) as well as by tomography and other imaging techniques.

These advances increase the already recognized advantages of *in situ* testing:

1 *In situ* tests can be performed in soil deposits in which undisturbed sampling is impossibile or unreliable.
2 They can be performed in difficult-to-operate conditions, such as continental shelf and sea floor.
3 Many tests produce a continuous record of the soil profile, which allows one to detect boundaries and macrofabric of different layers.
4 The volume of soil tested is larger than the volume of samples usually tested in laboratory.

In situ tests obviously also have limitations, and these include:

1 The drainage conditions are often poorly defined, so that the problem arises of defining whether the test is performed in virtually undrained, drained or partially drained conditions.
2 During *in situ* tests the soil is subjected to effective stress paths which may not reproduce design conditions.

In addition, it must be realized that the interpretation of an *in situ* test requires the solution of a complex boundary value problem. The result of a test depends on the combination of many factors, i.e. at least strength, stiffness and *in situ* stresses, so that soil parameters cannot be measured separately.

By considering these latter aspects, the subject in this book is addressed with the aim of identifing the problems associated with *in situ* testing and to provide a rational basis for the interpretation of these tests, in order to obtain the relevant soil parameters. Therefore, a description of the presently used testing devices and their capability is first given in the following; then, rather than dealing with the interpretation of each test separately, the matter has been organized in order to show how to assess specific parameters required to analyse geotechnical structures.

7.5.1 Standard penetration test (SPT)

The SPT (*Standard Penetration Test*) was developed in USA in 1927, and is a test performed during boring in order to obtain a measure of soil resistance to the dynamic penetration and a disturbed soil sample. The test is carried out by dropping a falling weight of 63.5 kg onto the drill rods from a height of 760 mm, and the number of blows (N_{SPT}) necessary for penetration of a predefinite distance of a standard sampler (see Figure 7.13) is the required measure of the state condition of the soil.

When making such measurements, the sampler is driven a total of 450 mm, but this penetration is divided in three separate advances of 150 mm: the initial seating drive of the sampler of 150 mm is disregarded, because it reflects the influence of the disturbed soil at the bottom of the hole; then the operator counts the number of blows

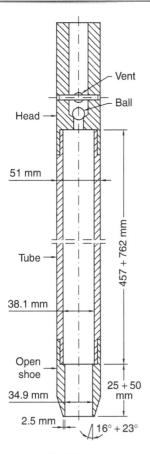

Figure 7.13 SPT split-spoon sampler.

to advance the sampler each of the remaining 150 mm and adds these last two counts to obtain the number of blows (N_{SPT}) necessary for penetration of 300 mm, which represents the required measure.

The test is usually performed at intervals ranging from 45 to 105 cm in a borehole, with a diameter ranging from 65 to 115 mm, as larger diameters can have a significant influence on the test results. The drilling operation must be interrupted each time the test is performed.

The mud required to support the borehole should be maintained up to the groundwater level or higher, to avoid an inflow of water, which would cause low test results.

The main advantages of the standard penetration test are that this test is relatively simple to perform and inexpensive. On the contrary the disadvantage is that the accuracy is dependent on the details of the procedure. As an example, at present there is wide variability in apparatuses and procedure and there are four methods of releasing the hammer, ranging from the manual procedure to the automatic mechanism, so that there is a variation on the delivered energy and a need to normalize the measured N_{SPT} to a standard energy ratio.

As a general indication, the SPT should not be used in soils containing coarse gravel, cobbles or boulders, as the sampler has the tendency to give high, erroneous and unconservative results. Furthermore, this test is not effective in fine-grained soils, especially soft and sensitive clays. It should be used mainly as a penetration index for sandy soils.

7.5.2 Cone penetration tests (CPT and CPTU)

The mechanical *cone penetration test* (CPT), initially introduced in 1917 by the Swedish State Railways, in 1927 by the Danish Railways, and in 1935 by the Department of Public Works in the Netherlands, is now widely used for geotechnical characterization. The relevant features of the *mechanical cone* are the following (see Figure 7.14): the cone has an apex angle of 60° and the base diameter is 35.7 mm; the friction sleeve has an area of 150 cm^2.

The test consists of first pushing the cone section at a constant rate of 20 mm/s by the inner rod, until the retaining sleeve engages with the friction jacket. The cone is advanced in this way 40 mm and the tip resistance (which is indicated in the following with the symbol $q_c [FL^{-2}]$) is measured. Both the cone and the jacket are then pushed together. This allows us to measure the sum of the tip resistance and the friction resistance. The tip resistance, previously measured, is then subtracted to give the side

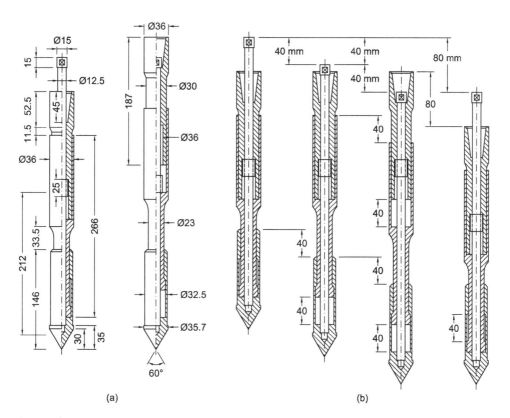

Figure 7.14 Mechanical cone penetrometer (from Begemann, 1953).

(or friction) resistance (indicated as $f_s[FL^{-2}]$), so that it is inherent in the process that the tip resistance and the frictional resistance are not evaluated at exactly the same depth. Finally, the outer rods are pushed and the procedure is repeated at intervals of about 200 mm.

The mechanical cones are cost-effective and simple to operate, but they have the disadvantages of a slow incremental procedure (every 200 mm) and poor accuracy in soft soils.

The *electrical cone* (Figure 7.15) has a friction sleeve and a tip of the same diameter, and is pushed with a single rod system. The tip resistance and the friction are measured by load cells, with a capacity of 10 to 150 kN for the tip resistance, and of 7.5 to 15 kN for the side friction, depending on the strength of the soil.

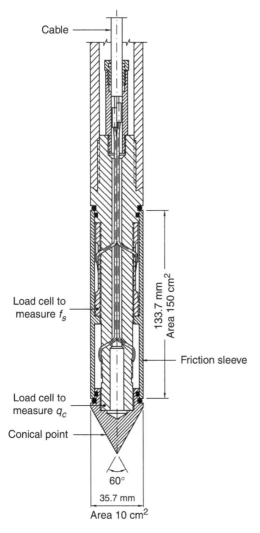

Figure 7.15 Typical design of electric friction cone penetrometer (Holden, 1974; Schaap and Zuidberg, 1982).

The main advantages of the electrical cones are the high repeatability, the high accuracy, the continuous record of data and the possibility of including additional sensors (Mitchell, 1988; Mayne, 2006). For example, the introduction of an inclinometer allows one to measure the deviation from the vertical, thereby preventing errors in detecting soil layers. Recent advances in data acquisition systems have also further improved the capability of the CPT, allowing one to obtain data in both graphical and digitized form. All these improvements in the equipment and testing procedures make the CPT the most frequently used tool for offshore investigations.

The importance of monitoring the pore pressure during cone penetration was recognized during the mid 1970s and almost at the same time it was followed by the development of the piezocone (Janbu and Senneset, 1974; Torstensson, 1975; Wissa *et al.*, 1975; Baligh *et al.*, 1980). The **piezocone** (CPTU) (Figure 7.16) measures simultaneously the cone resistance, the skin friction and the pore pressure generated during penetration. The measured pore pressure depends on the location of the pore pressure sensor, and three possible locations are (1) on the cone face, (2) just behind the cone and (3) behind the friction sleeve. However, there are reasonable arguments supporting that the best location is just behind the cone, because this is less subject to damage and abrasion, and offers the same stratigraphical details as the sensor located on the cone. The main advantages over the conventional CPT are that drained penetration can be distinguished from partially drained or undrained penetration, and cone resistance can be corrected for unbalanced water pressure load resulting from different areas at the front and back of the cone.

The CPTU has a wide range of applications in geotechnical engineering, including soil profiling, assessment of stress history and soil strength and soil permeability. In addition, further advances are related to techniques introduced in the area of

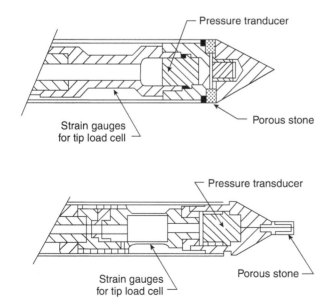

Figure 7.16 Schematic sections of piezocone tips.

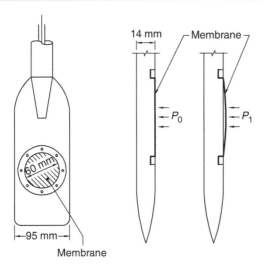

Figure 7.17 Marchetti dilatometer (Marchetti, 1975).

environmental geotechnics, mainly aimed at identifying contamination of polluted sites (Mitchell, 1988; Mitchell and Brandon, 1998; Pasqualini *et al.*, 2001; Kurup, 2006).

7.5.3 Marchetti dilatometer

The **flat dilatometer** was developed in Italy by Marchetti in the late 1970s (Marchetti, 1975; Marchetti *et al.*, 2001). A flat plate (Figure 7.17), 14 mm thick, hosts a flexible stainless steel membrane, 60 mm in diameter, located on one face. The plate is pushed down from the surface, by using the same technique used for CPT, and the test is performed at spaced intervals of 200 mm. This consists of inflating the membrane by using nitrogen pressure, and the pressure required to just lift the membrane off the sensing disc (*lift-off pressure*) and the pressure required to cause a 1.1 mm deflection are recorded. These values are used to detect soil properties, as discussed in Sections 7.7, 7.8 and 7.10. The main advantages of the DMT are the simplicity of the testing procedure, the high repeatability and the near continuous records. Furthermore, the capability of these tests has recently been enhanced through the introduction of seismic receivers, which allow one to perform shallow seismic tests (as discussed in 7.5.7 and Section 7.11).

7.5.4 Field vane test

Introduced in Sweden in 1911, the field vane test has become the most widely used test for *in situ* measurement of the undrained strength of soft clays in the late 1940s, mainly due to contributions of Carlson (1948), Skempton (1948), Cadling and Odenstat (1950) and Bjerrum (1972, 1973). The test provides measurements at intervals as small as 0.5 m, reproducible values and economy if compared to sampling and laboratory tests.

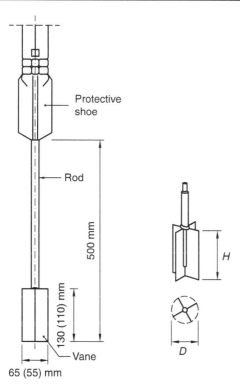

Figure 7.18 Vane test.

The test (Figure 7.18) consists of lowering into the soil a rotating *vane*, within an external casing. At the selected depth, the vane is pushed down 0.5 m and is rotated, and the measured torque applied through the rods allows one to obtain the undrained strength of the soil, as discussed in Section 7.8. This is a common approach, but as an alternative approach the vane can be pushed into the soil without casing. In this case the measured torque is the value needed to rotate both the vane and the rods, so that the additional measurement of the torque needed to rotate the rod alone is required.

Measurements are also made on the remoulded clay after 25 revolutions, to obtain the soil *sensitivity*. This parameter is defined as the strength ratio of the undisturbed versus remoulded soil.

The *vane* is made up of four cutting blades with a height-width ratio $H/D = 2$. The most widely used dimensions are $H = 130\ mm$ and $D = 65\ mm$, even though larger vanes are sometimes used in softer clays in order to increase the accuracy of torque measurements.

Soil disturbance during the insertion of the vane is mainly due to the volume of soil displaced by the blade thickness and it is advisable to have a blade thickness of about 2 mm and a ratio between the volume of the blade and the soil swept when rotating the vane by less than 15%.

The results are influenced by many factors such as the delay between the vane insertion and the testing (a rest period of 5 minutes is usual and appropriate) and the rate of

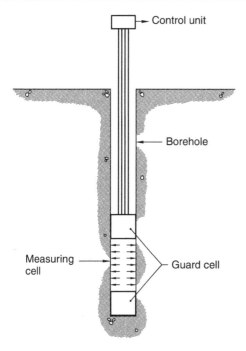

Figure 7.19 Menard's pressuremeter.

rotation (usually 6° to 12°/minute). A review of the current performance, interpretation and use of this test has been provided by Chandler (1988).

7.5.5 Pressuremeter tests

The pressuremeter test (Figure 7.19) is an *in situ* test made by the expansion of a cylindrical cell. Measurements are taken during this expansion of the pressure of the water injected to produce the expansion, and, assuming that the test can be modelled as a cylindrical cavity of infinite length, analytical solutions can be derived (see Sections 7.8 and 7.10), which allows us to obtain fundamental soil parameters (initial horizontal stress, strength, stiffness and consolidation parameters).

Two types of pressuremeter probes are currently in use. *Menard's pressuremeter* (Figure 7.19) is installed through a predrilled borehole, and the instrument is equipped with two guard-cells to minimize the end effects, ensuring the expansion of a cylindrical cavity (Menard, 1957, 1971).

The *self-boring technique* was developed at *Ecole Nationale de Ponts et Chausses* in 1967 (Baguelin *et al.*, 1972, 1978) and independently at *Cambridge University* (Wroth and Hughes, 1972), with the aim of minimizing the disturbance of the soil during the installation of the instrument.

Both these probes (Figure 7.20) apply the same concept: a volume of soil equal to the cell is slurried and washed up to the surface, without, in theory, disturbing the soil at the surface of the hole. The French probe (called *Pafsor*), an inflatable cell of 132 mm diameter and ratio of length to diameter of 4 or 2, is inserted into the ground by a drill

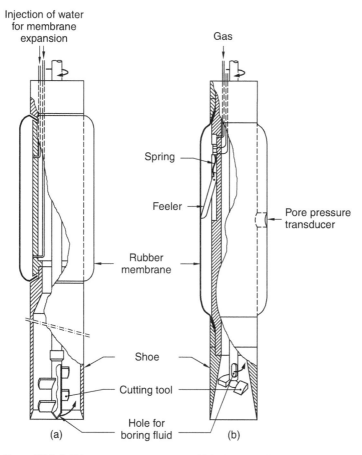

Injection of water
for membrane
expansion

Gas

Spring

Feeler

Pore pressure
transducer

Rubber
membrane

Shoe

Cutting tool

Hole for
boring fluid

(a)

(b)

Figure 7.20 Self-boring pressuremeters: (a) Pafsor; (b) Camkometer.

rod, and washing water is injected through the rod to the cutting tool which grinds the soil, which then flows to the surface passing through the inner part of the instrument. The rubber membrane (reinforced with longitudinal steel sheets) is deformed by water, and the measurements of the injected volume and the required pressure are taken.

In the *Camkometer* (English device), water is pumped down by means of a central rod, which also rotates the cutting tool and the soil is removed to the surface through a central rod. The membrane expansion is obtained by applying an internal nitrogen pressure, and the expansion is controlled by three radial feelers, located at the mid height of the cell.

Two small pore pressure transducers are also fixed to the rubber membrane and move together with it during the expansion in contact with the soil. The cell diameter is usually 80 mm and the length is 480 mm.

A new *in situ* testing device for the measurement of soil undrained strength and stiffness is the so-called **cone pressuremeter** (Jezequel *et al.*, 1982; Withers *et al.*, 1986). This test is intended to combine the operational advantages of a standard cone penetrometer with the capability of a pressuremeter cell for detecting strength and

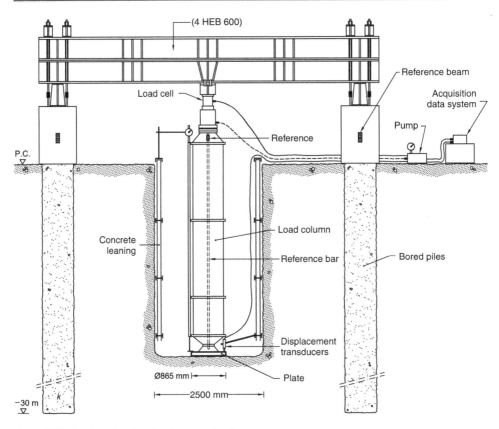

Figure 7.21 Design of a plate load test at depth.

stiffness parameter, and to achieve this a pressuremeter has been installed just behind the cone. Because the cone is pushed into the ground, the pressuremeter test must be considered as a *full displacement pressuremeter test*. Houlsby and Withers (1988) have provided an analysis of this test in clay, showing that parameters obtained from the interpretation of the unloading phase are in agreement with other measurements.

7.5.6 Plate load test

Plate loading tests can be considered as one of the first examples of *in situ* tests performed in order to obtain soil stiffness. This test can be performed at the surface or at the base of a borehole. In this latter case, the test arrangement is rather expensive (see Figure 7.21), but, if performed using plates of large diameter (of the order of 900 mm), these tests are of particular relevance for stiff fissured clays and in all other cases where there are no alternative procedures.

7.5.7 Shallow seismic exploration tests

Shallow seismic exploration tests of soils represent an important class of field tests, because of their non-invasive character. This allows to preserve the initial structure of

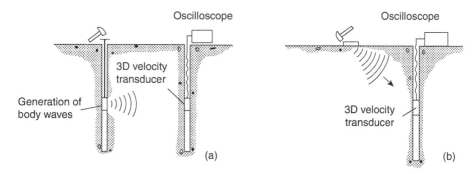

Figure 7.22 Cross-hole and down-hole tests.

soil deposits as well as the influence of all diagenetic phenomena (sutured contacts of grains, overgrowth of quartz grains, precipitation of calcite cements and authigenesis) contributing to a stiffer mechanical response. Soils are tested at small strain level, i.e. less than 10^{-3}%, and a reasonable assumption is to analyse the results of these tests by referring to the wave propagation theory in elastic materials.

The **cross-hole test** represents one of the most reliable methods of determining the shear modulus at small strain amplitude. In short, seismic energy is produced in a borehole (Figure 7.22a) and the time the seismic wave takes to reach another borehole is measured.

Two boreholes are usually sufficient to perform the test, but in order to eliminate the errors due to the triggering of the timing instruments, a preferable array is constituted of three or more boreholes. In such a way the wave velocity can be computed from the time intervals between pairs of holes.

As far as the borehole is concerned, the casing should provide a good coupling with the surrounding soil, the space between the casing and the soil being filled with cement grout or sand. PVC or aluminium casing are often specified because of their low impedence.

In addition to these recommendations, spacing between the boreholes should be selected in order to avoid problems with refracted waves, when operating in the presence of layers of different stiffness.

The arrangement for the **down-hole test** is shown in Figure 7.22b. In this case, the impulse is generated at the surface and the receivers are clamped to the borehole at different depths. This offers the advantage of being a more economical test arrangement, if compared to cross-hole tests, but care is required with data interpretation (as discussed in detail in Section 7.11).

Tests performed with the **seismic cone** (Robertson *et al.*, 1985; Hepton, 1988; Mayne, 2006) and the **seismic dilatometer** (SDMT) (Martin and Mayne, 1997; Mayne *et al.,* 1999) are conceptually similar to down-hole tests, the only difference is represented by the fact that the receivers are no longer clamped to a borehole, but they are inserted into a seismic module, placed above the tip of the cone penetrometer or the DMT blade (Figure 7.23).

The test configuration with two receivers avoids possible inaccuracy in the determination of the 'zero time' at the hammer impact; moreover, the couple of seismograms

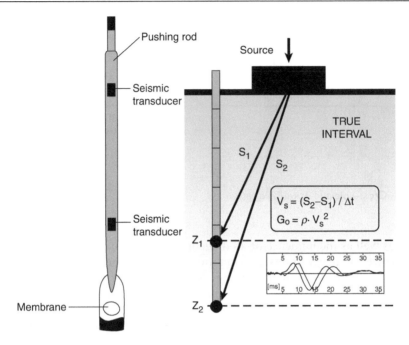

Figure 7.23 Seismic dilatometer.

recorded by the two receivers at a given test depth correspond to the same hammer blow and not to different blows in sequence, not necessarily identical. Hence the repeatability of wave velocity measurements is considerably improved.

Finally, the so-called **SASW method**, pioneered by Stokoe at the Texas University (see Stokoe *et al.,* 1994), is a non-invasive seismic technique, which represents the most significant recent development in shallow seismic exploration for geotechnical engineering application. In this method, a source at the surface generates Rayleigh surface waves, detected by at least two receivers (usually vertical velocity transducers with natural frequencies between 1 and 4.5 Hz) located at known distance apart from the source (Figure 7.24). If we consider a layered half-space, the velocity of propagation of a surface wave depends on the frequency (or wavelength) of the wave. This dependence is called *dispersion* and arises because waves of different wavelengths sample different portions of the soil: high frequency or short wavelengths sample the top layer, whereas low frequencies or long wavelengths sample deeper layers. Therefore, by using surface waves with a range of wavelengths or frequencies all layers can be sampled and the

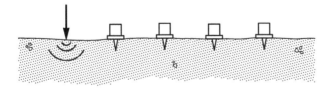

Figure 7.24 SASW test.

stiffness profile can be determined by using an inversion algorithm (see Section 7.11). Despite its simplicity, this method requires a high degree of expertise.

7.6 Soil profiling

As already mentioned, by using continuous penetration devices, such as the electrical cone (CPT) and the piezocone (CPTU), it is possible to obtain cost-effective detailed information on soil profile, including the identification of different layers and the location of drainage boundaries.

Originally, the cone penetrometer was used to measure just the tip resistance. Successively, the measurement of the sleeve friction was envisaged as an estimate of pile shaft resistance. Because in sands the tip resistance tends to be high and the friction ratio tends to be low, while the reverse is observed in clay soils, both these measurements have been used in recent times as indicators of the soil type (Robertson, 1990). However, there are limitations associated with the use of such empirical correlations (Mitchell and Brandon, 1998), and the reader is not encouraged to use these as a substitute of soil borings for detecting a site stratigraphy. In particular it must be observed that, despite the good repeatability of the tip resistance values, the repeatability of the sleeve friction may be poor, as it depends on several factors such as maintenance of the instrument, mechanical compliance with the cone and inclusions in the seals.

When using CPTU, simultaneous measurements of tip resistance and pore pressure make the detection of soil layering straightforward. This can be shown by referring to Figure 7.25, where the results of a test performed in the stratified deposit at

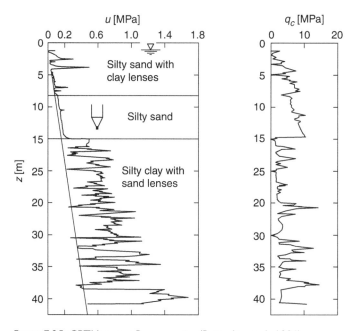

Figure 7.25 CPTU test at Ravenna site (Battaglio *et al.*, 1986).

Ravenna (Italy) is reported. In the first horizon, 15 m of thickness, of loose to medium dense sands, high values of the tip resistance are associated to pore pressure values close to the initial equilibrium values. On the contrary, in the lower soft to medium silty clay stratum, a large excess pore pressure develops.

Note that, since the resolution in terms of pore pressure is higher than that in terms of tip resistance, the CPTU is at present the most useful tool for detecting soil layering. However, it must be outlined that an important source of errors in the measured pore pressure may occur if the measuring system is not fully saturated (see Battaglio *et al.*, 1986). This usually leads to incorrected initial pore pressure values, lower capacity of detecting soil layering, and delayed recovery of the pore pressure level when restarting penetration phase after a dissipation test.

7.7 *In situ* horizontal stress

The *in situ* initial stress state is described by the vertical effective stress σ'_{vo} and the horizontal effective stress $\sigma'_{ho} = K_o \sigma'_{vo}$. Whereas it is relatively simple to compute the vertical effective stress, as already discussed in Chapter 4, the evaluation of the horizontal stress is a very complex task, because the value of the at rest earth pressure coefficient depends on the history of the deposit. In particular, it must be remembered that K_o can be estimated through empirical correlations in NC soil deposits; in OC soils, empirical correlations only apply if the overconsolidation ratio is due to simple unloading. If more than one cycle of unloading and reloading has occurred or there is the presence of other phenomena (cementation, ageing or capillarity), direct measurements are necessary.

7.7.1 *Pressuremeter tests*

The possibility of evaluating the *in situ* horizontal stress certainly represents one of the most interesting features of pressuremeter tests. These tests are *in situ* load tests made by the expansion of a cylindrical cell. During this expansion measurements of radial stress σ_r applied to the soil and the related radial displacements u_r of the membrane (Camkometer) or of the injected volume (Pafsor) are taken, and the results are usually represented as in Figure 7.26. The first curve (Figure 7.26a) refers to a Menard pressuremeter test. As a predrilled hole is required for the installation

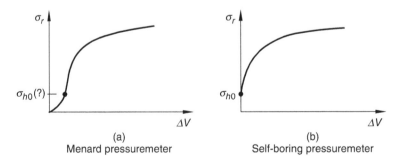

Figure 7.26 Examples of expansion curves from pressuremeter tests.

of the instrument, disturbance is introduced in the surrounding soil. Therefore, the first portion of the expansion curve shows the radial stress that must be applied to push the membrane into contact with the soil; the quasi-linear portion being the soil response until yield and the last part reflecting the elasto-plastic response. The second curve (Figure 7.26b) is obtained from a good quality self-boring pressuremeter test. In this case the curve starts without the reloading portion, because the self-boring technique is presumed to allow one to install the instrument without disturbance of the soil.

In practice there are many factors that influence the self-boring technique, including the penetration rate, the drilling mud pressure and flow rate, the cutter position with respect to the cutting shoe edge and the geometrical configuration of the cutter, so that some disturbance is likely to occur. The problem is even more complex in sands, and in this case even a small disturbance during the probe installation leads to a situation in which the so-called *lift-off pressure*, i.e. the horizontal stress measured at the beginning of the expansion phase, is far from matching the *in situ* horizontal stress.

Because of all these aspects, the experience gained at the Technical University of Torino (Ghionna *et al.*, 1981, 1983) was that, despite its potentiality, the self-boring pressuremeter test still requires more research in those areas where there is a real need to measure the *in situ* state of stress, that is in stiff clays and in sands.

7.7.2 Marchetti's dilatometer

The flat dilatometer is being used increasingly, because of its simplicity to operate. To date its interpretation has been largely performed using empirical correlations, and more insight has been gained only recently, by using strain path analyses or cavity expansion methods.

When evaluating the *in situ* horizontal stress, the empirical approach is based on the idea (Marchetti, 1980) that the *lateral stress index*, defined as the difference between the lift-off pressure and the equilibrium pore pressure u_o, normalized with respect to the effective overburden stress, i.e.:

$$K_D = \frac{p_o - u_o}{\sigma'_{vo}},\qquad(7.10)$$

can be correlated to the *coefficient of earth pressure at rest* by means of the following correlation:

$$K_o = \left(\frac{K_D}{\beta_K}\right)^{0.46} - 0.6.\qquad(7.11)$$

As expected, this correlation depends on the soil type, as reflected by different values of β_K on different soil deposits (see Figure 7.27).

Yu *et al.* (1993) have suggested to simulate the expansion of a flat cavity, showing (through numerical analysis, since no analytical solutions are available for the

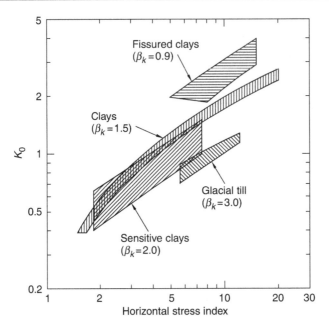

Figure 7.27 Correlation between K_o and horizontal stress index (Kulhawy and Mayne, 1990).

expansion of a flat cavity in soils) that the lateral stress index can be linked to the coefficient of earth pressure at rest, as follows:

$$K_D - K_o = \left(1.57 \ln I_R - 1.75\right)\frac{s_u}{\sigma'_{vo}}, \tag{7.12}$$

where $I_R = G/s_u$ is the rigidity index.

Similarly, when dealing with dilatometer tests in sands, Yu (2004) has recently shown that the ratio K_D/K_o increases with soil friction angle, but the influence of the soil stiffness is also important. Therefore, both these results corroborate the following conclusion (which is also common to the interpretation of CPT): strength, stiffness and *in situ* effective stress combine to give the soil response, so that is difficult to detect the initial horizontal stress without an appreciation of the soil strength and stiffness.

7.8 Undrained shear strength

7.8.1 Field vane test

Since the 1940s, the field vane test was the most popular test to measure the undrained strength in *soft clays*. In the early 1970s, Bjerrum (1972, 1973) greatly improved its capability through well-documented back analyses of failure embankments on soft clays, and further studies have been reported by Ladd (1975), Ladd *et al.* (1977), Azzouz *et al.* (1983), Chandler (1988) and Leroueil *et al.* (1990).

We briefly mentioned in Section 7.5 that the test (Figure 7.18) consists of lowering into the soil a rotating *vane*, within an external casing. At the selected depth, the vane

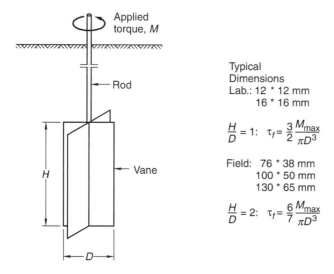

Figure 7.28 Interpretation of field vane test.

is pushed down 0.5 m and is rotated, and the measured torque applied through the rods allows, to obtain the undrained strength of the soil.

This requires the following assumptions: the test is performed in undrained conditions, and there is a uniform mobilisation of the undrained strength, which is supposed to be constant along vertical and horizontal surfaces. Then, by considering the equilibrium around the vertical axis (Figure 7.28), one obtains:

$$s_u(FV) = \frac{6M}{7\pi D^3},\qquad(7.13)$$

where M is the applied torque.

Measurements are also made on the remoulded clay after 25 revolutions, to obtain the soil *sensitivity*. This parameter is defined as the ratio of the undisturbed strength versus the remoulded soil strength.

The obtained results are influenced by many factors such as the delay between the vane insertion and the testing (a rest period of five minutes is usual and appropriate) and the rate of rotation (usually 6° to 12°/minute).

Furthermore it must be recalled that the undrained strength is not a soil property, but it rather represents the soil response, so that it depends on the test used and may be different when considering different modes of failure of soil elements along a potential failure surface. For this reason, Bjerrum (1972, 1973) first suggested that the undrained strength from field vane should be corrected to obtain the strength mobilised at failure, using an empirical correction of the form:

$$s_u(at\ failure) = \mu \cdot s_u(FV)\qquad(7.14)$$

where the coefficient μ is given in Figure 7.29 as a function of the plasticity index.

The scatter which characterize this empirical correction may reflect errors in measurements of soil strength as well as specific field conditions, i.e. strength profile and

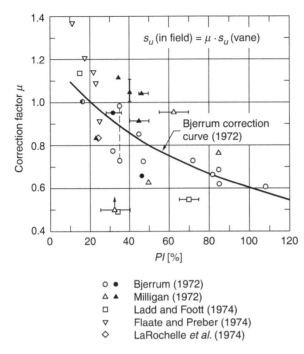

$$s_u \text{ (in field)} = \mu \cdot s_u \text{ (vane)}$$

Bjerrum correction curve (1972)

o • Bjerrum (1972)
Δ ▲ Milligan (1972)
□ Ladd and Foott (1974)
▽ Flaate and Preber (1974)
◇ LaRochelle *et al.* (1974)

Figure 7.29 Vane correction factor (from Ladd, 1975; Ladd et al., 1977).

mode of failure, so that from the data of Figure 7.29 it can be inferred that the Bjerrum correction can involve errors up to 30% in the computed safety factor (see Ladd, 1975; Ladd *et al.*, 1977; Azzouz *et al.*, 1983; Leroueil *et al.*, 1990 for further discussion of this topic).

7.8.2 Undrained strength from CPT

The approaches suggested to analyse the cone penetration problems in clays can be grouped into two main classes: those referring to bearing capacity theory and those referring to cavity expansion theories. The use of bearing capacity theory is not as simple as appears, and the suggested solutions often involve the assumption of boundary conditions which are not appropriate for the cone penetration problem (see Houlsby and Wroth, 1982). If we consider a shallow bearing capacity problem, the displaced soil moves essentially outwards and upwards to the free surface. On the contrary, a deep penetration requires that the displaced soil is essentially compensated by deformations of the soil. For this reason, cavity expansion theories have been suggested to investigate the deep penetration problem (Vesic, 1972). However, since the expansion of a cavity does not predict correctly the strain paths followed by soil elements, Baligh (1985) suggested the use of a more sophisticated approach, called the *strain-path method*. Further analytical studies by Levadoux and Baligh (1980) and Teh and Houlsby (1991) have shown that the soil strength, the stiffness and *in situ* stresses combine to give the cone resistance, so that the soil strength cannot be deduced in a rational way by the measured tip resistance without appreciation of the influence of soil stiffness and horizontal stress.

Due to the complexity of the penetration process and the difficulties of an accurate knowledge of soil stiffness and horizontal stress, the empirical approach still remains the most practised one.

In this approach, the undrained strength is obtained from the following relationship (which suggests a bearing capacity approach):

$$s_u = \frac{q_c - \sigma_{vo}}{N_c} \tag{7.15}$$

where σ_{vo} is the total overburden stress at the depth where the cone resistance is measured, and N_c is called the *cone factor*.

Taking into account the introductory considerations, it is not surprising that the information published in literature concerning the cone factor shows some scatter, also because the undrained strength used in the correlations depends on the test used for its determination.

(a) Lunne *et al.* (1976) and Baligh *et al.* (1980) have suggested that the undrained strength, as obtained by field vane, should be considered as the reference strength in soft to medium clays, as it has been extensively checked through evidence of foundation failures. Moving along this line, the collected data show an average value $N_c = 14$, with an uncertainty in the predicted strength of 33%.

(b) The problem is more difficult in overconsolidated clays. Kjekstad *et al.* (1978) indicate $N_c = 17 \pm 5$, for intact clays, using the value obtained from triaxial compression tests as a reference strength.

(c) Based on experience with large diameter (865 mm) plate loading tests performed in stiff fissured clays, Marsland and Powell (1979) suggest N_c ranging from 10 to 30, the upper limit being associated with the higher values of the undrained strength. However, in stiff clays it is difficult to establish to what extent the behaviour may be considered fully undrained.

More recently, Karlsrud *et al.* (2005) have attemped new correlations, comparing CPTU (**piezocone**) results against the undrained strength from triaxial compression tests on block samples of high quality (peak strength). By using the measured *excess pore pressure* during penetration, the suggested correlation now becomes:

$$s_u = \frac{u_2 - u_o}{N_{\Delta u}}, \tag{7.16}$$

where u_2 is the measured pore pressure just behind the cone and u_o is the initial pore pressure.

The cone factor $N_{\Delta u}$ depends on stress history, clay sensitivity and plasticity index, and on the basis of the influence of these factors the following correlations have been suggested:

$$N_{\Delta u} = 6.9 - 4 \cdot \log OCR + 0.07 PI, \tag{7.17}$$

for low sensitivity clays (sensitivity lower than 15), and for high sensitivity:

$$N_{\Delta u} = 9.8 - 4.5 \cdot \log OCR. \tag{7.18}$$

In (7.17), the plasticity index (PI) must be introduced in %.

7.8.3 Pressuremeter tests

Theoretical interpretation methods of pressuremeter tests are based on the fundamental assumption that the test can be simulated as the expansion of an infinite long cylindrical cavity. This simplifies the problem, which can be analysed as a plane strain problem; then, the further assumption of axial-symmetry gives the equilibrium equation to the following form (see Chapter 3):

$$\frac{\sigma_r - \sigma_\theta}{r} + \frac{d\sigma_r}{dr} = 0. \tag{7.19}$$

Strain components can be expressed in terms of displacement field through the relations:

$$\begin{aligned}
\varepsilon_r &= -\frac{\partial u_r}{\partial r} \\
\varepsilon_\theta &= -\frac{u_r}{r}
\end{aligned} \tag{7.20}$$

where the minus sign has been introduced according to the assumption that considers positive the compressive strain. Let us now assume that the soil behaves as an elastic perfectly plastic material. In the elastic phase, the introduction of the constitutive relationships allows one to express the equilibrium equation in terms of dispacement, i.e.:

$$\frac{d^2 u_r}{dr^2} + \frac{1}{r} \frac{du_r}{dr} - \frac{u_r}{r^2} = 0, \tag{7.21}$$

and the equation (7.21) is known as the *Cauchy-Euler equation*.

We can try to obtain a solution by setting $u_r = r^\alpha$, with $r > 0$, and we denote as a the current radius and by $\delta\sigma_r = \sigma_r - \sigma_{ho}$ the increment of radial stress, whose value at the cavity wall is $\delta\sigma_{r=a}$.

The boundary conditions to be taken into consideration are as follows:

$$\begin{aligned}
r \to \infty, &\quad u_r \to 0 \\
r = a, &\quad \delta\sigma_r = \delta\sigma_{r=a}
\end{aligned}$$

and the solution of the ODE (7.21) is expressed as:

$$u_r = \frac{a^2}{r} \frac{\delta\sigma_{r=a}}{2G}. \tag{7.22}$$

This can be substituted into the constitutive equations to obtain the stress distribution around the cavity, as a function of the applied radial stress:

$$\delta\sigma_r = \frac{a^2}{r^2}\delta\sigma_{r=a}$$
$$\delta\sigma_\theta = -\frac{a^2}{r^2}\delta\sigma_{r=a}$$

(7.23)

so that the elastic phase is completely investigated.

When considering the expansion of a cavity in clay soil, the assumption is usually introduced that the test is fast enough that the undrained condition holds. If according to this assumption the analysis is carried out in terms of total stresses, then a soil element reaches the plasticity condition when:

$$\sigma_r - \sigma_\theta = \delta\sigma_r - \delta\sigma_\theta = 2s_u,$$

and by considering that $\delta\sigma_\theta = -\delta\sigma_r$ (see equation 7.23) this occurs when the applied cavity stress reaches the value:

$$\delta\sigma_r = s_u.$$

(7.24)

The equilibrium equation (7.19) then writes:

$$\frac{2s_u}{r} + \frac{d\sigma_r}{dr} = 0,$$

which can be integrated with the following boundary conditions:

$$\delta\sigma_r = s_u \quad if \quad r = R_p,$$

to obtain:

$$\delta\sigma_r = s_u + 2s_u \ln\frac{R_p}{r}$$
$$\delta\sigma_\theta = -s_u + 2s_u \ln\frac{R_p}{r}$$

(7.25)

The radius R_p represents the boundary of the plastic zone, and its value:

$$R_p = a\exp\left(\frac{\sigma_{r=a} - \sigma_{ho} - s_u}{2s_u}\right)$$

(7.26)

can be obtained by considering that the condition $\delta\sigma_r = \sigma_{r=a} - \sigma_{ho}$ must hold at $r = a$. Therefore, the stress distribution within the plastic annulus is the following:

$$\sigma_r = \sigma_{r=a} + 2s_u \ln\frac{a}{r}$$
$$\sigma_\theta = \sigma_{r=a} - 2s_u\left(1 - \ln\frac{a}{r}\right)$$

(7.27)

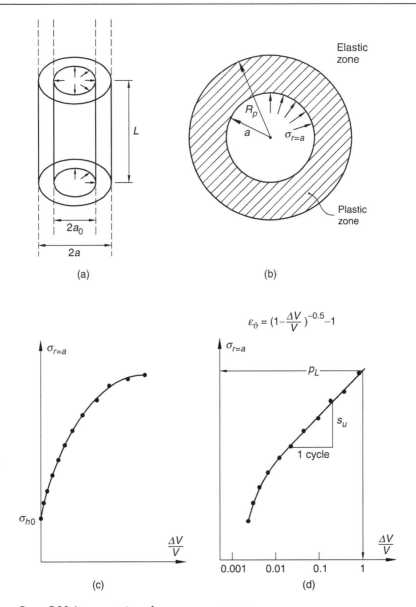

$$\varepsilon_\vartheta = (1 - \frac{\Delta V}{V})^{-0.5} - 1$$

Figure 7.30 Interpretation of a pressuremeter test.

In order to apply the derived theoretical solution to a pressuremeter test, we first observe that the radial displacement of a soil element at the interface between the plastic and the elastic zone (see Figure 7.30) can be obtained by using the elastic solution (7.22), which writes (assuming $r = a = R_p$):

$$\frac{u_r}{R_p} = \frac{s_u}{2G}.$$

Moreover, the constraint of no volume change implies:

$$\pi(R_p^2 - a^2) = \pi\left[\left(R_p - u_r\right)^2 - a_o^2\right],$$

and, by neglecting the term u_r^2, it follows:

$$u_r = \frac{a^2 - a_o^2}{2R_p}.$$

These expressions give:

$$\frac{s_u}{G} = \frac{a^2 - a_o^2}{R_p^2}$$

and then:

$$\frac{R_p^2}{a^2} = \frac{G}{s_u}\frac{\delta V}{V}$$

since: $\left(\frac{\delta V}{V} = \frac{a^2 - a_o^2}{a^2}\right)$.

By combining this latter expression with that of radius of the plastic zone (7.26) we finally obtain:

$$\delta\sigma_{r=a} = s_u\left(1 + \ln\frac{G}{s_u}\right) + s_u\ln\frac{\delta V}{V}, \tag{7.28}$$

where V is the current volume of the probe.

This is the expression first derived by Gibson and Anderson (1961), and Windle and Wroth (1977) suggested that if the test results (Figure 7.31) are plotted in terms of cavity radial stress against the logarithm of the volumetric strain (Figure 7.32), the slope of the plastic portion is equal to the undrained strength of the soil.

By defining as **limit pressure** the value p_L attained at $\delta V/V = 1$, the expression (7.28) is also referred in the form:

$$\sigma_{r=a} = p_L + s_u\ln\frac{\delta V}{V}. \tag{7.29}$$

More recently, a similar expression has been derived for the unloading stage by Jefferies (1988) within the small strain assumption, and Houlsby and Withers (1988) have provided a large strain analysis for the interpretation of the cone pressuremeter test.

Table 7.2 summarizes the experience gained at Technical University of Torino (Ghionna *et al.*, 1981, 1983) by comparing the undrained strength as obtained by

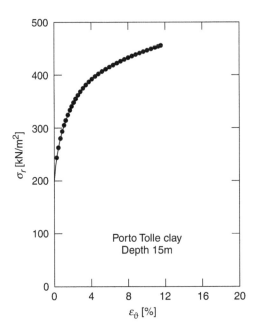

Figure 7.31 Example of expansion curve from a pressuremeter test.

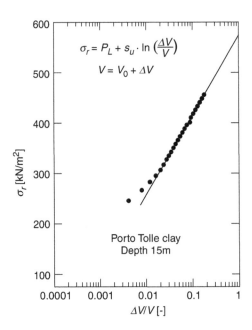

Figure 7.32 Interpreting a pressuremeter test according to Windle and Wroth (1977) procedure.

Table 7.2 Normalized undrained strength s_u/σ'_{vo} from SBP tests

Site	PI (%)	OCR	S_t	$TX - CK_oU$	DSS	FV	SBP
Porto Tolle	30	1.1–1.3	2–3	0.31	0.26	0.29	0.29±0.04
Trieste	47	1.0	2–4	0.32	0.28	0.35	0.33±0.11
Montalto di Castro	34	2.5–4	–	0.54	–	–	0.64±0.11
Bandar Abbas (Iran)	16–28	1.5–2.5	3–4	0.54	–	0.56	0.67±0.30
Onsoy	20–36	1.0–1.2	5–6	0.33	0.26	0.23	0.48±0.08
(Norvegia)	15–30	2	6–9	0.45	0.28	0.48	1.0–1.4
Drammen	10–15	1.15	7–8	0.34	0.22	0.25	0.65±0.08
(Norvegia)	25–30	1.5	7–8	0.40	0.32	0.34	0.65±0.08
Guasticce	63	1.1	–	0.32	–	0.30	0.46
Panigaglia	45–65	1	4–7	0.32	0.27	0.26	0.53±0.22
Taranto	22–30	22–30	–	2.28	0.94	–	2.3–3.2

S_t = soil sensitivity, as deduced from field vane tests; TX-$CK_oU = K_o$ consolidated-undrained tests, compression loading; DSS = direct simple shear; FV = field vane tests; SBP = self-boring pressuremeter tests.
Source: Ghionna *et al.*, 1983.

using the self-boring pressuremeter, the field vane and laboratory tests. As expected, the undrained strength depends on the test used for its measurement; therefore, the use of the undrained strength derived from pressuremeter tests in problems characterized by stress paths which deviate from those applied during this test can lead to a significant overestimation of the strength available in the field. This overestimation may also be induced by the non-cylindrical expansion of the probe, due to the limited value of the ratio *L/D*. In addition, it must be considered that the assumption of undrained condition seems to be reasonably accurate for soft clays, whereas it is unlikely to apply for stiff clays (Yu and Collins, 1998).

7.9 Shear strength of coarse-grained soils

The interpretation of cone penetration tests in sand is a more difficult problem than interpretation in clay and the currently used approaches are mainly based on tests performed in calibration chambers (see a summary of the researches provided by Lunne *et al.*, 1997).

The **calibration chamber** (Figure 7.33) is a double-walled cylinder, equipped with a base piston and a top lid. Sand specimens, to be tested within this cylinder, are usually obtained by *pluvial deposition* (raining), because this technique allows to obtain repeatable sample specimens (the rainfall is the main factor controlling the relative density of the pluviated sand specimen).

The sample is enclosed in a rubber membrane, and the applied radial stress is controlled through the pressure of water filling the annular space between the membrane and the wall of the chamber. The vertical stress is applied to the sample through the base piston. The hole in the centre of the top lid allows the introduction of the cone penetrometer for the simulation of the field test. During this stage, the condition of constant radial stress can be achieved by connecting the annular space to a pressure supply. On the contrary, a condition of zero radial movement can also be applied by preventing flow of water throughout the annular space.

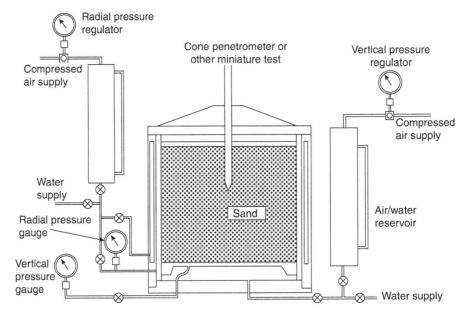

Figure 7.33 The calibration chamber (adapted from Bellotti et al., 1985, and Wesley, 2002).

Therefore, the calibration chamber offers the advantage that penetration tests and other tests (i.e. DMT, miniature plate loading tests, pressuremeter tests) can be performed under controlled conditions and the properties of the tested sands can be established by independent means, so that reliable correlation can be deduced.

Shortcomings derive from the fact that, even in reasonable large chambers, there are boundary effects in dense sands (Wesley, 2002), and the artificially prepared sand specimens cannot reproduce all diagenetic phenomena of natural sands.

In Chapter 5 we discussed that the critical state angle φ'_{cv} depends on mineralogy, roughness and grading of particles, but it is independent of the initial conditions. In silica sands it can range from 30° to 38° (higher values apply to angular particles), and in feldspar or carbonatic sands can reach values up to 40°.

On the contrary, the peak value of the angle of shear resistance is linked to the rate of dilation, it depends on soil state and cannot be regarded as a soil property. The rate of dilation depends on the specific volume and on the effective stress level, so that we reach the conclusion that peak value of the angle of shear resistance increases if the relative density increases, but decreases with higher level of effective stress.

Bolton (1986) supplied experimental evidences supporting these arguments and suggested the following relation, which expresses the influence on the angle of shear resistance of the combined effect of relative density and stress:

$$\varphi' - \varphi'_{cv} = mDI < 12°$$

$$DI = D_R \left(Q - \ln p'_f \right) - 1$$

(7.30)

where:

p'_f is mean effective stress at failure (expressed in kPa); m depends on deformation constraints (it is equal to 3 for triaxial strain conditions and equal to 5 for plane strain conditions); Q depends on grain crushing strength and ranges from 10, for quartz and feldspar sands, to 8 for limestone and 5.5 for chalk.

Therefore, this relation supports the idea that the assessment of the relative density can be converted into an estimate of the friction angle. This is the reason why in this section we concentrate on the assessment of relative density from penetration tests.

7.9.1 Relative density from SPT

Early work carried out by Terzaghi and Peck (1948) simply correlated the blowcount N_{SPT} with the relative density. Later, research by Gibbs and Holtz (1957) proved that the penetration resistance is also significantly affected by the overburden stress (Figure 7.34). However, subsequent researchers have definitively proved that any penetration resistance is mainly controlled by the effective horizontal stress. For this reason, the correlation suggested by Gibbs and Holtz (1957), shown in Figure 7.34, is appropriate for NC sands, but can lead to unconservative results if applied to overconsolidated

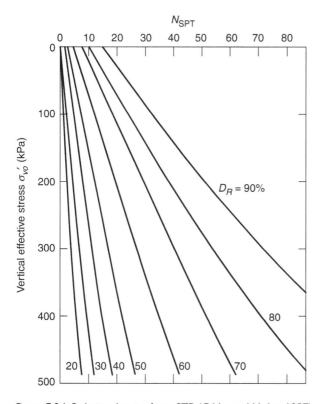

Figure 7.34 Relative density from STP (Gibbs and Holtz, 1957).

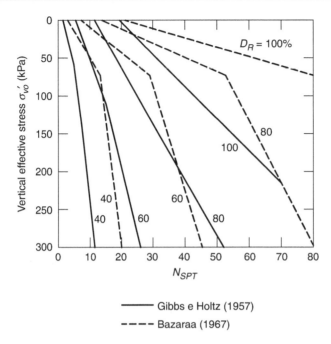

Figure 7.35 Correlations suggested by Gibbs and Holtz (1957) and Bazaraa (1967).

or artificially compacted materials. In the latter case the correlation in Figure 7.35, suggested by Bazaraa (1967), seems to be more appropriate.

Skempton (1986) reviewed the SPT data from Japan, UK, China and USA, and has suggested the following correlation:

$$D_R^2 = \frac{N_1}{60} \tag{7.31}$$

$$N_1 = C_N \cdot N_{SPT}$$

$$C_N = \begin{cases} \dfrac{2}{1 + \sigma'_{vo}/100} & \text{for fine sand} \\[2ex] \dfrac{3}{2 + \sigma'_{vo}/100} & \text{for coarse sand} \end{cases}$$

where N_1 is the SPT value that refers to an average energy ratio of 60% and to a reference stress of 1 atmosphere. The overburden effective stress σ'_{vo} is in kPa.

7.9.2 Relative density from CPT

When referring to cone penetration tests similar correlations have also been suggested, which consider the above-mentioned factors (see Bellotti *et al.*, 1985). The correlation

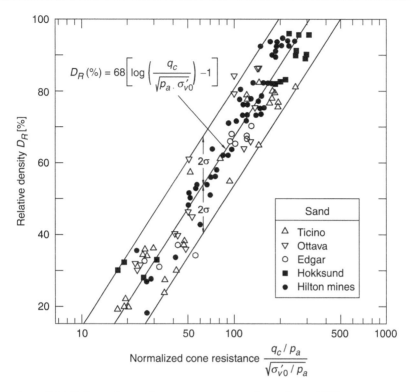

$$D_R(\%) = 68\left[\log\left(\frac{q_c}{\sqrt{p_a \cdot \sigma'_{v0}}}\right) - 1\right]$$

Figure 7.36 Relative density from cone resistance (Lancellotta, 1983).

shown in Figure 7.36 (Lancellotta, 1983), expressed in the form:

$$D_R = 68\left[\log\left(\frac{q_c}{\sqrt{p_a\sigma'_{v0}}}\right) - 1\right],\tag{7.32}$$

has the advantage of illustrating the range of actual data for five different NC sands, so that the confidence limits can easily be evaluated.

7.10 Stiffness

The evaluation of soil stiffness from *in situ* tests is of great practical interest in soil deposits where sampling of undisturbed samples is still unreliable, as is the case with cohesionless soils, or when laboratory samples, due to reduced dimensions, cannot be representative of the soil mass, as can be the case with stiff fissured soils. However, the interpretation of *in situ* tests is not so straightforward, because of two main factors: some tests are intrusive, so that the interpretation remains empirical; further, some tests require to solve a boundary value problem, because the stress path imposed to soil elements is different from point to point.

Therefore, it is necessary to recall these aspects in order to clarify the meaning of the parameters obtained from different tests, as well as their potential use in practice. In this section we first deal with stiffness related to medium and large strain levels, whereas seismic methods, used to access the small strain stiffness, will be discussed in Section 7.11.

7.10.1 Shear modulus from Marchetti's dilatometer

As already discussed in Section 7.5, the flat dilatometer tests (Marchetti *et al.*, 2001) uses a flat plate containing a flexible stainless steel membrane. The membrane is inflated using nitrogen pressure, and two basic measurements are performed: the first one, p_o, corresponds to the value required just to lift the membrane off the sensing disc (*lift-off pressure*); the second value, p_1, is that required to produce a 1.1 mm deflection at the centre of the membrane.

As is sketched in Figure 7.37, these values are used to compute the so called *dilatometer stiffness*:

$$E_D = 37.4 \left(p_1 - p_o \right),$$
(7.33)

which is empirically correlated to the constrained modulus (Marchetti *et al.*, 2001) or to the small strain shear modulus (Jamiolkowski *et al.*, 1988).

The potential of Marchetti's dilatometer has been recently increased, by means of geophones properly installed into the pushing rods (see Figure 7.23). This allows to also perform down-hole tests (with the so-called *true interval method*), so that direct measurements of the small strain shear modulus G_o can be done at the same depth the dilatometer modulus is being measured (see Foti *et al.*, 2006).

7.10.2 Plate loading tests

Plate loading tests can be considered as the first example of *in situ* tests performed in order to obtain soil stiffness. When performed using plates of large diameter (of the order of 900 mm), these tests are of particular relevance for artificially made fills, stiff fissured clays and in all other cases where there are no alternative procedures. One of the most comprehensive researches on this subject has been carried out by Marsland (1971, 1974) in the London stiff clay, using plates with a diameter of

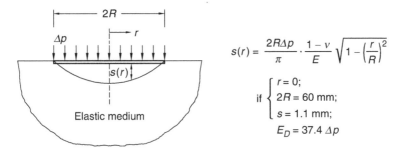

Figure 7.37 Interpretation of the dilatometer test (Marchetti, 1980).

865 mm. This investigation proved that the results depend on the time elapsed from the execution of the hole and the loading phase (as the delay allows swelling of the soil and opening of the fissures), on the dimension of the plate and on the boundary conditions of the hole.

When interpreting the result of a loading test, the assumption is made that the soil behaves as a linear, isotropic and homogeneous medium. With such an assumption the shear modulus is given by:

$$G = \frac{q}{w}(1 - \upsilon)R \cdot f(z),$$ (7.34)

where R is the radius of the plate, q is the load per unit surface, w is the measured settlement and $f(z)$ is a correction factor, ranging from 0.85 and 1, depending on the relative depth of the plate.

7.10.3 Shear modulus from pressuremeter tests

These tests simulate the expansion of a cylindrical cavity, and represent a boundary problem with well-defined boundary conditions. As already discussed in Section 7.8, if the loading path is such that the soil can be assumed to be elastic, as is the case when an unloading-reloading loop is performed (see Figure 7.38), then the radial displacement is given by:

$$u_r = \frac{a^2}{r}\frac{\delta\sigma_{r=a}}{2G},$$ (7.35)

where a is the actual radius of the probe.

It follows that, by considering a soil element at the probe interface, the shear modulus can be evaluated through the relation:

$$G = \frac{1}{2}\frac{\delta\sigma_{r=r}}{\varepsilon_\theta},$$ (7.36)

where the circumferential strain is defined as $\varepsilon_\theta = u_r/r$.

7.11 Seismic methods

Shallow seismic exploration tests of soils represent an important class of field tests, because of their non-invasive character. This allows to preserve the initial structure of soil deposits as well as the influence of all diagenetic phenomena (sutured contacts of grains, overgrowth of quartz grains, precipitation of calcite cements and authigenesis) contributing to a stiffer mechanical response. Soils are tested at small strain level, i.e. less than $10^{-3}\%$, and a reasonable assumption is to analyse the results of these tests by referring to the wave propagation theory in elastic materials. Accordingly, the soil stiffness G, indicated as G_o when referred to small strain level, is inferred from measurements of shear wave velocity, by using the relation (see Section 7.12 for

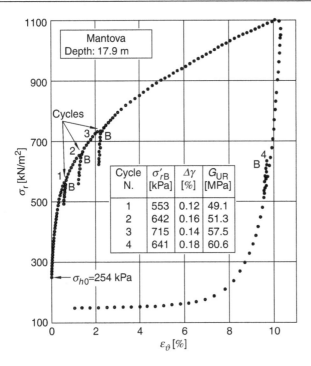

Figure 7.38 Example of a pressuremeter test in Po river sand (Bellotti *et al.*, 1986).

theoretical aspects of wave propagation):

$$G_o = \rho \cdot V_s^2 ,$$ (7.37)

where ρ is the soil density.

Further, as already discussed in Chapters 3 and 5, it is relevant to recall that soils during deposition usually experience one-dimensional deformation, so that there is a tendency for displacements to only occur vertically without any lateral component. The so-called *inherent anisotropy* reflects this depositional history and a rather realistic model is, in this case, the *cross-anisotropic medium* (see Figure 7.39): the soil response is different if the loading direction changes from vertical to horizontal, but it is the same when changes occur in the horizontal plane.

Consistently, when interpreting or comparing the results of cross-hole and down-hole, it should be clarified if the measured shear modulus is the one relevant to shearing in the horizontal plane, i.e. G_{hh} (deduced from waves propagating in the horizontal direction and polarized into the horizontal plane), or to shearing in the vertical plane, i.e. G_{vh} (deduced from waves propagating in the vertical direction and horizontally polarized).

7.11.1 Cross-hole and down-hole tests

The **cross-hole test** represents one of the most reliable methods of determining the shear modulus at small stain amplitude. When performing the test, seismic energy is

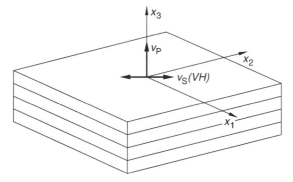

(a) waves propagating in x_3 direction

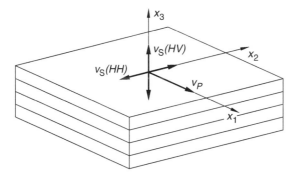

(b) waves propagating in x_1 direction

Figure 7.39 Wave propagation in a cross-anisotropic medium.

produced in a borehole (Figure 7.22.a) and the time the seismic wave takes to reach another borehole is measured, and depending on the generated polarized waves, the test allows to detect values of G_{hh} or G_{hv}.

The arrangement for the **down-hole test** is shown in Figure 7.22.b. In this case the impulse is generated at the surface and the receivers are clamped to the borehole at different depths. In this case, depending on the number of receivers used and on the specific testing procedures, three interpretation approaches can be followed.

In the so-called **direct arrival approach**, a travel time plot is made (Figure 7.40) of time versus depth as a single receiver is being lowered at different depths. The wave velocity is then determined from the slope of a best fit curve through the data points. By considering the obliquity of the seismic rays, the plotted travel time is a value corrected according to the relation:

$$t_c = \frac{z}{d}t,$$

where d is the distance between the source and the receiver.

In the **pseudo-interval** method, the time difference between wave arrivals at a single geophone at two successive depths is used as travel time for the

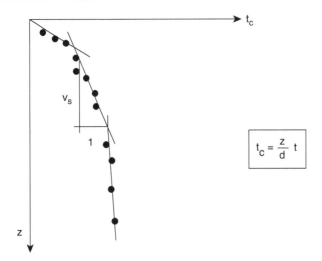

Figure 7.40 Down-hole test: direct arrival approach.

difference in depth, i.e.:

$$v_S = \frac{d_2 - d_1}{t_2 - t_1}.$$

Note that, with this arrangement, the repeatability of v_S measurements depends on the fact that the couple of seismograms recorded by the two receivers corresponds to different blows in sequence. This possible shortcoming is avoided in the **true-interval method**, the shear wave velocity v_S being now obtained as the ratio between the difference in distance from the source of two receivers $(d_2 - d_1)$ and the delay of the arrival of the impulse from the first to the second receiver $(t_2 - t_1)$.

Now we observe that, when dealing with down-hole tests or with seismic cone or SDMT tests, the measured velocity of propagation depends on both the shear moduli in the vertical and the horizontal planes (see Section 7.12), i.e.:

$$v_S = \sqrt{\frac{G_{HH} \sin^2 \alpha + G_{VH} \cos^2 \alpha}{\rho}}, \tag{7.38}$$

where the angle α is the direction of propagation with respect to the vertical. Presumed that this direction has a negligible deviation from the vertical, the velocity of propagation can be assumed to depend mainly on G_{VH}.

7.11.2 SASW method

This non-invasive seismic technique represents the most significant recent development in shallow seismic exploration for geotechnical engineering application, pioneered by

Stokoe (see Stokoe *et al.*, 1994). The key points of the SASW (*Spectral Analysis of Surface Waves*) method are the following.

A source at the surface generates Rayleigh surface waves, detected by at least two receivers (vertical velocity transducers with natural frequencies between 1 and 4.5 Hz) located at known distance apart from the source (Figure 7.24). Usually, practical consideration as attenuation and near field effects suggest to have a distance d such that:

$$\frac{\lambda}{2} < d < 3\lambda,$$

where λ is the wavelength.

Surface waves propagating in a non-homogeneous system are *dispersive*, i.e. the phase velocity depends on wavelength (or frequency). **Dispersion** arises from the fact that surface waves of different wavelength sample different depths (Figure 7.41). In fact, if we consider a uniform homogeneous half-space (Figure 7.42) the wave velocity is independent of wave length. On the contrary, if we consider a layer over a half-space (Figure 7.43), at high frequency or short wavelengths the velocity will be equal to the Rayleigh wave velocity of the top layer, whereas at low frequencies or long wavelength the velocity will tend towards the velocity of the half-space. By using surface waves with a range of wavelengths or frequencies and by assembling dispersion curves for all receiver spacings, it is then possible to define a dispersion curve for the tested soil deposit (Figure 7.44).

Recall that the *Fourier integral theorem* allows to express a transient wave as a superposition of steady-state components with a spectrum of frequencies. Therefore, when we consider a transient wave, each component propagates with the phase velocity corresponding to its frequency and the transient wave tends to spread or to disperse,

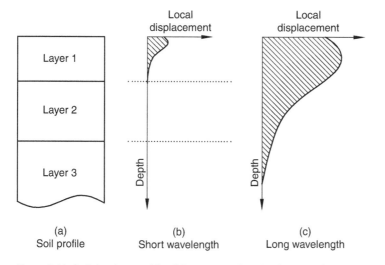

Figure 7.41 Soil depth tested by different wavelengths (adapted from Stokoe *et al.*, 1994).

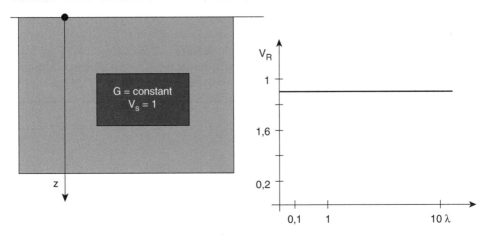

Figure 7.42 Rayleigh wave velocity in uniform homogeneous half-space (adapted from Stokoe *et al.*, 1994).

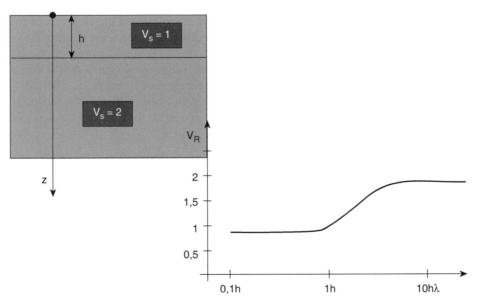

Figure 7.43 Rayleigh wave velocity in a layer overlying a half-space (adapted from Stokoe *et al.*, 1994).

as it propagates. For this reason we use the term *dispersion*, and, as a corollary, we cannot define a **phase velocity** for a transient wave (the term *phase velocity* is used because it refers to a point of constant phase, such as peak or trough). Provided the transient wave is characterized by a narrow spectrum of frequency, it propagates with a velocity called **group velocity** (see Figure 7.45).

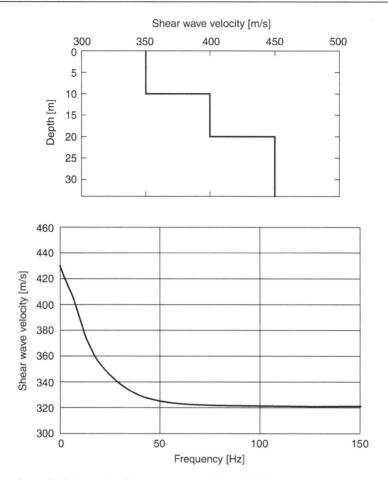

Figure 7.44 Example of dispension curve (Foti, 2000).

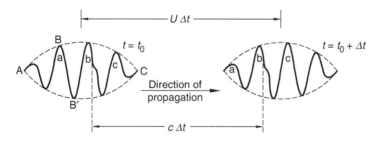

Figure 7.45 Difference between group and phase velocity.

Once the dispersion curve has been obtained, the **inversion process** allows to compute the shear wave velocity profile (Figure 7.46). In particular, *forward modelling* is one method of obtaining this profile. This procedure requires assuming a shear wave velocity profile and then determining the theoretical dispersion curve, which is

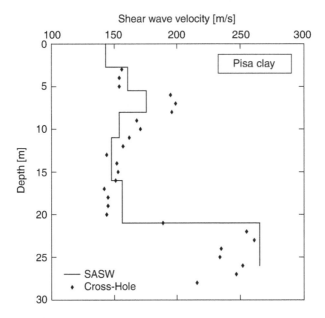

Figure 7.46 Profile of shear wave velocity in Pisa clay, as obtained from seismic tests (Foti, 2003).

compared with the experimental one. Based upon this comparison, the assumed profile is modified until the theoretical and the experimental dispersion curves closely match (see Nazarian, 1984; Sanchez-Salinero, 1987; Rix, 1988; Lai, 1998; Foti, 2000).

7.12 Wave propagation theory

Consider the motion equation (see Section 2.13):

$$\rho \ddot{u}_i = \sigma_{ij,j} + \rho b_i,$$

the costitutive equation:

$$\sigma_{ij} = \lambda \varepsilon_{kk} \delta_{ij} + 2G\varepsilon_{ij},$$

and the small strain tensor in terms of displacement field:

$$\varepsilon_{hk} = \frac{1}{2}\left(u_{h,k} + u_{k,h}\right).$$

By substituting the latter two expressions into the first equation we obtain the motion equation in terms of displacements:

$$\rho \ddot{u}_i = \rho b_i + \left(\lambda + G\right)u_{k,ki} + Gu_{i,kk}. \tag{7.39}$$

In order to highlight the *geometrical aspects* of wave propagation, consider now for sake of simplicity a *steady state one-dimensional wave*, i.e. a waves such that the dependent variable is a harmonic function of time:

$$u = A\cos(kx - \omega t).\tag{7.40}$$

The absolute value of the constant A is called *amplitude*, the constant k is called the *wave number* and ω is the *circular frequency*. The *phase velocity* of the wave is defined by:

$$c = \frac{\omega}{k},\tag{7.41}$$

and the distance required for the solution to undergo one complete oscillation is called *wavelength*:

$$\lambda = \frac{2\pi}{k}.\tag{7.42}$$

The number of oscillations per unit time at a fixed position is called *frequency*:

$$f = \frac{\omega}{2\pi},\tag{7.43}$$

and the time required for the solution to undergo one complete oscillation is called *period*:

$$T = \frac{1}{f} = \frac{2\pi}{\omega}.\tag{7.44}$$

By using the above definitions, the following relation holds between the phase velocity, the frequency and the wavelength:

$$c = f\lambda.\tag{7.45}$$

In the sequel, it is convenient to use, instead of (7.40) the complex representation:

$$u = Ae^{i(kx - \omega t)},\tag{7.46}$$

and it is understood that the real or the immaginary part of (7.46) has to be considered for a physical interpretation of the solution.

Moreover, when dealing with a *plane wave*, the relevant aspect is represented by its direction of propagation, given by the unit vector $\mathbf{n}$ perpendicular to the plane of the wave, whereas the orientation of the coordinate axes is a matter of convenience. In order to define a plane wave in a form independent of the orientation of the coordinate axes, let us assume $\mathbf{x}$ to be the *position vector* at any point in the wave front, so that $x = \mathbf{n} \cdot \mathbf{x}$ and equation (7.46) now writes:

$$u_i = a_i e^{i(\mathbf{k} \cdot \mathbf{x} - \omega t)},\tag{7.47}$$

where we have introduced the vector $\mathbf{k} = k\mathbf{n}$, pointing in the direction of propagation, called *propagation vector* or *wave vector*.

If we substitute (7.47) into (7.39) and we neglect the time independent part, we obtain the equation:

$$\rho c^2 a_i = (\lambda + G) a_k n_k n_i + G a_i. \tag{7.48}$$

This equation can be written in the more significant form:

$$\left(A_{ih} - \rho c^2 \delta_{ih} \right) a_h = 0, \tag{7.49}$$

which highlights that the squared velocities of propagation are the eingenvalues of the *acoustic tensor*:

$$A_{ih} = [\lambda + G] n_i n_h + G \delta_{ih}. \tag{7.50}$$

If the eigenvalue problem (7.49) is rewritten as:

$$\left(\frac{\rho c^2 - G}{\lambda + G} \delta_{ik} - n_i n_k \right) a_k = 0, \tag{7.51}$$

or in its matrix representation:

$$\begin{bmatrix} \dfrac{\rho c^2 - G}{\lambda + G} - n_1^2 & -n_1 n_2 & -n_1 n_3 \\[2mm] -n_2 n_1 & \dfrac{\rho c^2 - G}{\lambda + G} - n_2^2 & -n_2 n_3 \\[2mm] -n_3 n_1 & -n_3 n_2 & \dfrac{\rho c^2 - G}{\lambda + G} - n_3^2 \end{bmatrix} \begin{Bmatrix} a_1 \\ a_2 \\ a_3 \end{Bmatrix} = 0, \tag{7.52}$$

we can formulate the following observations.

Consider a wave propagating in the x_1 direction (Figure 7.47), represented by the unit vector $\mathbf{n}(1,0,0)$. Then, the velocity $c_P = \sqrt{(\lambda + 2G)/\rho}$ is associated to a

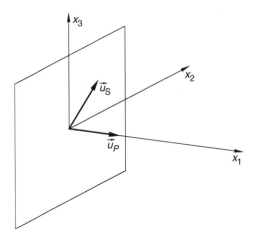

Figure 7.47 Displacement vectors of P and S waves propagating in x_1 direction.

displacement vector of components $a_1 \neq 0$; $a_2 = 0$; $a_3 = 0$, i.e. the displacement vector is collinear to the propagation vector, so that we are concerned with a **compression** or **P wave**.

If on the contrary we consider the solution $c_S = \sqrt{G/\rho}$, the displacement components are $(a_1 = 0;\ a_2 \neq 0;\ a_3 \neq 0)$, i.e. the particle movement is orthogonal to the propagation direction and we are concerned with a **shear** or **S wave**.

We also note that, since the acoustic tensor is symmetric and positive defined, the eigenvalues are real and positive and the eigenvectors, related to distinct eigenvalues, are orthogonal to each other.

All previous consideration can now be extended to the more interesting case of a *cross-anisotropic body* (Figure 7.39). To analyse the geometrical character of wave propagation, let's indicate, according to Love (1944), the non-vanishing components of the stiffness tensor C_{ijhk} as:

$$
\begin{aligned}
&C_{1111} = C_{2222} = A \\
&C_{3333} = C \\
&C_{3311} = C_{3322} = F \\
&C_{2323} = C_{1313} = L \\
&C_{1212} = N \\
&C_{1122} = C_{2211} = A - 2N
\end{aligned}
\qquad (7.53)
$$

To give the above elastic constant a physical meaning, we write the constitutive law in the following form:

$$
\begin{aligned}
&\sigma_{xx} = A\varepsilon_{xx} + (A - 2N)\varepsilon_{yy} + F\varepsilon_{zz} \\
&\sigma_{yy} = (A - 2N)\varepsilon_{xx} + A\varepsilon_{xx} + F\varepsilon_{zz} \\
&\sigma_{zz} = F\varepsilon_{xx} + F\varepsilon_{yy} + C\varepsilon_{zz} \\
&\sigma_{xy} = 2N\varepsilon_{xy};\ \sigma_{yz} = 2L\varepsilon_{yz};\ \sigma_{xz} = 2L\varepsilon_{xz}
\end{aligned}
\qquad (7.54)
$$

so that it appears that N represents the shear modulus in the horizontal plane, i.e. G_{HH}, and L is the shear modulus in the vertical plane, i.e. G_{VH}.

In addition, we also observe that, by substituting equation (7.47), as well as the constitutive equation into the equation of the motion, one gets:

$$
(C_{iklm}n_k n_l - \rho c^2 \delta_{im})a_m = 0. \qquad (7.55)
$$

Let us now consider the case where the normal to the plane wavefront (i.e. the direction of propagation) belongs to the vertical plane (x_1, x_3), i.e. $n_2 = 0$, the x_3 direction assumed to be the vertical one (Figure 7.39). In addition, suppose that the particle motion (direction of polarization) is given by the vector $\mathbf{a}(0, 1, 0)$. Then the system (7.55) reduces to:

$$
(C_{2112}n_1 n_1 + C_{2332}n_3 n_3 - \rho c^2)a_2 = 0. \qquad (7.56)
$$

If α is the angle between the direction of propagation and the vertical one, $n_1 = \sin\alpha$ and $n_3 = \cos\alpha$, so that, by accounting for (7.53) and (7.54), the wavefront propagates with a velocity equal to:

$$c = \sqrt{\frac{G_{HH}\sin^2\alpha + G_{VH}\cos^2\alpha}{\rho}}. \tag{7.57}$$

In particular, if the direction of propagation is coincident with the x_1 axis, then (7.57) gives $c = \sqrt{G_{HH}/\rho}$, a result which applies to cross-hole tests, when the induced shear waves is polarized in the horizontal plane. On the contrary, if the direction of propagation is coincident with the x_3 axis, then $c = \sqrt{G_{VH}/\rho}$.

Now we observe that, when dealing with down-hole tests or with SDMT tests, the measured velocity of propagation is the one given by equation (7.57), i.e. it depends on both shear moduli in the vertical and the horizontal planes, and only if the direction of propagation has a negligible deviation from the vertical, the velocity of propagation can be assumed to depend mainly on G_{VH}.

Interestingly, referring to (7.57), Foti *et al.* (2006) have suggested to perform SDMT tests at conveniently different distances between the source and the receiver, in order to change the direction of propagation α and therefore to obtain values of both G_{VH} and G_{HH}.

However, a relevant aspect to be outlined is that the direction of polarization must be coincident with the x_2 axis. If this is not the case, by using similar arguments, it can be proved that the velocity of propagation is a rather complicated function of four elastic constants, so that it is not easy to relate the measured wave velocity to soil parameters.

In fact, presumed both the directions of propagation and polarization belong to the plane (x_1, x_3), the system (7.55) writes:

$$\begin{bmatrix} A_{11} - \rho c^2 & A_{13} \\ A_{31} & A_{33} - \rho c^2 \end{bmatrix} \begin{Bmatrix} a_1 \\ a_3 \end{Bmatrix} = 0, \tag{7.58}$$

where (considering the non-vanishing terms):

$$\begin{aligned} A_{11} &= C_{1111}n_1^2 + C_{1331}n_3^2 = An_1^2 + Ln_3^2 \\ A_{33} &= C_{3113}n_1^2 + C_{3333}n_3^2 = Ln_1^2 + Cn_3^2 \\ A_{13} &= A_{31} = C_{1133}n_1n_3 + C_{1313}n_1n_3 = (F+L)n_1n_3 \end{aligned} \tag{7.59}$$

It follows that the speeds of propagation are:

$$\begin{aligned} c_1 \\ c_2 \end{aligned} = \left[(a \pm b)/2\rho \right]^{1/2}, \tag{7.60}$$

with:

$$a = A\sin^2\alpha + C\cos^2\alpha + L$$

$$b = \left\{ \left[(A - L)\sin^2\alpha - (C - L)\cos^2\alpha \right]^2 + 4(F + L)^2 \sin^2\alpha \cos^2\alpha \right\}^{1/2}. \qquad (7.61)$$

It is now proved that, if the direction of propagation and that of polarization do not coincide with a principal direction, displacements are neither truly longitudinal nor truly transverse, so that the motion cannot be decoupled anymore into P and S waves.

7.13 Soil porosity from wave propagation

Porosity is a fundamental parameter for the description of the natural state of a soil deposit and experimental evidences also prove that porosity affects the mechanical, hydraulic and electrical behaviour of soils.

In fine-grained soils, porosity is currently inferred from laboratory measurements on undisturbed samples, but in coarse-grained soils undisturbed samples require sophisticated freezing technique, which are far from being currently used. Therefore, there is a need to search for alternative procedures, and a promising one derives the porosity from measurements of seismic wave velocities, as it is shown in the sequel using concepts of poroelasticity (Biot, 1956).

Consider again the motion equation:

$$\rho \ddot{u}_i = \sigma_{ij,j} + \rho b_i, \qquad (7.62)$$

and assume all quantities as incremental with respect to an initial equilibrium configuration, so that we can neglect body forces.

When we take into consideration a saturated porous medium, the constitutive equation is expressed in terms of effective stresses, i.e.:

$$\sigma_{ij} - u\delta_{ij} = \lambda\varepsilon_{kk}\delta_{ij} + 2G\varepsilon_{ij}, \qquad (7.63)$$

and considering that the small strain tensor can be expressed in terms of displacement field:

$$\varepsilon_{hk} = \frac{1}{2}\left(u_{h,k} + u_{k,h}\right) \qquad (7.64)$$

the term $\sigma_{ij,j}$ writes:

$$\sigma_{ij,j} = \left(\lambda + G\right)u_{k,ki} + Gu_{i,jj} + u_{,j}\delta_{ij}. \qquad (7.65)$$

Taking into account the storage equation (see Chapter 6):

$$n\beta\frac{\partial u}{\partial t} + \nabla \cdot \mathbf{v} = \frac{\partial\varepsilon_v}{\partial t}, \qquad (7.66)$$

if we introduce the *assumption that there is no relative motion* between the water and the solid phase, i.e. $\nabla \cdot \mathbf{v} = 0$, it follows that:

$$n\beta \delta u = \delta \varepsilon_v , \tag{7.67}$$

and this allows to give the motion equation the following form:

$$\rho \ddot{u}_i = \left(\lambda + G + \frac{1}{n\beta} \right) u_{k,ki} + G u_{i,jj} . \tag{7.68}$$

This result, first proved by Biot (1956), tells us that the equation of the motion is similar to that of an elastic material provided that λ is replaced by $\lambda + 1/n\beta$ and, because there is no relative motion between water and solid matrix, the shear wave propagates at a speed equal to:

$$v_S = \sqrt{\frac{G}{\rho}} \tag{7.69}$$

and the compression wave propagates at a speed equal to:

$$v_P = \sqrt{\frac{\lambda + 2G + \frac{K^w}{n}}{\rho_s(1-n) + \rho_w n}} , \tag{7.70}$$

In equation (7.70) $K^w = 1/\beta$ is the bulk modulus of the fluid phase and ρ_s and ρ_w are the mass density of the soil particles and pore fluid.

If now we recall that the ratio $(\lambda + 2G)/G$ is a function of Poisson's ratio:

$$\frac{\lambda + 2G}{G} = f(v) = \frac{2(1-v)}{1-2v} , \tag{7.71}$$

by combining (7.69) and (7.70), i.e.:

$$\rho v_P^2 = \rho v_S^2 f(v) + \frac{K^w}{n}$$

$$\left[n(1-n)\rho_s + n^2 \rho_w \right] (v_P^2 - v_S^2 f(v)) - K^w = 0 ,$$

it is possible to obtain the following equation, which allows to estimate the soil porosity from measurements of shear and compression wave velocities in a saturated medium (Lancellotta, 2001; Foti *et al.*, 2002; Foti and Lancellotta, 2004):

$$n = \frac{\rho_s - \sqrt{(\rho_s)^2 - 4(\rho_s - \rho_w)\frac{K^w}{v_P^2 - f(v)v_S^2}}}{2(\rho_s - \rho_w)} . \tag{7.72}$$

Note that the use of (7.72) implies an assumption about the value of Poisson ratio, which must be arbitrarily introduced independently from the measurements of wave velocities. This can be accepted because numerical simulations prove that the solution (7.72) is fairly insensitive to this parameter.

Furthermore, as for the assumption of no relative motion between the two phases, this is reasonable provided that the saturated porous medium is tested at low frequency, since in this case the porous medium behaves as a closed system, i.e. in undrained conditions. In this respect, the following value of the frequency can be considered an upper bound for this assumption being retained (Miura *et al.*, 2001):

$$f_c = \frac{\mu n}{2\pi k \rho_w} = \frac{\rho_w g n}{2\pi K \rho_w} = \frac{g n}{2\pi K}, \tag{7.73}$$

where μ, k, K are the fluid viscosity, the permeability of the porous medium and its hydraulic conductivity.

7.14 Hydraulic conductivity

Most problems in geotechnical engineering have to deal with the recognition and evaluation of some aspects of drainage. In this respect, it is convenient to retain the distinction between the coarse and fine-grained soils, based on the hydraulic conductivity. No other parameter has such wide range, which may be explained by various factors, the most important being the macro and micro-structural features. In particular, the dominating influence of macrofabric features of the deposits leads to large differences between the values of flow and consolidation properties as derived from laboratory tests and those derived from *in situ* tests or back-analysis of full-scale structures. For these reasons, while the theory of water flow in porous media is founded on a robust theoretical basis, a reasonable determination of the *in situ* conductivity remains, in practice, difficult.

A further consequence is that the most reliable way of assessing the mass conductivity of coarse-grained soils ($K > 10^{-6}$ *m/s*) is to properly perform large-scale pumping tests. Borehole outflow and inflow tests are also used in these soils, but they yield, at best, only values representative of an order of magnitude of the *in situ* conductivity.

When assessing the flow and consolidation properties of fine-grained soils ($K < 10^{-6}$ *m/s*), by means of outflow or inflow tests performed with piezometers, it is essential to pay particular attention to the range of effective stress state within which each specific method is operating. The same applies to the self-boring devices, that offer the advantage of minimizing soil disturbance, as well as to the interpretation of piezocone dissipation tests.

7.14.1 Pumping tests

Pumping tests are performed to assess the characteristics of an aquifer. They consist of withdrawing the water, usually at constant rate, from one well and observing the water decline in one or more observation wells or piezometers. The water decline with time is indicative of the hydraulic properties of the aquifer.

Since the general set-up of the test depends on the soil profile and hydrological conditions, a sufficient number of boreholes should be executed before performing

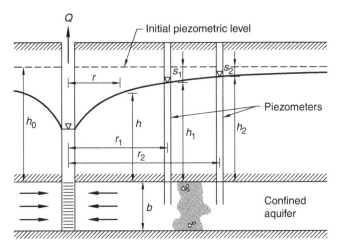

Figure 7.48 Radial flow in a confined aquifer.

a test in order to provide data on the main geological features of the aquifer, its lateral extent and the presence of recharge boundaries or barrier boundaries.

The methods of constructing pumping wells are described in detail by Todd (1980), so that only the main points concerning the length of the screen, the open area and the characteristics of the filter pack and the set-up of the piezometers or observation wells are here briefly recalled.

(a) When penetrating a **confined aquifer** (Figure 7.48), the well should be completed to its bottom, with the advantage that water flow can be assumed to be horizontal. Since flow lines are horizontal and the equipotential lines are vertical, the piezometers can be located even at short distances from the well and at any depth. From a practical point of view, a *screen* equal to 70% of the aquifer thickness can be considered to be equivalent to a fully penetrating well.

However, in many cases the confined aquifer is very thick and the well is necessarily only partially penetrating. The drawdown is affected in this case by the vertical flow component, and rather complex procedures must be applied to use the data from piezometers readings near the well. These difficulties can be avoided if the piezometers are located at distances greater than:

$$r \geq 1.5b\sqrt{\frac{K_r}{K_z}}, \tag{7.74}$$

because at such a distance the flow can be assumed to be horizontal. In (7.74) b is the thickness of the aquifer and K_r and K_z are the hydraulic conductivity for the radial and vertical flow respectively.

(b) In **unconfined aquifer** (Figure 7.49), it is sufficient to screen the lower half or third of the aquifer to obtain the maximum drawdown. But apart from any theoretical reason, when dealing with very thick aquifers the relative length of the screen is

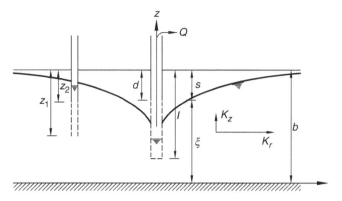

Figure 7.49 Pumping test in unconfined aquifer (Neuman, 1974).

reduced for economical reason. Therefore, if the length of the well is less than 70% of the aquifer thickness, the well must be considered partially penetrating, and the vertical flow component cannot be neglected when interpreting the experimental data. In practice, as suggested by Neumann (1974), the interpretation can be done by assuming the drawdown pattern is that of a fully penetrating well, provided that the piezometers are located at distances that satisfy the inequality (7.74) and the reading times exceed the value:

$$t = 10\frac{S_y r^2}{T}, \tag{7.75}$$

where S_y and T are the *specific yield* and the *transmissivity* (both these parameters are defined in the sequel).

In order to verify whether the effect of partial penetration has really dissipated at the selected radial distance, it is suggested installing two piezometers at the same radial distance but at substantial different depths. When the drawdown curves tend to merge, this is an indication that the vertical component of flow no longer plays a significant role and the conventional approach can be used.

Considering that most of the alluvial aquifers have a pronunced anisotropy, the pumping well can be assumed to be equivalent to a fully penetrating well in a saturated thickness equal to its length. In this way, a piezometer or an observational well that is fully screened over its depth can be used at shorter distances.

(c) As far as the characteristics of the well are concerned, the total open area of the screen should be such that the velocity of the entering water is low enough to minimize the friction losses. For practical purposes, a minimum open area of 15% is suggested. In addition, an artificial gravel filter should protect the screen and should be designed in order to fulfil the filter criteria discussed in Chapter 6.

(d) We now move to the interpretation of a *pumping test in confined aquifers*. This is a transient flow problem, the solution of which requires one to consider the conservation of mass, the state relation of the pore water, the equilibrium of the porous medium, Darcy's law and the constitutive equation of the solid phase.

If we introduce the definition of *specific storage* S_s as *the volume of water that a unit volume of aquifer releases under a unit decline of hydraulic head*, it can be proved that the following relation holds:

$$S_s = \gamma_w \left(n\beta + \frac{1}{K_{SK}} \right),$$
(7.76)

where K_{SK} is the bulk modulus of the solid phase.

If the aquifer thickness is equal to b, and we assume that the well fully penetrates the aquifer, the pumping rate Q is constant with time and the storage within the well can be neglected if compared to the volume of the aquifer, the initial-boundary problem can be formulated as follows:

$$\frac{\partial^2 h}{\partial r^2} + \frac{1}{r}\frac{\partial h}{\partial r} = \frac{S}{T}\frac{\partial h}{\partial t}$$

$$h(r, 0) = h_o, \; \forall r$$

$$h(\infty, t) = h_o, \; \forall t$$
(7.77)

$$\lim_{r \to o} \left(r\frac{\partial h}{\partial r} \right) = \frac{Q}{2\pi T}, \; if \; t > 0$$

The quantities:

$$S = S_s b$$

$$T = Kb$$
(7.78)

are defined as aquifer *storativity* (or *storage coefficient*) and *transmissivity*.

Having recognized the analogy between the heat conduction and groundwater flow, Theis (1935) first obtained a solution for the problem (7.77):

$$s = \frac{Q}{4\pi T} \int_u^\infty \frac{e^{-u}}{u} du,$$
(7.79)

where s is the *drawdown* (Figure 7.48) and the dimensionless variable u (which is not to be confused with the pore pressure) is defined as:

$$u = \frac{r^2 S}{4Tt}.$$
(7.80)

The integral in equation (7.79) can be expanded into a convergent series such as:

$$W(u) = \int_u^\infty \frac{e^{-u}}{u} du = -0.5772 - \ln u + u - \frac{u^2}{2 \times 2!} + \frac{u^3}{3 \times 3!} + \cdots.$$
(7.81)

Figure 7.50 shows the value of this integral, indicated as the *well function W(u)*, as a function of $1/u$, and the obtained curve is commonly called the *Theis curve*.

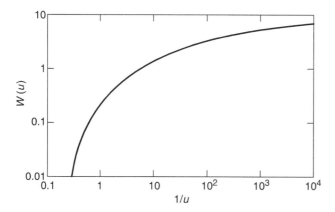

Figure 7.50 Theis curve.

However, the use of equations (7.79) and (7.81) is not so straightforward when the aim is to obtain the aquifer storativity and transmissivity from measurements of the drawdown *s* at the time *t* and at distance *r*. For this reason approximate procedures have also been developed, as discussed by Todd (1980). The approach suggested by Cooper and Jacob (1946) is based on the recognition than when *t* is large or *r* is small, so that the $u \leq 0.01$, the solution can be simplified to:

$$s = \frac{Q}{4\pi T}(-0.5772 - \ln u),$$ (7.82)

and by rearranging into decimal logarithms:

$$s = \frac{2.3Q}{4\pi T} \log\left(\frac{2.25Tt}{Sr^2}\right).$$ (7.83)

This suggests that by plotting the drawdown against the time on a semilogarithmic scale (Figure 7.51), a straight line is obtained, and the change Δs for a tenfold increase of time and the intersection t_o with the horizontal axis $s = 0$ are determined. Then, the transmissivity and the storativity are given by:

$$T = \frac{2.3Q}{4\pi \Delta s}$$ (7.84)

$$S = \frac{2.25Tt_o}{r^2}.$$ (7.85)

It is also of interest to observe that, when the Cooper and Jacob approximation holds, by considering two piezometers at distances r_1 and r_2, equation (7.83) can be rewritten as:

$$s(r_1) - s(r_2) = \frac{Q}{4\pi T} \ln\left(\frac{r_2}{r_1}\right)^2 = \frac{Q}{2\pi T} \ln\frac{r_2}{r_1}.$$ (7.86)

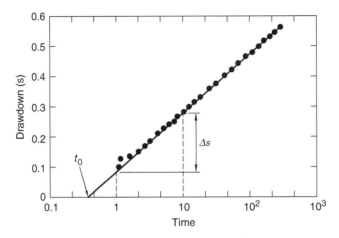

Figure 7.51 Interpreting a pumping test by the approximate method of Cooper and Jacob (1946).

This equation shows, as suggested by Neuman (1988), that the relative drawdowns change logarithmically with the radial distance in a time independent fashion. This means that the drawdown cone around the well is in a *pseudo-steady state condition* and equation (7.86) is analogous to the well-known equation for stationary conditions.

(e) When dealing with *unconfined aquifers* (Figure 7.49), the water produced by a well arises from three mechanisms: expansion of water due to reduction of water pressure, compaction of the aquifer due to the increased effective stresses (both these mechanisms are common to confined aquifers) and actual dewatering of the aquifer, i.e. the drawdown of the water table.

A complete analysis of this problem should recognize that the unconfined flow involves the saturated-unsaturated system to be taken into account. However, many studies have proved (see Freeze and Cherry, 1979) that the position of the water table is essentially unaffected by the unsaturated flow above the water table, so that this complete analysis is seldom used in practice.

The most widely used approach was suggested by Neuman (1972, 1974, 1975), and is based on *the concept of delayed water table response*. Referring to Figure 7.52, there are three phases that can be recognized in the pattern of a drawdown curve. During the first phase, which covers only a short period of time, the reaction of the unconfined aquifer is similar to that of a confined aquifer, in that water is released from expansion of water and compaction of the aquifer. Then there is a decrease in the slope of the drawdown versus time curve, because of gravity drainage effects. Finally, during the third phase, the drawdown curve tends to conform again to the Theis curve.

Note that the storage term of an unconfined aquifer is known as *specific yield* S_y and represents *the volume of water that an unconfined aquifer releases from storage per unit surface area per unit decline of water table*.

Accordingly, the specific yield of an unconfined aquifer, usually in the range of 0.01 to 0.3, is much higher than the storativity of a confined aquifer, ranging from 5.10^{-5} to 5.10^{-3}. This is because the water release in unconfined aquifers is produced

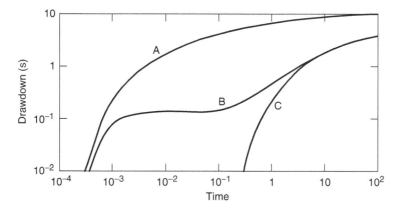

Figure 7.52 Delayed response of unconfined aquifer.

by an actual dewatering of soil pores, whereas water release in confined aquifers is the effect of water expansion and aquifer compaction produced by changes in water pressure.

The approach suggested by Neuman (1972) considers the unconfined aquifer as a compressible medium and the phreatic surface as a moving material boundary. The general solution depends on many parameters, so that it cannot be singled out on only one plot. However, if the pumping well and the observation well fully penetrate the saturated thickness of the aquifer, the general solution only depends on r if an average drawdown is considered. Therefore, the aquifer parameters can be derived by following the same procedure already seen for the confined aquifer, by observing that the first segment A of the response curve in Figure 7.52 will provide the value of the storativity S and the last segment C the value of the specific yield S_y.

When considering radial distances that satisfy the inequality (7.74) and observation intervals greater than (7.75), the vertical flow component can be neglected, so that the Dupuit assumption holds. In this circumstance, the drawdown cone around the well is in a *pseudo steady-state condition*, and the equation for stationary conditions applies (Neuman, 1988), so that:

$$h_2^2 - h_1^2 = \frac{Q}{\pi K} \ln \frac{r_2}{r_1}. \tag{7.87}$$

7.14.2 Permeability tests using piezometers

Permeability tests using piezometers are usually performed in low conductivity soils, and, since the test produces a seepage induced consolidation phenomenon, one needs to consider the transient character of the flow for a proper interpretation. Basic contributions have been given by Gibson (1963) and Wilkinson (1968) and are here summarized.

(a) *Falling head tests* are difficult to interpret, as the **soil** surrounding the piezometer is subjected to *non-monotonic* effective stress changes. In particular, if the water pressure in the piezometer is lowered below the ambient pressure, the clay surrounding the piezometer is initially subjected to consolidation and then to swelling. Since the compression index differs significantly from the swelling index, the theoretical solution, based on a single coefficient of consolidation, is only an approximation. This theoretical assumption is better satisfied if the water pressure is raised above the ambient pore pressure, since, in this case, a swelling process is followed by reconsolidation, with a consolidation coefficient that can be considered representative for both these phases. However, even in this case, the solution is far from being simple.

(b) Due to the above difficulties, *constant head tests* and in particular *outflow tests*, where the flow is directed from the piezometer towards the soil, are preferred. The interpretation is based on Gibson's (1963) solution for a spherical piezometer, extended by Wilkinson (1968) to the case of a cylindrical piezometer. By referring to the boundary conditions shown in Figure 7.53 and considering the high degree of anisotropy of natural clays, the basic solution reduces to that which corresponds to pure radial flow (Wilkinson, 1968)

$$Q = 2\pi b K_h \frac{\Delta u}{\gamma_w} \left(1 + \frac{2}{\pi \sqrt{T}} \right), \tag{7.88}$$

where:

$$T = \frac{c_h t}{a^2}. \tag{7.89}$$

If the rate of flow $Q(t)$ is plotted against $1/\sqrt{t}$, the experimental data fall on a straight line. The intercept of this line with the Q axis, denoted as Q_∞, corresponds to the flow rate at a stationary condition (attained as $t \to \infty$), so that

$$K_h = \frac{Q_\infty}{2\pi b \frac{\Delta u}{\gamma_w}}, \tag{7.90}$$

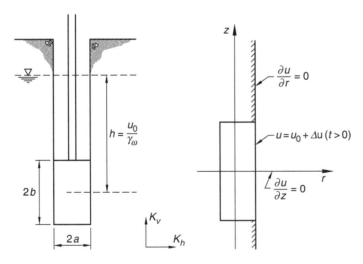

Figure 7.53 Boundary conditions for a piezometer test.

whereas the consolidation coefficient is related to the slope of the experimental line:

$$\beta = \frac{2Q_\infty a}{\pi \sqrt{c_h}}.$$ (7.91)

Note that the coefficients of consolidation and hydraulic conductivity obtained from outflow tests are representative of unloading-reloading conditions, so that they are appropriate for field problems where $(\sigma'_{vo} + \Delta\sigma'_{vo}) \leq \sigma'_p$.

Disturbance and remoulding effects (*smear*), due to piezometer installation, cause changes in the surrounding soil mass and the reliability of the measured parameters depends on the severity of these changes and the sensitivity of the soil.

Hydraulic fracture may also occur if an excessive increase of pore pressure is applied during the test.

7.14.3 Self-boring devices

The need to overcome effects due to soil disturbance and smear, related to the installation of piezometers and penetration of piezocones, renders the self-boring devices very attractive.

(a) The **self-boring permeameter** (Figure 7.54) (Jezequel and Mieussens, 1975) has two distinct parts: the filtering part is a porous cylinder, placed behind the cutting edge; the ward cells, placed on either side of the filtering section, consist of a rubber membrane that can be inflated by water or gas pressure. During the permeability test the cells are pushed into contact with the soil, to avoid leakage

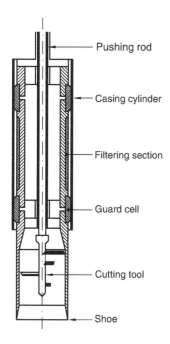

Pushing rod

Casing cylinder

Filtering section

Guard cell

Cutting tool

Shoe

Figure 7.54 Self-boring permeameter.

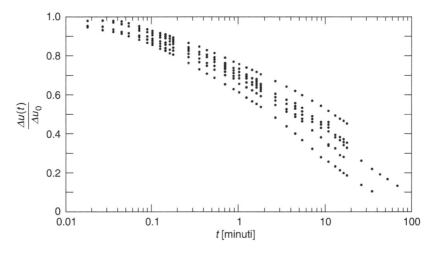

Figure 7.55 Examples of dissipation tests in Porto Tolle clay (Battaglio *et al.*, 1986).

of water between the permeameter and the soil. In this way, the boundary conditions (Figure 7.53) already seen for piezometers are realized, and all that has been stated concerning the piezometers also applies for the interpretation of this test.

(b) The **self-boring pressuremeter** offers the possibility of measuring the consolidation coefficient c_h by performing either a *strain holding test* (Clarke *et al.*, 1979) or a *pressure holding test* (Fahey and Carter, 1986). The basic aspects of these tests can be summarized as follows. If a pressuremeter cell is expanded in clay in undrained conditions, excess pore pressures are induced in the surrounding soil. If the diameter of the cell is held constant, the decrease of the excess pore pressure and the total radial stress can be measured, and this is called a *holding test*. Another possibility is to hold the total radial stress constant (pressure holding test), and the decrease of excess pore pressure and the increase in cavity diameter can be measured. The expansion of the pressuremeter is usually modelled as a cylindrical cavity expansion in a Tresca soil, and by assuming that the soil behaves elastically during the subsequent consolidation process, a closed form solution has been derived by Randolph and Wroth (1979). Futher aspects of this tests are discussed by Fioravante *et al.* (1994) and Yu (2004).

7.14.4 Piezocone dissipation tests

In recent years a lot of attention has been given to dissipation tests in clays, performed with the piezocone. The test consists of stopping the penetration of the cone and monitoring the decay of excess pore pressure with time (Figure 7.55). A simplified approach for the interpretation of dissipation records was first proposed by Torstensson (1975). The soil was assumed to be an elastic-perfectly plastic material subjected to isotropic initial stress. The initial excess pore pressure was estimated using the the cavity expansion theory (Vesic, 1972), and the consolidation process was studied using the linear uncoupled one dimensional theory.

At present, the most comprehensive investigation of this problem has been carried out by Baligh and Levadoux (1980), based on the strain-path method (Baligh, 1985), and the relevant conclusions of this work are the following.

(a) The effect of coupling between the total stresses and the pore pressure is small, except at the early stage of consolidation, so that the *uncoupled solution* provides a reasonably accurate prediction of the dissipation process.

(b) A two-dimensional analysis around the cone shows that the dissipation rate is mainly controlled by c_h and that even a tenfold change of c_v has a negligible influence on the shape of isochrones. Hence the test yields c_h values.

(c) The cone penetration process produces undrained shearing of the soil with excess pore pressure. When these excess pore pressures start to dissipate, the soil surrounding the cone is subjected to an increase of effective stresses under reloading conditions, and only after some dissipation has been taken place do the effective stresses equal those existing before the cone penetration. Therefore, the c_h obtained from early stage of consolidation (less than 50% of consolidation) is relevant for reloading conditions.

Based on this conclusion, Baligh and Levadoux (1980) suggest to compute the consolidation coefficient from:

$$c_h = \frac{TR^2}{t},$$ (7.92)

where R is the radius of the pushing rod and T is the time factor, depending on the cone geometry and the location of the filter stone (Figure 7.56). The values of c_h for a dissipation of 50% of the excess pore pressure are representative of horizontal flow in the OC range. In order to obtain the consolidation coefficient in the NC range the following rule is suggested:

$$c_h(NC) = c_h(OC)\frac{C_r}{C_c}.$$ (7.93)

Besides all the above examined theoretical aspects, the successful interpretation of dissipation records is controlled by the requirements of a rigid and well de-aired pore pressure measuring system.

7.15 Summary

Field investigation is aimed at assessing enough information to select the most appropriate foundation solution, to highlight problems that could arise during construction, and, more in general, to highlight potential geological hazards in the examined area.

The importance of field investigation should always be firmly in mind to authorities charged to make official decisions, by considering that most failures have been caused by lack of field investigation or undetected essential features, as well as that an unsatisfactory site investigation is a matter of litigation.

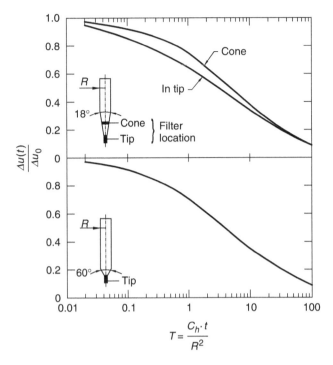

Figure 7.56 Pore pressure decay versus time factor, as predicted by Baligh and Levadoux (1980).

The best result is obtained if the programme is developed in stages and it is adapted during the investigation, when preliminary data start to be available.

A starting point is represented by desk studies, i.e. the collecting of all information sources such as geological maps and relative sections, airphotographs and reports on previously done investigations.

Preliminary investigations are aimed at confirming or reviewing the findings of desk studies, through trial pits, some preliminary borings, exploratory geophysical seismic tests and investigations aimed at defining groundwater conditions. The final step consists of a detailed investigation, including borings, sampling, *in situ* and laboratory tests.

Any carefully done investigation programme should be aimed at answering questions linked to the environment of sedimentation, synsedimentary and post-sedimentation events, including tectonic events, uplift, erosion and weathering. This outlines the multidisciplinary nature of the approach, requiring cooperation between geologists and geotechnical engineers, and there is no need to emphasize that answering the above-mentioned questions is the most fascinating aspect of field investigation.

In recent years, the capability of the existing techniques has been increased due to the introduction of new devices and to the development of more appropriate theoretical approaches.

Examples of new testing methods are related to techniques introduced in the area of environmental geotechnics, mainly aimed at identifying contamination of polluted sites.

Recent advances have also been made in the area of geophysical testing, as represented by shallow geophysical testing (SASW method is one of the examples) as well as by tomography and other imaging techniques. These advances increase the already recognized advantages of *in situ* testing:

1   *In situ* tests can be performed in soil deposits in which undisturbed sampling is impossible or unreliable.
2   They can be performed in difficult to operate conditions, such as continental shelf and sea floor.
3   Many tests produce a continuous record of the soil profile, which allows one to detect boundaries and macrofabric of different layers.
4   The volume of soil tested is larger than the volume of samples usually tested in laboratory.

Note that the interpretation of an *in situ* test requires the solution of a complex boundary value problem. The result of a test depends on the combination of many factors, i.e. at least strength, stiffness and *in situ* stresses, and, therefore, it is important to recognize the simplifying assumptions introduced by current interpretation approaches.

7.16 Further reading

Further reading on subsurface exploration includes the following:

M.J. Hvorslev (1949) *Subsurface Exploration and Sampling of Soils for Civil Engineering Purposes.* Waterways Experiment Stations.
A.J. Weltman and J.M. Head (1983) *Site Investigation Manual*, CIRIA, London.
British Standards BS 5930. Code of practice for Site Investigation.

Innovation on field testing is the subject of the second part of Theme Lecture of M. Jamiolkowski, C.C. Ladd, J.T. Germaine and R. Lancellotta (1985): *New Developments in Field and Laboratory Testing of Soils*, XI ICSMFE, San Francisco, 1, 57–153.

Key aspects of field tests interpretation are discussed by H.S. Yu (2004) *In situ Soil Testing: From Mechanics to Interpretation*, J.K. Mitchell Lecture, Proc. ISC-2 on Geotechnical and Geophysical Site Characterization, Millpress, Rotterdam.

Finally, the following are some relevant ASTM Standards for soil investigation:

ASTM D420. *Guide to Site Characterization for Engineering Design and Construction Purpose.*
ASTM D1587-00. *Standard Practice for Thin-Walled Tube Sampling of Soils for Geotechnical Purposes.*
ASTM D3441. *Cone Penetration Test.*
ASTM D5778. *Electrical Cone Penetration Test.*
ASTM D6066-96. *Standard Practice for Determining the Normalized Penetration Resistance of Sands for Evaluation of Liquefaction Potential.*
ASTM D2573. *Field Vane Test.*
ASTM D4719-87. *Standard Method for Pressuremeter Testing in Soils.*

Chapter 8

The collapse of soil structures

In recent years the design in geotechnical engineering has come to be considered in terms of *limit states*. The introduction of this concept can be so viewed as to remind the designers that a soil structure must satisfy a certain number of requirements. Some of these requirements must be fulfilled not only during the life of the structure, but also during its construction, and when a structure or parts of the structure fail to satisfy one or more of the above requirements, one can say that a *limit state* has been attained. In a rather wide perspective, the three major requirements are those of *strength*, *stiffness* and *stability*, that means that the structure must safely carry the imposed loads, it must not deflect reaching critical displacements and it must not develop unstable displacements.

In this chapter we concentrate on the *ultimate limit state*, at which a collapse mechanism takes place in the ground or in some parts of the structure due to interaction with the ground, while the *serviceability limit state*, at which deformations in the ground produce a loss of serviceability of the interacting structure, will be addressed in Chapter 9.

8.1 Theorems of limit analysis

The analysis of both serviceability and ultimate limit states is far from being simple, because of the non-linear, anisotropic, time-dependent soil behaviour, so that the need arises to introduce simplified assumptions as far as the constitutive laws are concerned.

In particular, in this chapter it will be shown how we can obtain solutions of engineering interest in a rather simple way, if the assumption is made that conditions at the instant of collapse can be analysed independently of the previous loading history, so that soil behaviour, at its ultimate state, can be thought as a *rigid perfectly plastic material* (see Figure 8.1a).

With such an assumption, we abandon the quest for the actual state of a soil structure and, instead, we concentrate on the way the structure might collapse. Obviously, it is not envisaged that the structure will actually collapse; rather the analysis is based on loads increased by a hypothetical factor, so that if the structure is loaded by smaller working loads it will not collapse.

Plasticity theorems help us to define bounds of collapse load, by using two approaches, conveniently referred to as the *static approach* and the *kinematic approach*.

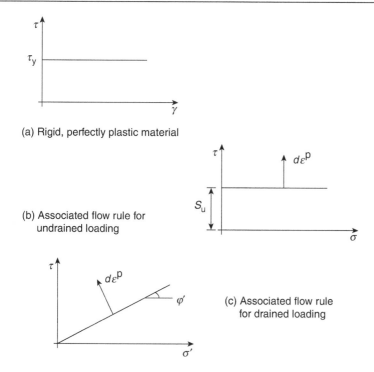

Figure 8.1 Yield criterion and associated flow rule.

The **lower bound theorem** (static approach) states that *if we find a stress field such that equilibrium equations and traction boundary conditions are satisfied and the stresses do not exceed the yield condition, the collapse cannot occur.*

Therefore, the stresses applied at the boundary are a lower bound of the collapse stresses and the stress field that satisfies these requirements is called a *stress admissible field*.

The **upper bound theorem** (kinematic approach) states that *if for any compatible plastic mechanism (i.e. a mechanism for which no gaps, separations or overlaps occur) the rate of work of external loads equals the rate of internal dissipation then the collapse must occur.*

The kinematic approach has the advantage of being intuitively simpler, but the predicted collapse load is always an *upper bound* of the exact solution (if not the correct one), so that it gives an *unsafe estimate* of the collapse load.

On the contrary, the collapse load predicted by the static approach, which is more difficult to master, is always a *lower bound* of the exact solution (if it is not equal to it), so that it is a *safe estimate* of the true collapse load.

It follows that the answers provided by the two approaches are different from each other, but usually the upper and the lower bounds are fairly close to one another, so that the true collapse load is then closely bracketed.

Proofs of these theorems, as well as of other auxiliary theorems, are beyond the scope of this book and therefore are not given, but may be found in the book of Chen (1975) or in the suggested reading.

However, it is here relevant to recall that the following conditions are required to prove these theorems:

1 The yield surface must be convex (recall that a closed surface is defined convex if through every point on it there is a plane such that all points interior to the surface lie on the same side of the plane).
2 The yield function plays both the role of specifying when plastic flow occurs and the role of plastic potential (i.e. it identifies the direction of the plastic incremental vector), so that we are in the presence of an *associated flow rule* (Figures 8.1b and 1c).

On the other hand, it is well recognized that the associated flow rule leads to unrealistic predictive behaviour when applied to frictional materials. In particular, the use of an associated flow rule with Coulomb's yield criterion leads to prediction of dilational rates which are much greater than those observed in experiments. For this reason, many efforts have been devoted to investigate the circumstances under which the bounding approaches can be applied to materials which do not obey associated flow rules. In this context it is of interest to recall the following two theorems of Radenkovic (1961, 1962), related to the assumption of a *non-associated flow rule* (we make reference to the book of J. Salençon: *Théorie de la plasticité pour les applications à la Mécanique des Sols*, Eyrolles, 1974, pages 139–140 and 161–163):

(a) If a static solution is obtained for a medium obeying an associate flow rule, this solution is still a lower bound with respect to the solution obtained considering a non associate flow rule;
(b) If a kinematic solution is obtained with the assumption of an associate flow rule, this solution is still an upper bound with respect to the solution obtained considering a non associate flow rule.

Further remarks are concerned with the mechanism of collapse. In structural engineering, when in a section of a frame the bending moment reaches the yield value, we say that in this section a plastic hinge is forming. In a statically undeterminate structure, any plastic hinge progressively reduces the degree of undeterminacy, until the frame doesn't exhibit enough constraints and merges into a mechanism.

In a continuum medium, forming a mechanism requires the introduction of a number of planes or surfaces of slipping, which divide the material into rigid blocks.

When considering a plane strain problem the intersections of these surfaces with the representative section appear as **slip lines,** that we can consider as a band of small thickness, within which the material is thought perfectly plastic and obeys the associated flow rule. Note that by localizing sliding on a single band simplifies the problem, since the internal dissipation all takes place within the band.

Are there kinematic requirements that such slip lines must fulfil? The assumption of an **associated flow rule** requires the increment of plastic displacement vector to have an obliquity equal to φ' with respect to the slip line, because any distortion is coupled with volume changes (see Figure 8.2). If r is the radial distance of

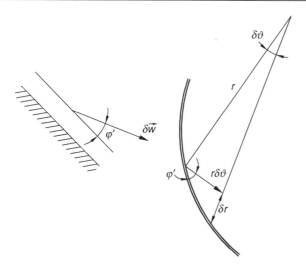

Figure 8.2 Shape of slip lines.

any point from the centre of instantaneous rotation, the geometry of Figure 8.2 tells us:

$$\frac{dr}{rd\theta} = \tan\varphi'$$

so that:

$$r = r_o e^{\theta\tan\varphi'}.$$

Therefore the following conclusion is reached: *in drained conditions, the slip lines must be straight lines or logspiral lines; in undrained conditions (i.e. $\tau_R = s_u$; $\varphi = 0$), they must be straight lines or circular arches.*

8.2 Yield criteria

As stated in Chapter 4, the term **yield** indicates the onset of irreversible (or plastic) deformations. Accordingly, the *yield criterion* is the condition for the development of irreversible deformations and, in mathematical terms, we can imagine this criterion as represented by a scalar function of stress $f(\sigma_{ij})$, such that:

$$f(\sigma_{ij}) = 0 \tag{8.1}$$

corresponds to yielding.

In order to give the yield criterion a geometric representation, we would need a six-dimensional stress space (because of six independent stress components) in which to represent a stress point, but our visual experience is confined to a three-dimensional space. An alternative is to make reference to the three principal stresses, plus information related to the principal directions. However, if we introduce the assumption of *isotropy*, i.e. there is no dependence of material behaviour on a given direction, then the six independent components σ_{ij} can be replaced, with no loss of generality, by the three principal stresses $\sigma_1, \sigma_2, \sigma_3$, and equation 8.1 can be rewritten as:

$$f(\sigma_1, \sigma_2, \sigma_3) = 0. \tag{8.2}$$

Therefore, if the principal stresses $(\sigma_1, \sigma_2, \sigma_3)$ are used as coordinates to define a point in a three-dimensional space, called *principal stress space*, in this space the equation (8.2) represents a surface, called the **yield surface**.

Unless we add other attributes to the function (8.2), this surface is fixed in the stress space, i.e. it neither moves nor expands, as it was the case for hardening materials discussed in Chapters 3 and 5.

The next aspect is about the shape of the yield surface in the principal stress space, but this point first requires further digress on theoretical aspects (for more insight into the subject presented in the following paragraphs see Harr (1966) and Davis and Selvadurai (2002)).

8.2.1 The π-plane

The line $\sigma_1 = \sigma_2 = \sigma_3$, equally inclined to all three principal axes is called **space diagonal**, and all planes perpendicular to this line have equation $\sigma_1 + \sigma_2 + \sigma_3 = $ constant and are called deviatoric planes. The special plane for which the constant is set zero is called the **π-plane** (Figure 8.3).

Let us now introduce a *local coordinate system* with unit vectors $\mathbf{n}_1, \mathbf{n}_2, \mathbf{n}_3$, oriented along and perpendicular to the space diagonal. If $\mathbf{n}_2$ is along the space diagonal and $\mathbf{n}_1$ is aligned with the projection of σ_1 on the π-plane (Figure 8.3), then these unit vectors

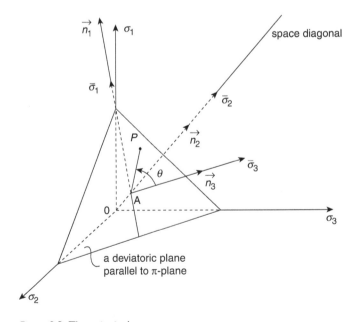

Figure 8.3 The principal stress space.

have the following components:

$$n_1 = \frac{1}{\sqrt{6}}\left\{\begin{matrix} 2 \\ -1 \\ -1 \end{matrix}\right\} ; \qquad n_2 = \frac{1}{\sqrt{3}}\left\{\begin{matrix} 1 \\ 1 \\ 1 \end{matrix}\right\} ; \qquad n_3 = \frac{1}{\sqrt{2}}\left\{\begin{matrix} 0 \\ -1 \\ 1 \end{matrix}\right\}. \qquad (8.3)$$

If a stress point in the principal space is now thought as represented by the vector $t(\sigma_1, \sigma_2, \sigma_3)$, the components of this vector in the n_1, n_2, n_3 directions are given by the scalar product of the vector t with the three unit vectors, i.e.:

$$\overline{\sigma}_1 = t \cdot n_1 = \frac{1}{\sqrt{6}}(2\sigma_1 - \sigma_2 - \sigma_3)$$

$$\overline{\sigma}_2 = t \cdot n_2 = \frac{1}{\sqrt{3}}(\sigma_1 + \sigma_2 + \sigma_3) \ . \qquad (8.4)$$

$$\overline{\sigma}_3 = t \cdot n_3 = \frac{1}{\sqrt{2}}(\sigma_3 - \sigma_2)$$

These can be inverted to give the relations that will be later used to represent the yield locus on the deviatoric plane:

$$\sigma_1 = \frac{1}{\sqrt{3}}\overline{\sigma}_2 + \sqrt{\frac{2}{3}}\overline{\sigma}_1$$

$$\sigma_2 = \frac{1}{\sqrt{3}}\overline{\sigma}_2 - \frac{1}{\sqrt{2}}\overline{\sigma}_3 - \frac{1}{\sqrt{6}}\overline{\sigma}_1 . \qquad (8.5)$$

$$\sigma_3 = \frac{1}{\sqrt{3}}\overline{\sigma}_2 + \frac{1}{\sqrt{2}}\overline{\sigma}_3 - \frac{1}{\sqrt{6}}\overline{\sigma}_1$$

Equations (8.4) show that the component $\overline{\sigma}_2$ is proportional to the mean stress (i.e. $\overline{\sigma}_2 = 3p/\sqrt{3}$), whereas the components $\overline{\sigma}_1$ and $\overline{\sigma}_3$ lie on the *deviatoric plane*. These can be combined to give the radial distance AP from the spatial diagonal (Figure 8.4), and it can be proved that this distance is linked to the deviatoric stress q, as defined in Chapter 2.

Finally, the radial direction with respect to the n_3 vector is defined by the angle θ, called the **Lode angle**, because it was introduced by Lode in 1926:

$$\tan\theta = \frac{\overline{\sigma}_1}{\overline{\sigma}_3} = \frac{1}{\sqrt{3}}\frac{2\sigma_1 - \sigma_2 - \sigma_3}{\sigma_3 - \sigma_2}. \qquad (8.6)$$

8.2.2 Yield criteria

Developments of the theory of plasticity are linked to the studies of ductile metals. The French engineer Tresca (1864) first suggested that yield would occur when the maximum shear stress on any plane reaches a limiting value. Referring to the case where $\sigma_1 \geq \sigma_2 \geq \sigma_3$, this criterion writes:

$$\sigma_1 - \sigma_3 = k, \qquad (8.7)$$

and in metal plasticity the constant k is the tensile strength.

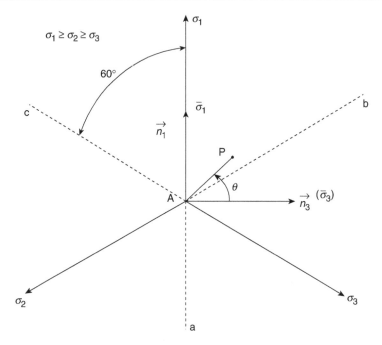

Figure 8.4 A view down the space diagonal.

In the context of soil mechanics, by assuming $k = 2s_u$, Tresca's criterion is well suited to represent the yield criterion of fine-grained soils when sheared in *undrained conditions*. Since this criterion does not depend on the mean stress, in the *principal stress space* equation (8.7) represents an infinitely long prism, whose axis is the space diagonal. Its intersection with the deviatoric plane (i.e. a view down the line $\sigma_1 = \sigma_2 = \sigma_3$) is a regular hexagon (Figure 8.5), as can be proved by substituting equations (8.5) into (8.7). When doing this exercise, it must be observed that the yield surface must be symmetric to the lines a, b, and c in Figure 8.4, that divide the deviatoric plane into six 60° segments. The linear equation obtained by substituting (8.5) into (8.7) applies only over the 60° segment for which $\sigma_1 \geq \sigma_2 \geq \sigma_3$, so that to describe the complete yield surface we need to consider all other possible combinations of principal stresses, i.e $\sigma_2 \geq \sigma_3 \geq \sigma_1$ and so on.

Later, von Mises (1913) suggested that yielding occurs when the second invariant of the deviator stress tensor reaches a critical value, i.e.:

$$f = \sqrt{J_2} - k = 0. \tag{8.8}$$

In the space diagonal, equation (8.8) represents an infinitely long circular cylinder (Figure 8.6), the intersection of which with the deviatoric plane is a circle.

For frictional materials, neither Tresca's nor von Mises's criterion is adequate, since neither of them depends on the mean stress. For this reason, the most popular and

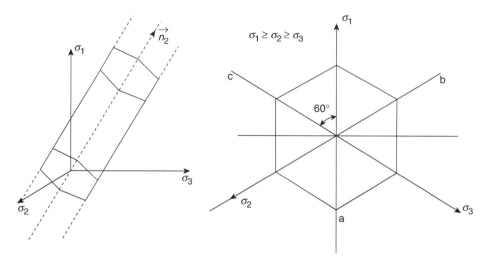

Figure 8.5 Tresca hexagon.

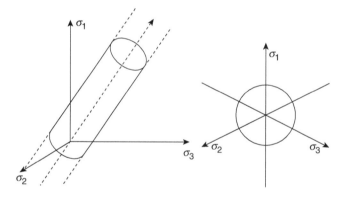

Figure 8.6 von Mises yield criterion.

useful criterion in soil mechanics is the one suggested by Coulomb (1773):

$$\tau_f = c' + \sigma'_f \tan \varphi',$$ (8.9)

according to which the shear strength τ_f depends on the intercept c' and on the friction component $\sigma'_f \tan \varphi'$. When expressed in terms of principal stresses, the same criterion for the combination $\sigma_1 \geq \sigma_2 \geq \sigma_3$ assumes the form:

$$\sigma'_1 - \sigma'_3 - (\sigma'_1 + \sigma'_3) \sin \varphi' - 2c' \cos \varphi' = 0,$$ (8.10)

which does not depend on the intermediate stress. In the principal stress space equation (8.10) represents an irregular pyramid, and its intersection with the deviatoric plane is an irregular hexagon (Figure 8.7), as it can be proved by substituting (8.5) into (8.10). Note that, in this case, the size of the irregular hexagon depends on

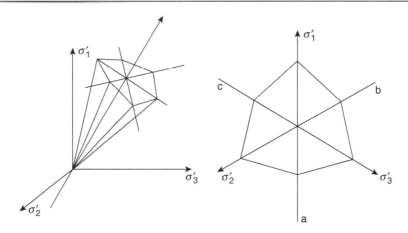

Figure 8.7 Coulomb yield criterion.

the mean stress p', so that it varies with distance along the diagonal space, although its characteristic shape is retained.

The Coulomb criterion is widely used in two dimensional problems, but for more general loading conditions Drucker and Prager (1952) suggested modifying this criterion by introducing a dependence on the mean stress:

$$J_2^{0.5} - \alpha I_1 - k = 0, \tag{8.11}$$

where I_1 is the first invariant of the stress tensor and J_2 is the second invariant of the deviator stress tensor. The two parameters α and k are often selected by requiring that this criterion matches with the Coulomb criterion at major vertices, so that:

$$\alpha = \frac{2 \sin \varphi'}{\sqrt{3}(3 - \sin \varphi')}$$

$$k = \frac{6c' \cos \varphi'}{\sqrt{3}(3 - \sin \varphi')}$$

In the principal stress space this criterion is a circular cone and its intersection with the deviatoric plane is a circle. However, the experimental research suggests that this circular shape does not agree with the available data, so that this criterion should be used with care in geotechnical engineering.

Since the Coulomb criterion contains corners, a feature which requires special numerical treatment, Matsuoka and Nakai (1974) have more recently suggested the following criterion:

$$I_3 - \alpha I_1 I_2 = 0, \tag{8.12}$$

which offers the mathematical advantages of the Drucker-Prager criterion and is in better agreement with experimental data of coarse-grained soils. In the deviatoric plane (Figure 8.8) it has a smooth curvilinear shape and is a convenient modification of the Coulomb criterion to describe tests other than triaxial.

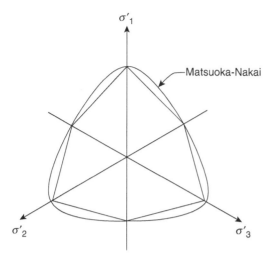

Figure 8.8 Matsuoka-Nakai criterion.

8.3 Rankine limiting states of stress

The problem of a soil mass which is in equilibrium and has everywhere attained the failure condition was first considered by Rankine (1857), and for this reason is referred as the **Rankine limiting stress field**.

In the context of plasticity theory, the construction of this stress field corresponds to a lower bound solution and therefore provides a safe estimate of earth thrust acting against a retaining wall. As it will be shown in the sequel, procedures can be in general developed for materials that exhibit both cohesion and internal friction. But in practice soil may be either considered as a purely cohesive material ($\tau_f = s_u$), an assumption appropriate for total stress analyses under undrained conditions, or purely frictional ($\tau_f = \sigma'_f \tan \varphi'$), an assumption appropriate for effective stress analyses under drained conditions, applicable to coarse-grained soils in almost all cases and to fine-grained soils in the long term.

For this reason we first concentrate on a cohesionless soil deposit, homogeneous in the horizontal direction, bounded by a traction free surface and characterized by the following parameters $c' = 0$; $\varphi' \neq 0$; $\gamma \neq 0$ (Figure 8.9a).

Because any vertical section is a section of symmetry, the vertical and the horizontal planes are principal planes. The integration of the indefinite equilibrium equation in the vertical direction, with the above specified boundary condition of traction free surface allows to compute the vertical effective stress acting at any depth on the horizontal plane, so that $\sigma'_{vo} = \gamma z$. On the contrary, the horizontal effective stress can not be univocally determined from the equilibrium equation in the horizontal direction, and without introducing further assumptions an infinite number of Mohr circles can be drawn (see Figure 8.9b), provided the stress state is far from failure.

If *the whole soil mass is subjected to lateral extension*, by progressively reducing the horizontal effective stress, while the vertical effective stress remains unchanged, a limiting stress condition will be reached, when the Mohr circle touches the failure

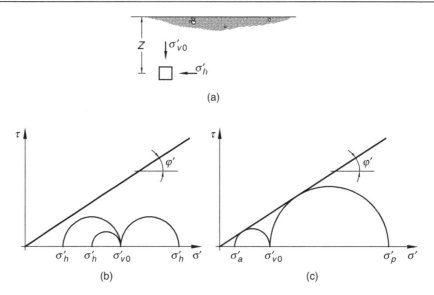

Figure 8.9 Limiting stress field.

envelope (consider the small circle in Figure 8.9c). The minimum value of the horizontal stress, attained in this way, is defined as the **active earth pressure**, and its expression can be easily derived by considering the Mohr-Coulomb criterion in terms of principal stresses:

$$\sigma'_a = \frac{1 - \sin\varphi'}{1 + \sin\varphi'}\sigma'_{vo} = \tan^2\left(\frac{\pi}{4} - \frac{\varphi'}{2}\right)\sigma'_{vo} = K_a\sigma'_{vo}, \tag{8.13}$$

where K_a is defined as the **coefficient of active earth pressure**.

It is important to realize that, since the whole soil mass has been subjected to a uniform lateral extension, a **uniform stress field** exists. Moreover, by considering the Mohr circle of stress at failure (Figure 8.10a), we also reach the conclusion that there are two families of planes on which the strength is fully mobilized, inclined at $\alpha = (\pi/4 + \varphi'/2)$ to the direction of the minimum principal stress. In the context of plasticity theory these directions are called *stress characteristics*.

There is, however, another possibility of finding an admissible uniform stress field, such that equilibrium equations and traction boundary conditions are satisfied and the soil has everywhere attained the failure condition.

In fact, if *the whole soil mass is subjected to lateral compression*, by progressively increasing the horizontal effective stress, while the vertical effective stress remains unchanged, a limiting stress condition will also be reached when the Mohr circle touches the failure envelope (consider the largest circle in Figure 8.9c). The maximum value of the horizontal stress, attained in this way, is defined as the **passive earth resistance**, and its expression can again be easily

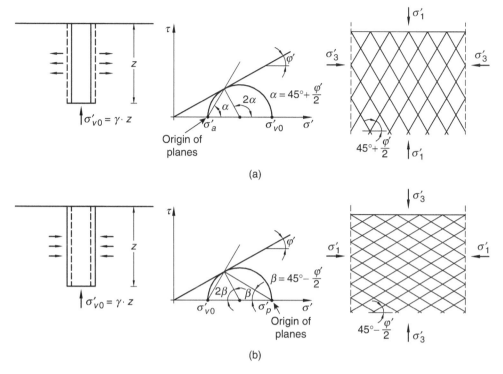

Figure 8.10 Rankine limiting stress field.

derived by considering the Mohr-Coulomb criterion in terms of principal stress:

$$\sigma'_p = \frac{1 + \sin\varphi'}{1 - \sin\varphi'}\sigma'_{vo} = \tan^2\left(\frac{\pi}{4} + \frac{\varphi'}{2}\right)\sigma'_{vo} = K_p\sigma'_{vo} \tag{8.14}$$

where K_p is the **coefficient of passive earth pressure**.

The previous considerations still apply: a uniform stress field exists and there are two families of planes of maximum stress obliquity on which the shear strength is fully mobilized, inclined at $\beta = (\pi/4 - \varphi'/2)$ to the direction of the major principal effective stress (Figure 8.10b).

8.3.1 Conjugate stresses within an infinite slope

We now consider the problem of investigating the Rankine states of stress in a semi-infinite soil mass, bounded by a surface at an angle $i < \varphi'$ to the horizontal (Figure 8.11). A slope of infinite extent, with constant properties at any given depth below the surface irrespective of the considered vertical, is designated as an *infinite slope*.

This definition requires that strata of different soil properties must be parallel to the surface and implies that the stress state is independent of the coordinate x_1 along the slope (see Figure 8.12 and the Example 8.1). Therefore, the state of stress acting on

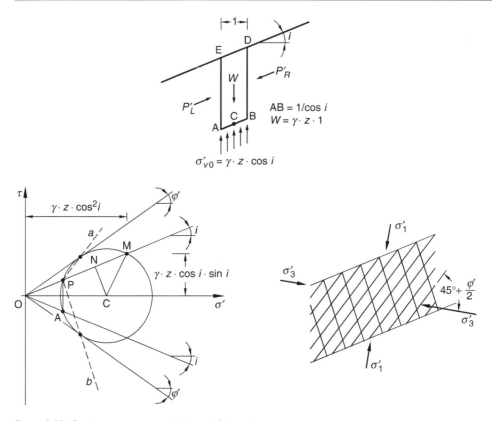

Figure 8.11 Conjugate stresses within an infinite slope.

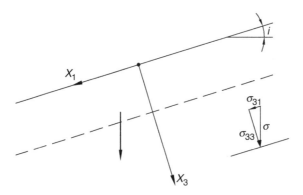

Figure 8.12 Stress vector on a plane parallel to the sloping surface.

any vertical section, such as EA or DB, must be the same, so that the total forces on these sections have the same magnitude as well as parallel lines of action. It can also be deduced that this line of action is the same if one considers that the summation of the moments about point C must be zero. The resultant of stresses at the base AB must balance the weight of the soil column (Figure 8.11), so that:

$$\sigma'_{vo} = \gamma z \cos i \tag{8.15}$$

and acts vertically.

A first conclusion is that stresses on vertical planes or on planes at the same slope of the surface are each parallel to the plane of the other. When such a property holds, we say that the planes are **conjugate planes** and stresses are **conjugate stresses**. Here again, while the stress vector acting on a sloping plane is defined by (8.15), the stress acting on the vertical plane is indeterminate and we are interested to define its limiting value.

In Figure 8.11 the segment OM represents the vertical stress σ'_{vo} acting on a section that has the same obliquity as the slope. The Mohr circle at failure must pass through point M and touch the failure envelope. The origin of planes is represented by point P and the directions of failure planes are given by straight lines a and b. The *minimum lateral pressure (active pressure)* acting on vertical planes can be derived by considering that the vertical through the origin of planes intersects the Mohr circle at point A. Therefore:

$$OM = \sigma'_{vo} \qquad\qquad OA = \sigma'_a$$

$$ON = OC \cos i \qquad NC = OC \sin i \qquad MC = OC \sin \varphi'$$

$$\frac{\sigma'_a}{\sigma'_{vo}} = \frac{ON - MN}{ON + MN}$$

so that:

$$\frac{\sigma'_a}{\sigma'_{vo}} = \frac{\cos i - \sqrt{\cos^2 i - \cos^2 \varphi'}}{\cos i + \sqrt{\cos^2 i - \cos^2 \varphi'}} \cdot \tag{8.16}$$

The same considerations also apply when one is interested to find the maximum lateral resistance (*passive resistance*), so that the ratio between σ'_p and σ'_{vo} is given by:

$$\frac{\sigma'_p}{\sigma'_{vo}} = \frac{\cos i + \sqrt{\cos^2 i - \cos^2 \varphi'}}{\cos i - \sqrt{\cos^2 i - \cos^2 \varphi'}} \cdot \tag{8.17}$$

If the obtained limiting values of active and passive stresses are used for the calculation of earth thrust against vertical retaining walls, it is important to realize that the Rankine state of stress imposes an implicit assumption to the mobilized friction between the wall surface and the soil. For the active state, the line of action of the resultant active pressure must be parallel to the sloping surface (Figure 8.13), and in the case of horizontal surface the Rankine theory assumes that the wall friction is zero. More in general, we say that the condition of conjugate planes must be fulfilled.

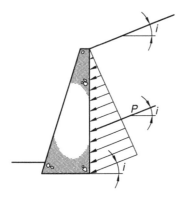

Figure 8.13 The active thrust as predicted by Rankine solution.

Example 8.1
Proof of equation (8.15)

With reference to Figure 8.12, let us assume a reference frame with the x_1 axis parallel to the sloping surface. If the soil properties are assumed to be constant on any plane orthogonal to the x_3 axis, the partial derivatives of any quantity with respect to x_1 and x_2 must be zero. Therefore, the indefinite equilibrium equations assume the form:

$$\frac{\partial \sigma_{31}}{\partial x_3} - \gamma \sin i = 0$$
$$\frac{\partial \sigma_{33}}{\partial x_3} - \gamma \cos i = 0 \qquad (8.18)$$

The integration of (8.18) (subjected to the boundary condition of traction free surface) and the change of coordinate $x_3 = z \cos i$ gives:

$$\sigma_{31} = \gamma z \sin i \cos i$$
$$\sigma_{33} = \gamma z \cos^2 i$$

and this result proves that the stress vector on any plane parallel to the sloping surface acts vertically and its magnitude is equal to:

$$\sigma_{vo} = \gamma z \cos i.$$

8.4 Coulomb critical wedge analysis

8.4.1 Active pressure

Well before Rankine's approach, Coulomb solved in 1773 the problem of soil thrust against a retaining wall, by using a procedure that nowadays is referred as the

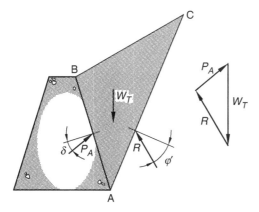

Figure 8.14 Coulomb's method.

limit equilibrium method. This procedure basically consists of devising an arbitrary mechanism of collapse and the collapse load is found from the equilibrium of forces that act at boundaries.

When applied to the problem of deriving the **active earth pressure**, the Coulomb theory considers the limit equilibrium of a soil wedge ABC bounded by the wall face and a planar failure surface (Figure 8.14).

It is assumed that on the failure plane AC the Coulomb criterion applies, so that the resultant force R has the obliquity φ' with respect to the normal to the plane. If δ is the friction angle at the soil-wall interface, the resultant force P_A has the obliquity δ with respect to the normal to the wall. The weight W_T of the sliding wedge is known in both direction and magnitude; on the contrary, neither the line of action and the magnitude of R and P_A are known. Therefore, there are four unknown and only three equations of equilibrium. The force equilibrium equations allow to compute the magnitude of the active thrust P_A and the force R (see the force polygon in Figure 8.14), but their lines of action remain unknown.

Note that the value of the computed thrust depends on the location of the failure surface, therefore more than one assumption is usually required in order to find the worst mechanism, which gives the maximum value of the active thrust. Provided the wall and the soil surface are planar (Figure 8.15), the following analytical expression for the active pressure is obtained:

$$P_A = \frac{1}{2}\gamma H^2 K_a, \tag{8.19}$$

where:

$$K_a = \frac{\cos^2(\varphi' - \beta)}{\cos^2\beta \cdot \cos(\beta + \delta)\left[1 + \sqrt{\dfrac{\sin(\delta + \varphi')\sin(\varphi' - i)}{\cos(\beta + \delta)\cos(\beta - i)}}\right]^2}. \tag{8.20}$$

For the special case where $\beta = i = 0$, the coefficient of active pressure assumes the simplified expression:

$$K_a = \frac{\cos^2 \varphi'}{\cos \delta \left[1 + \sqrt{\dfrac{\sin (\delta + \varphi') \sin \varphi'}{\cos \delta}} \right]^2} , \qquad (8.21)$$

and, if in addition we assume $\delta = 0$, the Coulomb wedge analysis gives the same results as does Rankine's method.

As previously observed, the line of action of P_A is unknown, unless further assumptions are introduced. A convenient one is to assume that at any depth down along the wall surface we can locate a potential sliding surface, and, provided the wall and the soil surface are planar, this assumption gives a triangular distribution of the earth pressure.

If the backfill has an irregular shape (Figure 8.16), it can be assumed that the point of application of P_A is given by the intersection of the wall surface and the line drawn through the centroid parallel to the sliding surface.

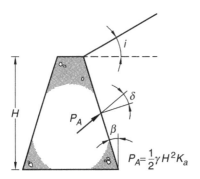

Figure 8.15 Active thrust predicted by Coulomb's method.

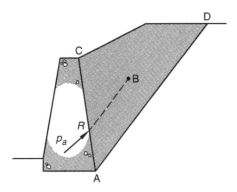

Figure 8.16 Point of application of the active thrust.

8.4.2 Passive resistance

The assumption of a planar sliding surface is no longer consistent when friction forces act at the interface between the wall and the soil. For this reason, while the Coulomb theory gives reasonable values of the active thrust, for the passive case this theory gives unsafe results, especially when $\delta > \varphi'/3$. As in practice a realistic value of δ is close to φ', a more refined analysis is required to estimate the passive earth pressure. Examples of these analyses include the numerical solution obtained by Sokolowski (1965), based on the *method of characteristics*, and the solution obtained by Caquot and Kerisel (1948), based on *curved (log-spiral) failure surface*. Within this context, in Section 8.5 we provide an analytical solution, based on the lower bound theorem of plasticity theory.

8.4.3 The Coulomb wedge analysis in terms of plasticity theory

The Coulomb wedge analysis does not consider the kinematic compatibility conditions, as required by the upper bound theorem (we did not sketch any velocity diagram when we applied this procedure). However, the obtained solution can be related to the upper bound limit analysis (the kinematic theorem of the plasticity theory). Heyman (1972) proved in fact that, in the case of a smooth wall, the Coulomb solution is identical to that obtained by applying the upper bound theorem of plasticity.

Later, Collins (1973) proved that, in the case of a rough wall also, Coulomb's solution can be interpreted in terms of upper bounds, by using a generalized limit theorem which applies in the presence of rough boundaries (Collins, 1969).

Therefore, since the upper bound theorem overestimates the collapse load, the Coulomb solution gives an unsafe estimate of the earth thrust against a retaining wall (this is the reason why we try to find the worst mechanism, which gives the maximum value of the active thrust and the minimum value of the passive resistance).

Despite this, experience has shown that the wedge analysis gives reasonable results of the active thrust on which one can rely in engineering practice, with the advantage that it can be easily adapted to layered soil conditions and irregularly shaped boundaries. For this reason, the limit equilibrium method is the most widely used procedure for examining the stability of soil structures and will be further examined when dealing with slope stability analysis, in Section 8.19.

8.5 Stress discontinuities

In Section 8.4 we reached the conclusion that, if the whole soil mass is subjected to *lateral extension*, by progressively reducing the horizontal effective stress, while the vertical effective stress remains unchanged, a limiting stress condition will be reached, when the Mohr circle touches the failure envelope (Figure 8.17a). The minimum value of the horizontal stress, attained in this way, was defined as the *active earth stress* and, since the whole soil mass has been subjected to a uniform lateral extension, a **uniform stress field** exists. In particular, because of the relatively simple boundary condition, represented by a *smooth wall*, it was an easy task to construct the stress field.

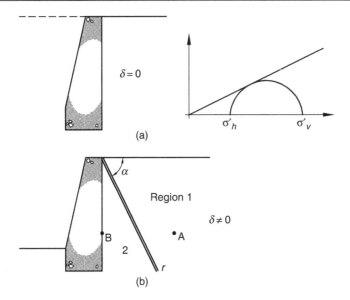

Figure 8.17 Stress discontinuity.

If we consider the problem of a rough wall (Figure 8.17b), we can argue that the average effective stress and the direction of principal stresses at points A and B will be different, so that the construction of an admissible stress field is now a little more complex.

Stresses can change from place to place in a rather smooth way or we can devise the presence of jumps or discontinuities when moving from one region to another, in which we can imagine to subdivide the stress field. These discontinuities are called **stress discontinuities**, and the objective of this section is to show how the lower bound theorem of plasticity can be applied to the determination of the passive earth resistance, by separating the stress field into zones of uniform stress through the introduction of a convenient number of stress discontinuities.

Referring to Figure 8.18, consider a discontinuity between two states of stress relative to regions A and B.

In order to satisfy equilibrium, the Mohr circles of these regions must have a common point at X, and a relevant result that can be deduced from Figure 8.18 is that there is a jump in the direction and magnitude of the major principal stress across the discontinuity.

By assuming a small rotation, the shift in centre of the Mohr's circles is related to the rotation $d\vartheta$ of principal direction through the equation:

$$\frac{ds'}{s'} = 2d\vartheta \tan\varphi', \tag{8.22}$$

where φ' is the angle of shear strength and s' is the abscissa of the centre of the Mohr circle.

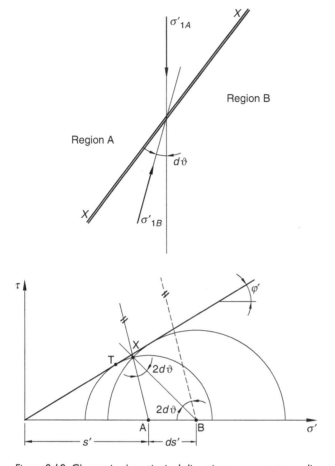

Figure 8.18 Change in the principal direction across a stress discontinuity.

Equation (8.22) can be proved by observing that, if $ds' \rightarrow 0$, $\sin 2d\vartheta \cong 2d\vartheta$, the common point $X \rightarrow T$ and $\overline{BX} \cong \overline{AX} = s' \sin \varphi'$. By applying the sine theorem to the triangle ABX, one gets:

$$\frac{\overline{AX}}{\cos \varphi'} = \frac{ds'}{\sin 2d\vartheta}$$

and the equation (8.22) is proved.

If, in order to smooth the overall stress field, we consider a fan of stress discontinuities (Figure 8.19), across which the rotation of the principal directions assumes the finite value ϑ, then the shift between the two extreme Mohr circles is defined by:

$$\frac{s'_1}{s'_2} = e^{2\vartheta \tan \varphi'}. \tag{8.23}$$

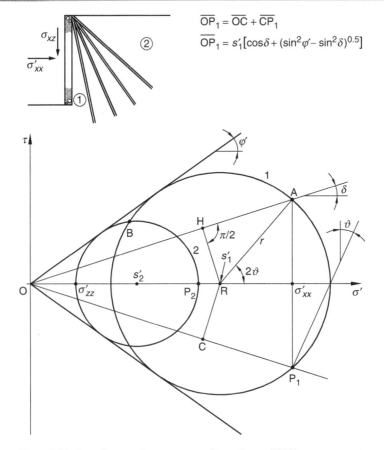

Figure 8.19 Fan of stress discontinuities (Lancellotta, 2002).

Consider now the **passive earth resistance** problem sketched in Figure 8.19. In order to obtain the value of σ'_{xx} at the wall interface as a function of the overburden stress $\sigma'_{zz} = \gamma z$ in region 2, the following relations apply:

$$\overline{OP_1} = \overline{OC} + \overline{CP_1}; \qquad \overline{OP_1} = s'_1\left[\cos\delta + \sqrt{\sin^2\varphi' - \sin^2\delta}\right], \tag{8.24}$$

where δ is the friction angle at the wall-soil interface.

Since $\sigma'_{xx} = \overline{OP_1}\cos\delta$, by observing that $\sigma'_{zz} = \gamma z$, and:

$$s'_2 = \frac{\sigma'_{zz}}{1 - \sin\varphi'}, \tag{8.25}$$

it follows that:

$$\sigma'_{xx} = \left[\frac{\cos\delta}{1 - \sin\varphi'}\frac{s'_1}{s'_2}\left(\cos\delta + \sqrt{\sin^2\varphi' - \sin^2\delta}\right)\right] \cdot \gamma z. \tag{8.26}$$

By inserting equation (8.23) into (8.26), we obtain the solution for **the passive earth resistance** (see Lancellotta, 2002):

$$K_p = \left[\frac{\cos\delta}{1-\sin\varphi'}\left(\cos\delta + \sqrt{\sin^2\varphi' - \sin^2\delta}\right)\right]e^{2\vartheta\tan\varphi'} \tag{8.27}$$

where:

$$2\vartheta = \sin^{-1}\left(\frac{\sin\delta}{\sin\varphi'}\right) + \delta. \tag{8.28}$$

Equation (8.28) is derived from the Mohr's of region 1 in Figure 8.19, by observing that:

$$\overline{HR} = r\sin(2\vartheta - \delta) = \overline{OR}\sin\delta = \frac{r}{\sin\varphi'}\sin\delta. \tag{8.29}$$

The obtained solution (8.27) is of value in engineering practice, as it is a conservative estimate of the exact solution and it is in agreement with Sokolowski (1965) solution, based on the method of characteristics (see Figure 8.20). It can further be inferred that, when the boundary condition of a smooth wall applies ($\delta = 0$), then equation (8.27) merges into Rankine's solution.

The corresponding expression for the **active earth pressure** can be derived by observing that the following relations hold (see Figure 8.21):

$$s_1' = \frac{\sigma_{zz}'}{1+\sin\varphi'}; \quad \overline{OA} = \overline{OH} - \overline{AH}$$

$$\overline{OH} = s_2'\cos\delta; \quad \overline{AH} = s_2'\sqrt{\sin^2\varphi' - \sin^2\delta},$$

so that:

$$\sigma_{xx}' = \left[\frac{\cos\delta}{1+\sin\varphi'}\frac{s_2'}{s_1'}\left(\cos\delta - \sqrt{\sin^2\varphi' - \sin^2\delta}\right)\right]\gamma z. \tag{8.30}$$

By inserting equation (8.22) into (8.30) we obtain:

$$K_a = \left[\frac{\cos\delta}{1+\sin\varphi'}\left(\cos\delta - \sqrt{\sin^2\varphi' - \sin^2\delta}\right)\right]e^{-2\vartheta\tan\varphi'}, \tag{8.31}$$

where:

$$2\vartheta = \sin^{-1}\left(\frac{\sin\delta}{\sin\varphi'}\right) - \delta. \tag{8.32}$$

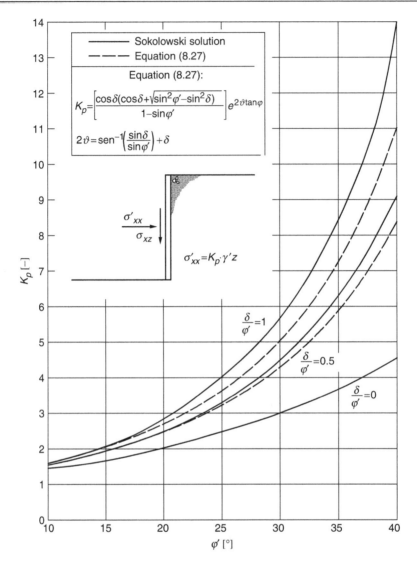

Figure 8.20 Passive earth resistance solution based on lower bound theorem of plasticity (Lancellotta, 2002).

Here again, equation (8.32) is deduced by observing that:

$$\overline{HR} = r\sin\alpha = \overline{OR}\sin\delta = \frac{r}{\sin\varphi'}\sin\delta$$

$$O\hat{R}H = \frac{\pi}{2} - \delta; \quad A\hat{R}H = \frac{\pi}{2} - \delta - 2\vartheta; \quad \alpha = 2\vartheta + \delta$$

$$\sin(2\vartheta + \delta) = \frac{\sin\delta}{\sin\varphi'}.$$

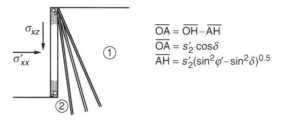

$$\overline{OA} = \overline{OH} - \overline{AH}$$
$$\overline{OA} = s_2' \cdot \cos\delta$$
$$\overline{AH} = s_2'(\sin^2\varphi' - \sin^2\delta)^{0.5}$$

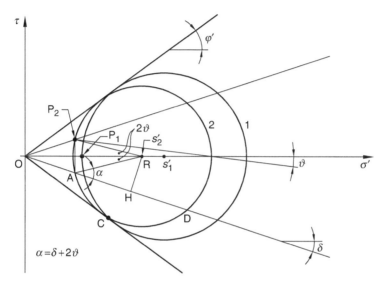

Figure 8.21 Active earth thrust based on lower bound theorem of plasticity (Lancellotta, 2002).

Finally, we remark that equations (8.27) and (8.32) can be expressed as a single equation:

$$K_{p,a} = \left[\frac{\cos\delta}{1 \mp \sin\varphi'} \left(\cos\delta \pm \sqrt{\sin^2\varphi' - \sin^2\delta} \right) \right] e^{\pm 2\vartheta\,\tan\varphi'}, \tag{8.33}$$

with:

$$2\vartheta = \sin^{-1}\left(\frac{\sin\delta}{\sin\varphi'} \right) \pm \delta, \tag{8.34}$$

where the upper sign applies for the passive coefficient and the lower sign for the active coefficient.

The coefficient of passive earth resistance for the case of a vertical wall against a soil with **sloping ground surface** at an angle $i > \varphi'$ can be obtained as a special case of a more general solution, derived in Section 8.11. Here we simply quote the solution of interest:

$$K_p = \frac{\sigma'_{xx}}{\gamma z \cos i} = \left[\frac{\cos \delta}{\cos i - \sqrt{\sin^2 \varphi' - \sin^2 i}} \left(\cos \delta + \sqrt{\sin^2 \varphi' - \sin^2 \delta} \right) \right] e^{2\vartheta \tan \varphi'}$$

(8.35)

where:

$$2\vartheta = \sin^{-1} \left(\frac{\sin \delta}{\sin \varphi'} \right) + \sin^{-1} \left(\frac{\sin i}{\sin \varphi'} \right) + \delta + i.$$

(8.36)

8.6 Earth retaining structures

8.6.1 Introduction

Earth retaining structures are designed to provide lateral support to non-stable soil masses. Examples of retaining walls constructed of stone masonry are very common to find; however, at present the usual practice is that of concrete or reinforced concrete gravity walls.

The term **gravity wall** (see Figure 8.22a) usually indicates a masonry or concrete wall whose stability depends entirely on its weight. Dimensions are established in such a way that no tension stresses are produced by the resultant R in any section.

The **semigravity wall** (Figure 8.22b) is a slender structure that requires some reinforcement at the inner face and foundation.

The **cantilever wall** (Figure 8.24c) consists of a concrete foundation slab and a vertical stem, fully reinforced because of the reduced dimensions. The weight of the soil on the slab significantly contributes to the overall stability.

The **counterfort wall** (Figure 8.22d) differs from the previous solutions in that the front face slab is supported by the inner counterforts.

A rather different solution is represented by the so-called **crib walls** (Figure 8.22e). These walls consist of an assembly of structural units composed in such a way as to form a cell or crib, that is successively filled with soil.

Finally, **diaphragm walls** or **flexible earth retaining structures** (Figure 8.22f) consist of vertical slabs or sheet-piles supported at various levels by anchors or struts and by the contrast offered by the soil in the embedded part.

8.6.2 Computing earth thrust in practice

As already discussed in Section 8.3, the **Rankine theory** (see equation 8.16) is one of the approaches used for the calculation of earth thrust against retaining walls. However, this requires to assume that vertical planes and planes at the same slope of the surface are **conjugate planes** and stresses acting on these planes are **conjugate stresses**. Therefore, for the active state, the line of action of the active thrust must be parallel to the sloping surface (Figure 8.13), and in the case of horizontal surface the Rankine theory implies that the wall friction is zero.

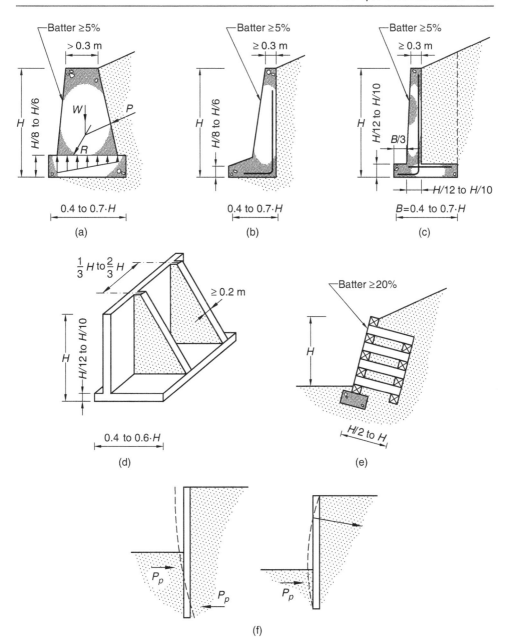

Figure 8.22 Earth retaining structures.

Further, we have to consider the constraints that the soil below the wall applies to the overlying soil mass. These constraints imposes shear stresses and violate the assumptions of Rankine's theory, but the effects are thought to be small and are disregarded.

An alternative, much used in practice for computing the active earth thrust on rough retaining walls, is provided by the **Coulomb theory**, which considers the limit equilibrium of a soil wedge ABC bounded by the wall face and a planar failure surface (Figure 8.14).

When the wall and the soil surface are planar (Figure 8.15), the analytical expression for the active pressure is given by equation (8.19) with the active earth pressure coefficient provided by equation (8.20).

We recall that the assumption of a planar sliding surface is no longer consistent when friction forces act at the interface between the wall and the soil. For this reason, while the Coulomb theory gives reasonable values of the active thrust, for the passive case this theory gives unsafe results, especially when $\delta > \varphi'/3$. As in practice a realistic value of δ is close to φ', a more refined estimate of the passive earth resistance is provided by equation (8.27), based on the lower bound plasticity theorem.

8.6.3 Tension cracks

In Chapter 5 we analysed the peak strength envelope of stiff clays, that is curved and passes close to the origin. In fact, possible effects of dilation and destruction of bonding at low confining pressures give rise to the pronounced non-linearity of the failure envelope near the origin.

For this reason care is required when the actual peak envelope is approximated with a straight envelope described by the equation:

$$\tau = c' + \sigma' \tan \varphi',$$

because this may result unsafe if the stress level in the field is smaller than the effective stress range applied in the laboratory tests.

When referring to the straight envelope approximation, it is common practice to consider the theoretical case in which both the strength contributions are present. To solve this problem in a rather elegant and concise manner, Caquot (1934) suggested that, *in the presence of both cohesive c' and friction φ' components, the solution can be obtained by the one relative to the cohesionless case, by simply adding an isotropic state of stress equal to $c'\cot\varphi'$*. Then, the limiting state of stress can be expressed as (see Figure 8.23a):

$$\sigma_3' + c' \cot \varphi' = \frac{1 - \sin \varphi'}{1 + \sin \varphi'} \left(\sigma_1' + c' \cot \varphi' \right), \tag{8.37}$$

and rearranging:

$$\sigma_3' = \sigma_1' \cdot \frac{1 - \sin \varphi'}{1 + \sin \varphi'} - 2c' \cdot \frac{\cos \varphi'}{1 + \sin \varphi'}. \tag{8.38}$$

Since:

$$\frac{\cos \varphi'}{1 + \sin \varphi'} = \frac{\sqrt{1 - \sin^2 \varphi'}}{1 + \sin \varphi'} = \frac{\sqrt{(1 + \sin \varphi')(1 - \sin \varphi')}}{1 + \sin \varphi'} = \sqrt{\frac{1 - \sin \varphi'}{1 + \sin \varphi'}},$$

equation (8.38) assumes the form:

$$\sigma_3' = K_a \sigma_1' - 2c' \sqrt{K_a}. \tag{8.39}$$

For a level ground surface and a smooth vertical wall, the corresponding values of active and passive stress then became:

$$\sigma_a' = K_a \sigma_{vo}' - 2c' \sqrt{K_a}$$
$$\sigma_p' = K_p \sigma_{vo}' + 2c' \sqrt{K_p} \tag{8.40}$$

where:

$$K_a = \frac{1}{K_p} = \tan^2 \left(\frac{\pi}{4} - \frac{\varphi'}{2} \right) = \frac{1 - \sin \varphi'}{1 + \sin \varphi'}. \tag{8.41}$$

It can now be observed, by inspection of Figure 8.23b, that the largest Mohr circle at failure, that can be drawn with $\sigma_a' = 0$, gives the largest value of σ_{vo}' that can be supported and, since $\sigma_{vo}' = \gamma z$, the first expression of (8.40) allows to compute the height of the unsupported portion of soil (assuming that the water level is at depth):

$$z_0 = \frac{2c' \tan \left(\dfrac{\pi}{4} + \dfrac{\varphi'}{2} \right)}{\gamma}. \tag{8.42}$$

Within the depth z_0, the attainment of failure conditions would require the presence of tensions, but stability analyses have to take into account the lack of tensions

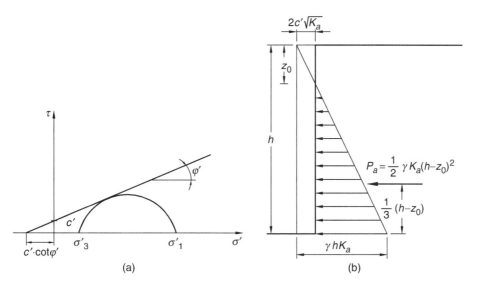

(a)   (b)

Figure 8.23 The tension cracks problem.

between the wall and the soil as well as the possible pre-existence of cracks or the possibility for their development. For this reason, the active force is estimated as (see Figure 8.23b):

$$
\begin{aligned}
P_a &= \frac{1}{2}\left[(h - z_0)\left(\gamma h K_a - 2c'\sqrt{K_a}\right)\right] \\
&= \frac{1}{2}\left[(h - z_0)(\gamma h K_a - z_0 \gamma K_a)\right] = \frac{1}{2}\gamma K_a (h - z_0)^2
\end{aligned}
\tag{8.43}
$$

Note that, although the above conventional analysis provides a theoretical estimate of the depth of tension cracks, research is still required on mechanisms involved in their development, and the depth of such cracks depends on factors such as seasonal effects of droughts and vegetation and the amount of surface water. Furthermore, it is usual in the design to assume that these cracks can fill with water in times of heavy rainfall, and this provides an additional hydrostatic pressure against the wall.

8.6.4 Short-term analysis

When dealing with the stability analysis of an earth retaining structure supporting fine-grained soft soils (see Figure 8.24), the stress removal due to excavation induces negative excess pore pressure and the effective stress path during undrained unloading ($\delta\sigma_h < 0; \delta\sigma_v = 0$) is represented by the path from A to B. In time, consolidation takes place and the effective stress path goes towards point D, eventually reaching the failure envelope. It therefore follows that the safety margin available in undrained conditions is higher than in the long term, when excess pore pressure has been completely dissipated, and the stability analysis is then performed by referring to the final drained or steady state flow conditions in terms of effective stresses.

When considering a diaphragm wall (Figure 8.25), the excavation would produce a total stress path for a soil element A given by the straight line $A - B$ (*extension-unloading*); on the contrary, movements of the diaphragm push the same element along the total stress path $A - D$ (*extension-loading*). As a consequence it is quite difficult to define a priori what would be in reality the total stress path (presumed to be $A - F - G$ or alternatively $A - F' - G'$), so that one cannot say if during consolidation the effective stress path will move from point E towards G (with an increase of safety margin) or towards G' (with a reduction of safety margin). The engineer in this case must consider both the events, by performing the stability analysis in undrained conditions, as well as the stability analysis in the long term.

When performing the **short-term analysis** (i.e. undrained conditions) in terms of total stresses, the limiting active and passive values of the total lateral stress are:

$$
\begin{aligned}
\sigma_a &= \sigma_{vo} - 2s_u \\
\sigma_p &= \sigma_{vo} + 2s_u
\end{aligned}
\tag{8.44}
$$

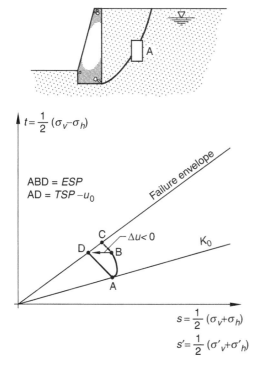

Figure 8.24 Stability analysis of an earth retaining wall.

If we also consider the wall adhesion c_w, the corresponding modified expression for the active and passive pressure are the following:

$$
\begin{aligned}
\sigma_a &= \sigma_{vo} - 2s_u\sqrt{1 + \frac{c_w}{s_u}} \\[1em]
\sigma_p &= \sigma_{vo} + 2s_u\sqrt{1 + \frac{c_w}{s_u}}
\end{aligned}
\tag{8.45}
$$

and, because the installation of the diaphragm produces some smear of the soil in contact with the wall, it is suggested to consider the adhesion c_w to be only a fraction of the undrained strength S_u.

As already seen for the long-term analysis, also in this case we have to consider that uncemented intact clay soils can deliver tension only by virtue of suction in the pore water, but the interface between the soil and the wall cannot transmit tensile stresses. Therefore, in the absence of a surface surcharge, a no tension condition should be applied up to a depth $z_o = 2s_u/\gamma$, if we consider a dry open interface, or to a depth $z_o = 2s_u/(\gamma - \gamma_w)$, when considering a flooded interface, and the design should allow full hydrostatic pressure in this zone, in the latter case.

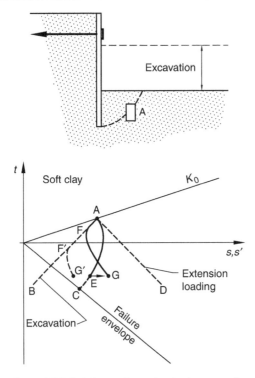

Figure 8.25 Stability analysis of a diaphragm wall.

8.6.5 Mobilization and distribution of earth pressure

Studies on earth thrust mobilization (see Fang and Ishibashi, 1986; Potts and Fourie, 1986; Paik and Salgado, 2003) have definitively shown that the nature of wall movement, whether translation or rotation, has a major effect on the magnitude and distribution of earth pressure. In particular, if a retaining wall is rotating about the top (as may be the case with bridge abutment), it is proved (Fang and Ishibashi, 1986) that the stress distribution is highly non-linear, it increases near the top beyond the active value due to arching effects and the magnitude of arching stresses increases with increasing soil density. Therefore, the magnitude of the total active thrust may be higher (by about 17%) than the value predicted by Coulomb theory and the location of the thrust resultant rises at about 0.46 the wall height from the base. Simple translation or rotation about the toe gives a total thrust which is nearly the same as that computed from Coulomb theory. However, simple translation gives a non-linear stress distribution, roughly parabolic, so that the point of application of the resultant is higher than one-third of the wall height.

As far as the movements needed to mobilize the limit values of active thrust and passive resistance, this relation has been traditionally based on the work of Terzaghi (1934) and is sketched in Figure 8.26. From this figure it is apparent that the deformations needed to mobilize the full passive resistance are much greater than those required for the active thrust. However, in stiff clays, characterized by high

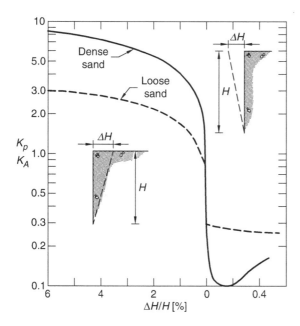

Figure 8.26 Mobilization of active thrust and passive resistance.

values of the coefficient of earth pressure at rest K_o and of soil stiffness, active and passive values can be mobilized at comparable displacements (Potts and Fourie, 1986).

As expected, these conclusions have relevance when evaluating bending moments under serviceability conditions, since they depend on the mode of wall deformation and, because of the mechanisms involved, a set of springs can hardly reproduce the stress distribution.

8.6.6 Surcharges

If a uniformly distributed surcharge q acts over the entire soil surface (Figure 8.27), then the vertical stress at any depth increases up to the value $(\gamma z + q)$ and an additional active stress $K_a q$ will result. The corresponding thrust on the vertical wall face will be $K_a q H$, acting at mid-height.

Other cases of surcharge, namely line, point or strip surcharge are more diffi-cult to analyse. Magnitude and distribution of the induced earth pressure is not well understood, and for this reason several calculation methods are quoted in literature.

8.6.7 Compaction

In order to guarantee that the backfill has adequate strength and stiffness, compaction to a minimum dry density is usually specified. The compaction process induces horizon-tal earth stresses, which can vary significantly in magnitude and distribution, depending on the compaction process.

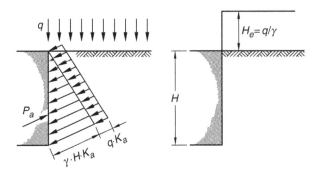

Figure 8.27 Active thrust in presence of a uniform surface load.

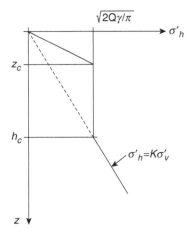

Figure 8.28 Idealized distribution of horizontal stresses in a backfill compacted in layers.

A detailed analysis has been given by Broms (1971) and Ingold (1979), and a simplified approach can be summarized as follows (see Figure 8.28).

Let us assume that the intensity of the *line load*, imposed by compaction, be Q. This should be the weight of the roller divided by its width, when considering dead weight rollers, while the centrifugal force generated by the vibrating mechanism should be added when considering vibratory rollers.

The maximum horizontal pressure induced by compaction should be evaluated as:

$$\sigma'_h = \sqrt{\frac{2Q\gamma}{\pi}}.$$ (8.46)

If this value is presumed to be attained for all successively compacted layers, then a locus of maxima can be identified, ranging from a minimum z_c and a maximum h_c critical depth, as shown in Figure 8.28, being:

$$z_c = K\sqrt{\frac{2Q}{\gamma\pi}} \qquad h_c = \frac{1}{K}\sqrt{\frac{2Q}{\gamma\pi}}.$$ (8.47)

In the above expressions, the earth pressure coefficient K ranges from the active value, if the wall can move to attain the active condition, to the at rest value for unyielding walls.

8.6.8 Drainage provisions

Unsatisfactory performance of retaining walls is very often caused by a poorly designed or constructed drainage system. A drainage system is designed to prevent excessive water pressure to act against the wall, and to fulfil this requirement various types of provisions can be devised, depending on the retaining wall, soil conditions and possible frost effects.

Figure 8.29 considers the simplest case when water ponds behind the wall, resting upon an impermeable stratum. In this case, the resultant of the *effective stresses* against the wall is computed by using equation (8.16) or (8.19), provided only that the **buoyant unit weight** γ' is used, i.e.:

$$P'_a = \frac{1}{2}\gamma' H^2 K_a,$$

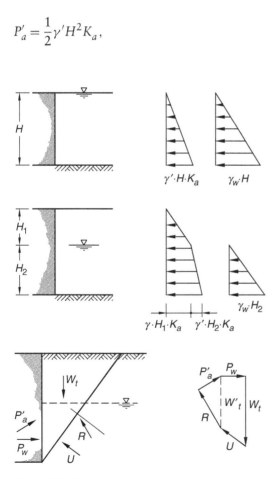

Figure 8.29 Hydrostatic pore pressure.

and the thrust exerted by the pore pressure must be added in order to obtain the total thrust:

$$P_a = \frac{1}{2}\gamma' H^2 K_a + \frac{1}{2}\gamma_w H^2 .$$

When we consider the more common situation in which the phreatic surface is below the soil surface (see the second case sketched in Figure 8.29), we must observe that the effective stress against the wall increases γK_a per unit depth above the water table and $\gamma' K_a$ per unit depth below the water table.

Weepholes are quite usual (see Figure 8.30a) and effective with pervious backfill, when the hydraulic conductivity is so high that rain intensity is unlikely to produce a continuous flow pattern. Weepholes should have at least a 100 mm diameter, the spacing should range from 1.50 to less than 3.00 m both horizontally and vertically and filter fabric should also be used in order to prevent plugging.

With semipervious backfill, weepholes can be clogged, and in addition the outlets can be closed by ice in freezing conditions, so that a continuous **vertical drain** at the back of the wall is a better solution (Figure 8.30b). A flow-net can be drawn for this case and the pore water pressure at any point such as A on the potential failure surface $C - B$ can be found (we also suggest the reader refer to the closed-form solution presented by Barros, 2006). The influence of seepage forces can then be accounted for, by considering the total weight of the sliding wedge and the pore

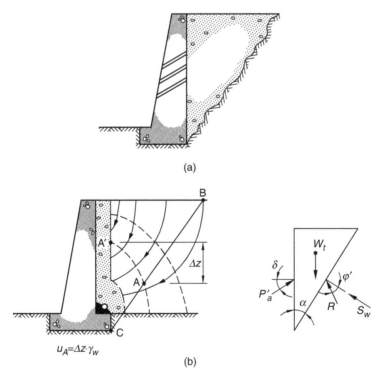

Figure 8.30 Drainage provisions and pore water pressure.

pressure acting on the boundary *CB*. Note that, since the hydraulic conductivity of the drain is higher that that of the backfill, the pore pressure within the drain is atmospheric. It follows that at any point at the interface between the drain and the backfill the total head is equal to the elevation head and therefore equipotentials must be spaced at equal vertical intervals. This boundary is neither a flow line nor an equipotential.

We outline that the vertical drain prevents water pressure on the wall, but the effect of seepage is to increase the overall active thrust.

When general conditions during construction permit the disposal of a **sand filter along the slope** of a natural soil, before the placement of the backfill (see Figure 8.31a), this represents an even better solution. In this case, flow lines and equipotential lines are respectively vertical and horizontal, and, if the hydraulic conductivity of the backfill is constant, the pore pressure is zero at any point on the failure surface.

In presence of **clay backfill**, Terzaghi and Peck (1967) suggest a drainage system as shown in Figure 8.31b, in order to prevent water content changes of the clay material producing cracking and swelling.

Different levels of water on the sides of **diaphragm walls** will cause seepage and unbalanced net water pressure on the structure (see Figure 8.32). The pore water pressure needs to be evaluated according to the principles discussed in Chapter 6,

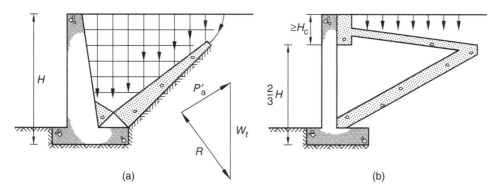

(a)   (b)

Figure 8.31 Drainage systems.

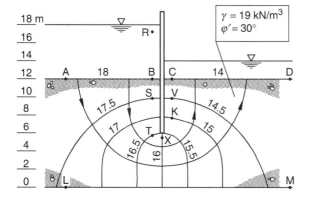

Figure 8.32 Flow under a diaphragm wall.

and it must be observed that seepage conditions increase the active thrust behind the diaphragm, whereas conditions of upwards seepage in front of the structure are associated to a decrease of passive resistance, if compared to the case where no seepage occurs (see the illustrative Example 8.2).

Example 8.2

Given the diaphragm wall sketched in Figure 8.32, *find* the limiting values of the active stress behind the wall and passive stresses in front of the wall.

Solution. The following boundary conditions are relevant for the confined flow under consideration:

$\overline{AB}, \overline{CD}$ are equipotential lines, respectively behind and in front of the wall;

$\overline{LM}$ and the wall surface $\overline{BTXC}$ are flow lines.

These boundary conditions allow to draw the flow net shown in Figure 8.32, and this gives the possibility to evaluate the hydraulic head and the pressure head at all points of interests.

Point	Hydraulic head (m)	Elevation head (m)	Pressure head (m)
B	18	12	6
S	17.5	10	7.5
T	16.5	6	10.5
X	16	6	10
K	15	8	7
V	14.5	10	4.5
C	14	12	2

Once the pressure head has been computed, the vertical effective stress can be determined. If the assumption of a smooth wall is introduced, the Rankine limiting state of stress provides the following earth pressure coefficients $K_a = 0.33$ and $K_p = 3.00$, and, accordingly, the active and passive effective stresses assume the values given in the following table.

Point	σ_{vo} (kPa)	u (kPa)	σ'_{vo} (kPa)	σ'_a (kPa)	σ'_p (kPa)
B	58.86	58.86	0.00	0.00	–
S	96.86	73.58	23.28	7.68	–
T	172.86	103.01	69.85	23.05	–
K	95.62	68.67	26.95	–	80.85
V	57.62	44.15	13.47	–	40.41
C	19.62	19.62	0.00	–	0.00

The total thrust on both sides is obtained by summing up the effective active (behind the wall) and passive (in front of the wall) normal stresses and the water pressure.

8.7 Design of retaining walls

Retaining walls do not usually succeed in fulfilling stability requirements because very often drainage provisions are ineffective or because the wall foundation has not been properly designed (Terzaghi and Peck, 1967). Therefore, a careful investigation of the soil beneath the wall foundation is needed and the design of a retaining wall should consider the following steps (Figure 8.33).

The active thrust P_a is first estimated and it is convenient to consider its vertical and horizontal components, i.e. P_V, P_H. Then the weight W of the wall is computed by subdividing its actual shape into simpler shapes to facilitate the computation (see Figures 8.35 and 8.36).

(a) The requirement that the wall must be safe against **sliding** is expressed by the following inequality:

$$\frac{(W + P_V)\tan\delta}{P_H} \geq 1.5, \tag{8.48}$$

where a global factor of safety of 1.5 has been introduced, to take into account uncertainties linked to the estimate of the active thrust.

Note that the passive earth resistance on the front side of the foundation should be disregarded when computing the safety factor against sliding, because it is always doubtful to rely on its presence and because the seasonal changes of water content render this contribution rather unreliable.

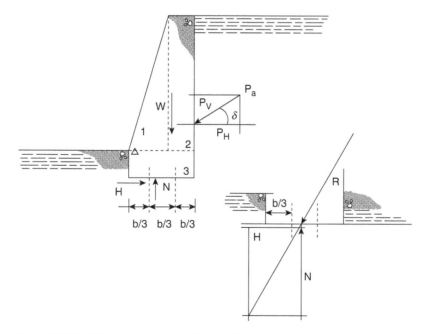

Figure 8.33 Stability analysis of a retaining wall.

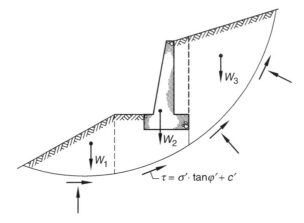

Figure 8.34 Stability analysis of a wall retaining a sloping soil.

(b) The eccentricity of the resultant force, measured from the centre line of the base, should not exceed one-sixth of the base width and should be close to the mid point of the base if the wall is founded on high compressibility soils.

This requirement is aimed at avoiding excessive tilting of the wall, as well as to limit the maximum bearing stress. An alternative way of expressing this requirement is to suggest that the ratio of resisting to overturning moments should be at least 1.5, a criterion that is usually interpreted as a safeguard against *toppling*, if the retained wall is *supported by rock*.

(c) The contact normal stress should not exceed the allowable **bearing capacity**. This is a very important step, that will be examined in detail in Section 8.12.

(d) In addition to the above mentioned requirements, it must be remembered that when the wall has been designed to resist the thrust due to an unstable slope (Figure 8.34), the overall stability of the whole soil mass with the inserted wall needs to be carefully investigated, by using the procedures developed for *slope stability analyses* in Section 8.18.

In order to provide all information needed to perform the above-mentioned analyses, soil investigations should reach a depth at least equal to the wall height. Furthermore, the foundation of the wall should be placed at depth where frost action and moisture changes can be excluded.

8.8 Design of diaphragm walls

Diaphragm walls are cast *in situ* structures, obtained by filling a trench with concrete. During the excavation, the stability of the trench is assured by the pressure of a slurry of bentonite and arching effects that develop around the hole. If the slurry pressure is higher than the water pressure in the soil, water tends to flow from the slurry into the soil, so that the bentonite will form a cake on the excavation faces and a full hydrostatic pressure is developed by the slurry.

The density of the slurry is a key factor: high densities are advisable because of the trench stability, but a slurry of low density is more easily removed from steel reinforcements.

Example 8.3
Examine the stability conditions of the retaining wall shown in Figure 8.35.

Let us start by computing the active thrust as predicted by the Coulomb theory. Assuming a wall friction angle $\delta = 2\varphi'/3 = 20°$, the active earth pressure coefficient is estimated as $K_a = 0.275$ (by using equation 8.20) and the resultant thrust assumes the value:

$$P_a = \frac{1}{2}\gamma h^2 K_a + qhK_a = 386.72 + 34.38 = 421.10\,\text{kN/m}$$

(note that a full height of 12.50 m, as composed of sandy silt, has been considered, for sake of simplicity). It is convenient to decompose this thrust into its horizontal and vertical components:

$$H = 395.83\,\text{kN/m}; \qquad V = 144.02\,\text{kN/m},$$

so that, by assuming a friction coefficient for the concrete wall base, cast against the ground, equal to 0.55 (as suggested by Terzaghi and Peck, 1967), the safety factor against sliding is equal to:

$$F_H = \frac{(W+V)\tan\delta_s}{H} = \frac{(1110+144.2)\cdot 0.55}{395.83} = 1.74.$$

The overturning moment is equal to:

$$M_R = (386.72\cdot 0.94)\cdot\frac{12.5}{3} + (34.38\cdot 0.94)\cdot\frac{12.5}{2} - 114.02\cdot 7.00$$
$$= 708.53\,\text{kN}\cdot\text{m/m}.$$

and the moment of the wall weight (see details in Figure 8.35) is equal to:

$$M_S = 5450.10\,\text{kN}\cdot\text{m/m},$$

so that the eccentricity of the vertical component of the resultant, with respect to the centre line of the base, will be within the middle third of the base, being:

$$e = \frac{b}{2} - \frac{M_S - M_R}{N} = 3.50 - \frac{5450.10 - 708.53}{1110.00 + 144.02} < \frac{b}{6}.$$

In order to answer the requirement related to the bearing capacity (i.e. can the foundation soil safely support this vertical load?) we need to learn about this subject, so that the related solution will be discussed in the worked Example 8.8.

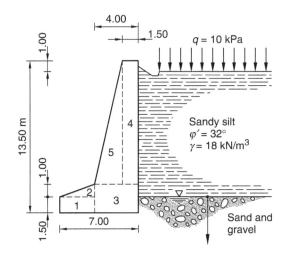

$$P_a = \frac{1}{2}\gamma h^2 K_a + qhK_a = 386.72 + 34.38 = 421.10 \text{ kN/m} \quad \begin{cases} V = 144.02 \text{ kN/m} \\ H = 395.83 \text{ kN/m} \end{cases}$$

$$\delta = 20° \; ; \; \beta = 0° \; ; \; h = 12.50 \text{ m} \; ; \; K_a = 0.275$$

$$K_a = \cfrac{\cos^2 (\varphi' - \beta)}{\cos^2\beta \cdot \cos (\beta + \delta)\left[1 + \sqrt{\cfrac{\sin (\delta + \varphi') \cdot \sin (\varphi' - i)}{\cos (\beta + \delta) \cdot \cos (\beta - i)}}\right]^2} \qquad \text{Equation (8.20)}$$

$$F_H = \frac{(W + V)\tan\delta_s}{H} = \frac{(1110 + 144) \cdot 0.55}{395.83} = 1.74$$

Wall element	Weight [kN/m]	Lever arm [m]	Ms [kN·m/m]
1	108.0	1.50	162.00
2	36.00	2.00	72.00
3	240.00	5.00	1200.00
4	396.00	6.25	2475.00
5	330.00	4.67	1541.10
$\Sigma = 1110.00$			$\Sigma = 5450.10$

$$M_R = 363.52\frac{12.5}{3} + 32.32\frac{12.5}{2} - 144.02 \cdot 7.00 = 708.53 \text{ kN·m/m}$$

$$F_R = \frac{M_S}{M_R} = \frac{5450.10}{708.53} = 7.69$$

Figure 8.35 Stability analysis of a retaining wall.

Example 8.4
Examine the stability of the cantilever wall shown in Figure 8.36.

Consider the *virtual back A − B* of the wall and assume that the Rankine active conditions apply on this virtual section. Then, the use of equation (8.16) gives $K_a = 0.287$ and $P_a = 34.34$ kN/m.

Note that this assumption implies that vertical planes and those parallel to the sloping surface are conjugate planes, so that $\sigma'_{vo} = \gamma z \cos i$ and the vector P_a has an obliquity equal to the sloping surface. If the active thrust is decomposed into its vertical and horizontal components, we get $V = 5.96$ kN/m and $H = 33.81$ kN/m.

The weight of the wall is computed by subdividing its actual geometry into a number of simpler shapes (as illustrated in Figure 8.36) so that we obtain a total weight 101.05 kN/m, which includes the backfill overlying the base heel. By assuming a friction angle for the foundation base against the soil equal to 0.50, the safety factor against sliding is equal to:

$$F_H = \frac{(101.05 + 5.96) \cdot 0.50}{33.81} = 1.58 \,.$$

By computing the overturning moment of the soil thrust and the stabilizing moment of the total weight of the cantilever wall (including the backfill acting on the heel), the eccentricity of vertical component of the resultant with respect to the centre line of the base is obtained as:

$$e = \frac{b}{2} - \frac{108.22 + 5.96 \cdot 1.80 - 33.81 \cdot 1.19}{101.05 + 5.96} = 0.165 < \frac{b}{6} \,.$$

Assuming that the contact normal stress is distributed linearly, the maximum normal stress is equal to:

$$p_{max} = \frac{N}{b} \left(1 + \frac{6e}{b}\right) = 92.15 \, \text{kPa},$$

and this can be considered to be a tolerable bearing stress.

However, a more refined analysis concerned with bearing capacity is presented in the illustrative Example 8.8.

One advantage of diaphragm walls over sheet piles is that they can be used as structural elements to be included in a construction.

(a) A **cantilever** (or unpropped) **diaphragm wall** (Figure 8.37) is used as a temporary support and when the expected soil movements around the excavation are not of concern for existing structures. If this is not the case, the diaphragm wall can be supported near the top by a tie-back or prop.

The failure mechanism of a cantilever wall is a rotation around point O, near the base, and its stability completely depends on the passive resistance mobilized in front of the wall above point O and on the back below the same point (Figure 8.37b). A simplified analysis considers the earth pressure distribution

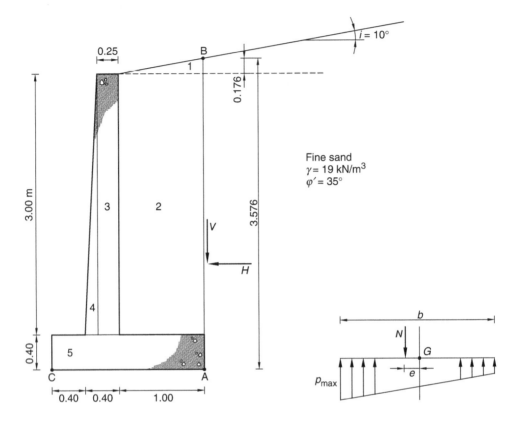

$$K_a = \frac{\cos i - \sqrt{\cos^2 i - \cos^2 \varphi'}}{\cos i + \sqrt{\cos^2 i - \cos^2 \varphi'}} = 0.287 \qquad \text{(Equation 8.16)}$$

$$P_a = 34.34 \text{ kN/m} \begin{cases} V = 5.96 \text{ kN/m} \\ H = 33.81 \text{ kN/m} \end{cases} \qquad F_H = \frac{(101.05 + 5.96)\, 0.50}{33.81} = 1.58$$

Wall element	Weight [kN/m]	Lever arm [m]	M_c [kN·m/m]
1	1.67	1.487	2.45
2	57.00	1.30	74.10
3	18.75	0.675	12.66
4	5.63	0.50	2.81
5	18.00	0.90	16.20
$\Sigma = 101.05$			$\Sigma = 108.22$

$$e = \frac{b}{2} - \frac{108.22 + 5.96 \cdot 1.80 - 33.81 \cdot 1.19}{101.05 + 5.96} = 0.90 - \frac{78.65}{107.01} = 0.165 < \frac{b}{6}$$

$$P_{max} = \frac{N}{b}\left(1 + \frac{6e}{b}\right) = \frac{107.01}{1.80}\left(1 + \frac{6 \cdot 0.165}{1.80}\right) = 59.45 \cdot 1.55 = 92.15 \text{ kPa}$$

Figure 8.36 Stability analysis of a cantilever wall.

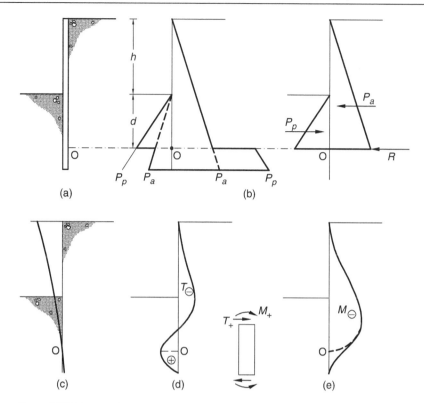

Figure 8.37 Design of a cantilever diaphragm wall.

shown in Figure 8.37b, and the net soil resistance below point O is represented by force R.

The depth of embedment d is obtained by equating moments about O and this value is increased (usually by 20%) in order to allow the development of the passive resistance below O, because the equilibrium is possible only when involving the passive resistance on the back of the wall. Therefore, it must always be checked that the estimated additional length can provide the required resistance.

Because of the relevance of the passive resistance in assuring the overall equilibrium, it is usual practice to apply a safety factor F_p (of the order of 2) to the passive resistance in front of the wall (also see the following Section 8.9, where the way safety factors may be introduced is discussed in a more general context).

The above procedure also applies to short-term analysis, performed in terms of undrained shear strength. Earth pressures are expressed in terms of total stresses, but the usual practice of factoring the passive resistance should be avoided, since in this case consistent results are only obtained by factoring the undrained shear strength (see Burland *et al.*, 1981).

(b) When dealing with a **propped diaphragm wall** (Figure 8.38), the stability of the diaphragm relies on the passive resistance in front of the wall and on the horizontal force provided by the prop or anchor.

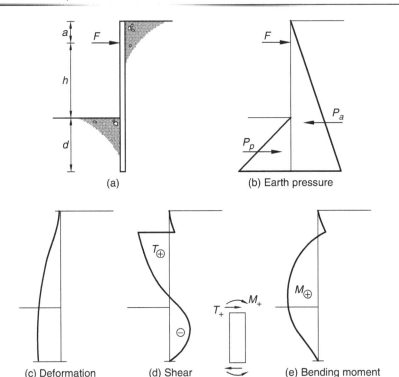

(a)

(b) Earth pressure

(c) Deformation   (d) Shear   (e) Bending moment

Figure 8.38 Design of a propped diaphragm wall (free earth support).

The simplest method of analysis, the **free earth support**, is based on the assumption that the depth of embedment or the wall rigidity prevents an inverted curvature near the base, so that the base is free to rotate. The failure mechanism envisaged in this case involves a rotation around the anchor.

The depth of embedment d is first obtained by considering equilibrium of rotation around the application point of the anchor force F; the equilibrium of horizontal translation then gives the force of the anchor.

The passive resistance must be here again divided by a factor of safety F_p and the anchor force is increased by 25% (see Section 8.9).

(c) Bending moments under serviceability conditions depend on the mode of wall deformation, the stress distribution being influenced by these modes. Limit analysis does not consider these aspects, because soil behaviour has been assumed rigid perfectly plastic and wall flexibility has not entered the analysis.

The influence of wall flexibility can be anticipated by observing that a flexible wall tends to bend towards the dredge level, so that, under serviceability conditions, little passive resistance is mobilized near the toe and the line of action of the passive force moves towards the dredge surface. This will result in a reduced lever arm and a reduced bending moment.

In order to account for the wall flexibility, Rowe (1952, 1955) introduced a flexibility number $\rho = H^4/EJ$, where H is the total height of the wall and EJ is the wall flexural rigidity, and showed that the ratio between the actual moment and that computing assuming a limiting stress distribution depends on this flexibility number.

An alternative procedure is suggested by Padfield and Mair (1984), based on the consideration that at the excavation level the actual stress distribution is likely to be the fully mobilized passive resistance. Therefore, when designing cantilever diaphragm walls, the maximum bending moment may be computed by using unfactored soil strength parameters. Since the depth of embedment is designed by using factored parameters, its actual value is greater than that required to give limit equilibrium with unfactored parameters. This requires that the additional length is not considered when computing the bending moment.

The same procedure also applies to propped walls: on the basis of the free-earth support method, the maximum bending moment is again computed with unfactored soil parameters, disregarding the additional depth of embedment. Prop loads, computed at limiting conditions, should be increased by 15 to 25%, and in any case always conservatively assessed, because of serious consequences of any failure.

8.9 Choice of strength parameters and factor of safety

Conventional design of embedded retaining structures is performed on the basis of limit equilibrium stress distribution. However, despite the widespread use of this approach, there is still debate about the way the safety factor should be applied, as well as about the values of shear strength and wall friction which should be adopted (see Bolton and Powrie, 1987, 1988; Burland *et al.*, 1981; Carder and Symons, 1989; Padfield and Mair, 1984; Powrie, 1996; Simpson, 1992). Most of these difficulties arise from the fact that when conventional procedures were set out, differences between peak and critical state strength were not recognized. In addition, it is not so simple to forecast an appropriate value of wall friction, since it depends on the direction and amount of wall movements, so that lower values of the interface friction may be mobilized and different values may be required, on either sides, to satisfy vertical equilibrium.

Traditionally, a reduction factor F_p (of the order of 2) is applied to the passive resistance, as a consequence of the observation that limiting active conditions are attained more rapidly than limiting passive conditions. This was the route followed in the previous Section 8.8 and related illustrative examples.

A scrutiny of this approach reveals that the application of the factor of safety to the passive earth pressure coefficient implies an equivalent coefficient, directly applied to the soil strength, which reduces as far as the angle of shear resistance increases. Therefore, to some extent this approach may be revealed misleading if interpreted in terms of mobilized shear strength.

Alternatively, Eurocode 7 (1995) suggests the application of a partial factor of 1.25 to $tan\varphi'_k$, where φ'_k is named the *characteristic value* of the angle of shear strength, defined as cautious estimate of the soil strength governing the occurrence of a limit state. Similarly, a partial factor of 1.4 to the characteristic value of the undrained shear strength is applied when reference is made to short-term analysis. In this case,

Example 8.5
Given the excavation in Figure 8.39, *find* the depth of embedment and the anchor force of the diaphragm wall.

1   The active earth pressure K_a, by making use of Coulomb critical wedge solution and assuming a wall friction $\delta = 2\varphi'/3 \cong 22°$, is equal to 0.254. The horizontal component of the active pressure is then given by $\sigma'_h = K_a \sigma'_{vo} \cos \delta = 0.24 \sigma'_{vo}$.

The passive earth resistance coefficient K_{PH}, based on the lower bound solution (equation 8.27), is equal to 5.66, assuming $\delta/\varphi' = 0.5$. Note that in this case the solution already provides the horizontal component.

2   In order to compute the depth of embedment, reference is made to the free earth support method, so that the required depth is obtained by considering the equilibrium of rotation around the anchor point:

$$\frac{1}{2} \frac{K_{PH}}{F} \gamma d^2 \cdot \left(\frac{2}{3} d + h\right) - \frac{1}{2} \gamma K_{AH} (a + h + d)^2 \cdot \left[\frac{2}{3}(a + h + d) - a\right] = 0.$$

If the passive resistance is factored by a factor of safety equal to 2, the above equation gives $d = 2.50$ m.

3   The equilibrium of translation gives $T_h = P_a - P_p = 251.37 - 168.31 = 83.06$ kN/m.

If an inclined anchor (10° from the horizontal) is used, the required force is 84.34 kN/m and a load factor of 1.25 must be applied when designing the anchor. It must also be considered that, in order for the anchor be fully effective, its foundation must be behind the critical active wedge, as shown in Figure 8.39c. Where an anchor plate should be used, it must be checked that critical active and passive wedges do not intersect (see Figure 8.39d).

4   Note that we did not consider in this illustrative Example unplanned events, such as surcharges on the retained side and unexpected additional excavation, as discussed in Section 8.9.

the factor of safety applied to soil strength can be considered a *strength mobilization factor*, since its main purpose is to limit wall movements under working conditions (see British Standards BS 2002).

In both cases, the fundamental question of what angle of shear resistance (i.e. peak or critical state?) should be used still remains. In this respect, analyses provided by Powrie (1996) and Padfield and Mair (1984) are of particular value. Based on centrifuge and laboratory tests, as well as on the behaviour of real walls, Powrie (1996) suggests that the collapse of either embedded cantilever or propped walls can be well or slightly conservatively predicted by using the critical state value φ'_{cs}, assuming on both sides the soil-wall friction $\delta = \varphi'_{cs}$ and using earth pressure coefficients which are consistent with the presence of wall-soil rough interface (i.e. Caquot and Kerisel (1948) or those derived by using lower bound solutions of plasticity). As a consequence, in order to

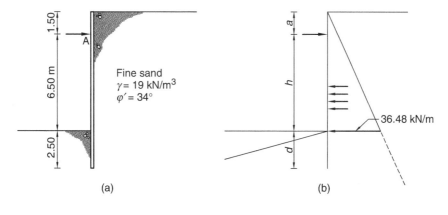

(a)   (b)

(1) $K_a = 0.254$ (equation 8.22 with $\delta = 22°$); $K_{AH} = 0.24$

(2) $K_{PH} = 5.665$ (equation 8.27 with $\delta = 17°$); $\dfrac{K_{PH}}{F} = \dfrac{5.665}{2} = 2.83$

(3) $\Sigma M_A = 0$

$$\frac{1}{2}\frac{K_{PH}}{F}\gamma\left(\frac{2}{3}d+h\right)d^2 - \frac{1}{2}\gamma K_{AH}\left[\frac{2}{3}(d+h+a)-a\right](h+a+d)^2 = 0$$

$$d = 2.50 \text{ m}$$

(4) $\Sigma H = 0$

$$T_h = 251.37 - 168.31 = 83.06 \text{ kN/m} \qquad T = \frac{T_h}{\cos 10} = 84.34 \text{ kN/m}$$

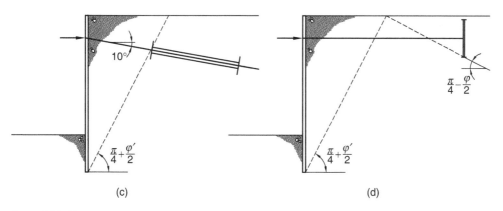

(c)   (d)

Figure 8.39 Illustrative example of a tie-back diaphragm wall.

guarantee a satisfactory performance of the wall, the design should be based on limit equilibrium stress distribution computed with the mobilized soil strength $\tan\varphi'_{mob} = \tan\varphi'_{cs}/F_{\varphi'}$, being $F_{\varphi'} \geq 1.25$, and $\delta = \varphi'_{mob}$.

This procedure can be expected to be slightly less conservative than that suggested by Padfield and Mair (1984) for permanent works in stiff clays, based on a mobilizing factor 1.2 applied to φ'_{cs}, with $\delta = \varphi'_{mob}/2$ in front of the wall and $\delta = 2\varphi'_{mob}/3$ behind.

Furthermore, note that it is also suggested (see Eurocode 7 and British Standards BS 2002) that the design should consider unplanned events, such as a surcharge on the retained side and additional excavation. For the additional unplanned excavation, it is suggested a value equal to 10% of the wall height above excavation for cantilever walls, and equal to 10% of the distance between the lowest support and the excavation level for propped walls, with a maximum of 0.5 m in both cases, *unless the excavation level is reliably controlled throughout the work*. These unforeseen additional excavation values have large influence on safety against passive failure and on bending moments.

8.10 Strutted excavations

The term temporary or provisional structure indicates all the supports needed to make an excavation feasible. In this respect, the use of sheet piles or diaphragm walls, supported by multiple rows of props or anchors, represents only one possibility, considering that technological progresses today offer alternative procedures linked to the use of soil improvement or treatment techniques (see for example freezing, injections and grouting). More classical techniques of installing props or anchors at different levels, as the excavation proceeds, are dealt with in this section. These techniques have received a lot of attention, due to the recent increase in the use of the 'top down' construction method.

It may be of interest to briefly recall the main characteristics of these structures: (a) they do not generally represent the final construction, even though, when dealing with diaphragm walls, the designer tends to consider them as a part of the construction (typically a multipropped retaining structure is appropriate when the wall is part of the structure of a deep basement, the floor slabs acting as props); (b) the economy and safety margin are often linked to the provisional character of these structures; (c) quite often, it is difficult to have a definite design in advance, and some flexibility is advisable in order to modify the choices according to the real soil conditions encountered during work; (d) this latter statement should be accepted on the basis of the **observational method** suggested by Peck (1969), which requires a careful and continuous inspection of the performance of the structure during work, and an inventory of possible alternatives if the original assumptions deviate from reality.

The main difficulties encountered when designing strutted excavations are linked to the fact that the earth pressure distribution acting on the wall and prop loads strongly depends on details of the excavation programme. For this reason, the current practice relies on empirical approaches based on case histories, and in order to proceed with such an analysis, let's start by considering the most relevant mechanisms of failure.

Failures of *strutted excavations in sands* above the water table have occurred mainly due to *buckling of struts*. The phenomenon has sometimes been preceded by crippling of the wall, but no failures due to heave of the bottom of the excavation have been reported. On the contrary, in *sands below the water table*, instability may arise due to *upward seepage* into the excavation, or a *blowup of the bottom* can occur if a pervious stratum embedded in clay or silty clay remains at its original hydrostatic pressure.

Excavations in clays usually proceed rapidly, therefore undrained conditions prevail and if the depth of excavation is excessive compared to the strength of the clay, a *bearing capacity failure* at the bottom can occur due to the soil outside the structure acting as a load.

Collapses have also been reported due to buckling of struts, in absence of bottom heave or failure.

To avoid the danger of buckling of the struts, they should be designed on the basis of the maximum load that each strut has to bear during all stages of the excavation.

Deformations of the structure, as the excavation proceeds, are different from those associated to classical earth pressure theories and loads on the upper struts are higher than those predicted by these theories, due to stress distribution.

To solve this problem, the measured loads on the struts have been converted into a diagram of **apparent pressure** (Flaate and Peck, 1973; Peck, 1969; Terzaghi and Peck, 1967).

In this view, when dealing with **excavations in sand**, it has been suggested to refer to a constant with depth diagram (Figure 8.40a), the earth pressure being:

$$p_{app} = 0.65\gamma H \frac{1 - \sin\varphi'}{1 + \sin\varphi'}. \tag{8.49}$$

When dealing with **excavations in clays**, the first step is to evaluate the stability number:

$$N = \frac{\gamma H}{s_u}. \tag{8.50}$$

If this number is close to 8, a bottom failure is likely to occur. On the contrary, if N is less than 4, the behaviour of the clay is expected to be essentially elastic and the apparent pressure diagram is that of Figure 8.40c. The lower value, equal to $0.25\gamma H$, is associated to relatively shallow excavations in stiff clays.

Finally, if the stability number is higher than 4, the apparent pressure of Figure 8.40b should be used, where the value of m is often taken to be equal to 1, even though a value as low as 0.4 would be appropriate for soft clays.

Note that these diagrams are not intended to give the distribution of loads on the wall: they represent an **artifice** for calculating only prop loads, whereas bending moments and shear forces on the wall cannot be calculated by apparent envelopes. In this respect, these diagrams are subjected to be refined, as more field data are available.

Further observations are the following: when dealing with soft to medium clays, the apparent pressure represents the total pressure, whereas in sands water pressure has to be added to give the total pressure envelope. Prop loads for clays have been developed for short-term conditions, so that some care is required when considering long-term conditions.

In order to evaluate bending moments and shear forces on the wall, advanced numerical approaches allow to account the complex soil-structure interaction. However, since such methods are relatively new, simplified methods still appear in codes of practice or are used as a preliminary design tool.

When using such simplified procedures it must always be recognized that their reliability is open to debate, and that the analyses should be performed by considering each stage of the construction sequence.

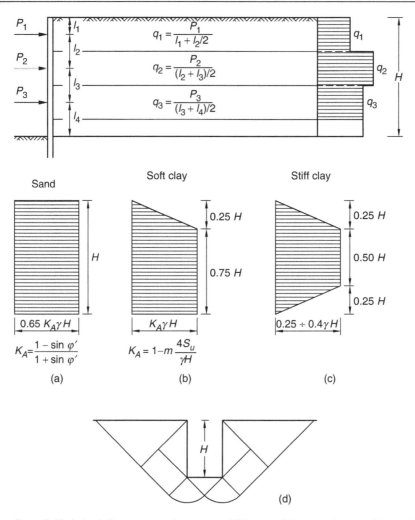

Figure 8.40 (a, b, c) Apparent earth pressure; (d) bottom failure mechanism (adapted from Flaate and Peck, 1973)

8.11 Earth pressure in presence of seismic actions

Even though there has been considerable debate in the literature about the use of force-based approaches (see Steedman and Zeng, 1990a; Zeng and Steedman, 1993), current practice for computing earth thrust on earth retaining structures in the presence of seismic action still relies on an extension of the Coulomb solution, due to Okabe (1924) and Mononobe and Matsuo (1929), and referred to in the literature as the **Mononobe-Okabe approach** (see Kramer, 1996).

8.11.1 Seismic active earth thrust

The approach suggested by Okabe (1924) and Mononobe and Matsuo (1929) considers the Coulomb wedge (Figure 8.41) acted upon by horizontal and vertical

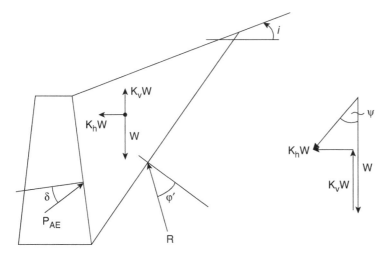

Figure 8.41 Mononobe-Okabe method.

pseudo-static forces (in addition to forces acting under static conditions), whose magnitude is related to the mass of the wedge, through the pseudo-static accelerations:

$$
\begin{aligned}
a_h &= k_h g \\
a_v &= k_v g
\end{aligned}
\tag{8.51}
$$

Then, by following the same line of reasoning of Section 8.4, the most adverse active thrust can be expressed in the form:

$$
P_{AE} = \frac{1}{2} K_{AE} \gamma H^2 (1 - k_v),
\tag{8.52}
$$

where the seismic active pressure coefficient is given by:

$$
K_{AE} = \frac{\cos^2 (\varphi' - \beta - \psi)}{\cos \psi \cos^2 \beta \cos (\delta + \beta + \psi) \left[1 + \sqrt{\dfrac{\sin (\delta + \varphi') \sin (\varphi' - i - \psi)}{\cos (\delta + \beta + \psi) \cos (i - \beta)}} \right]^2},
\tag{8.53}
$$

and:

$$
\psi = \tan^{-1} \left[\frac{k_h}{1 - k_v} \right].
\tag{8.54}
$$

This is a conceptually simple approach, and most researchers agreed that the esti-mated total lateral thrust is approximately correct. On the reverse, the distribution of the dynamic increment requires further scrutiny, due to the fact that the Mononobe-Okabe approach assumes a uniform acceleration field through the backfill, whereas phase changes can have an influence on the actual distribution of the earth thrust.

Tajimi (1973) first investigated this point, showing that the earth stress distribution is a function of a dimensionless parameter $\omega H/v_s$, where v_s is the shear wave velocity and ω is the the angular frequency of the base shaking. Steedman and Zeng (1990a) have explored further this point, reaching the conclusion that the resultant thrust acts above one-third of the height of the wall. Furthermore, since the maximum thrust and the maximum moment are in phase with the acceleration at the mid height, this acceleration may be the most appropriate value for the design.

The **presence of water** in the backfill also influences the seismic loads, depending on whether the pore water moves or not relative to the soil during earthquakes. This aspect is discussed by Matsuzawa *et al.* (1985) and Kramer (1996), and accordingly when the hydraulic conductivity of the soil is low enough (i.e. $K \leq 5 \cdot 10^{-4}$ m/s) that no relative movement of the water is expected, then the inertial forces can be assumed to be proportional to the *total unit weight* of the soil. This is equivalent to assume (see Ebelin and Morison, 1992; Dakoulas and Gazetas, 2008; Kramer, 1996):

$$\psi = \tan^{-1}\left[\frac{\gamma}{\gamma'(1-r_u)}\frac{k_h}{1-k_v}\right],\tag{8.55}$$

where r_u is the excess pore pressure ratio, i.e. $r_u = \Delta u/\sigma'_{vo}$.

Then, the active thrust can be computed using equation (8.52), where the unit weight $\gamma = \gamma'(1-r_u)$ should be introduced, and an equivalent hydrostatic pore pressure thrust should be added, based on a fluid unit weight $\gamma_{wd} = \gamma_w + r_u\gamma'$.

On the contrary, presuming the water is free to move, the vertical body force is given by $\gamma'(1\pm k_v)$, while the horizontal body force is contributed by the solid fraction and is equal to $\gamma_{dry}k_h$. Therefore the angle ψ assumes the expression:

$$\psi = \tan^{-1}\left[\frac{\gamma_{dry}}{\gamma'}\frac{k_h}{1-k_v}\right].\tag{8.56}$$

Furthermore, the hydrodynamic pore pressure must be added to the hydrostatic water pressure.

The resultant hydrodynamic pressure is commonly estimated by using the Westergaard (1931) solution:

$$P_{wd} = \frac{7}{12}\frac{a_h}{g}\gamma_w H^2.\tag{8.57}$$

Since the distribution with depth is given by:

$$u_d = \frac{7}{8}\frac{a_h}{g}\gamma_w\sqrt{z_w H},\tag{8.58}$$

the resultant P_{wd} may be assumed to act at $0.4H$ above the base of the wall.

Note that the pseudo-static approach allows to estimate the seismic thrust on the retaining wall, so that the ultimate limit state can be investigated in a rather simple and convenient way. But, as expected, more information is required when the overall performance of the wall has to be assessed, and this is related to permanent displacements that the wall can suffer. These displacements can be predicted by using the approach pioneeered by Richards and Elms (1979), based on the sliding block procedure originally proposed by Newmark (1965), successively improved by Whitman and Liao (1985).

8.11.2 Seismic passive earth resistance

We have already discussed that, when wall friction is present, a non-uniform stress field arises as well as a non-planar failure surface. This renders the problem of computing exact values of passive earth resistance rather complex. In particular, when dealing with passive earth resistance, standard codes suggest solutions provided by limit equilibrium methods with a curved (typically log spiral) surface (see Caquot and Kerisel, 1948). However, since these procedures are essentially of a kinematical nature, they are not conservative. In fact, should the assumed mechanism be admissible in kinematic terms, these solutions represent an upper bound to the exact solution. For this reason, it is of interest to rely on solutions based on the lower bound theorem of plasticity, as already discussed in Section 8.5. In the following we present an approach (Lancellotta, 2007), the novelty of which lies in a transformation of axes that allows the problem of seismic passive resistance to be solved using the same stress field equations as the usual static case.

Consider the problem shown in Figure 8.42: a soil surface, sloping at an angle i with respect to the horizontal axis x is subjected to the vertical body force γ, due to gravity, and to the horizontal body force $k_h\gamma$, which represents the seismic action, the coefficient k_h being the horizontal seismic coefficient (positive assumed if the inertia force is towards the backfill). In order to compute the passive resistance on a vertical wall of roughness δ (i.e. $\sigma'_{xz} = \sigma'_{xx}\tan\delta$), imagine transforming the problem geometry through a rigid rotation ψ, being:

$$\psi = \tan^{-1}k_h. \tag{8.59}$$

According to equation (8.59), ψ represents the obliquity of the body force per unit volume in presence of seismic action and it is also noted that the presence of a vertical component of the inertia forces could be taken into account by assuming:

$$\psi = \tan^{-1}\left(\frac{k_h}{1 \pm k_v}\right). \tag{8.60}$$

By referring to this transformed geometry we now deal with the problem of deriving the passive resistance acting on a rough wall, tilted from the vertical by the angle ψ, and interacting with a backfill of slope $\beta = i - \psi$. The resulting vertical body force is represented by the vector $\overline{\gamma} = \gamma\sqrt{1+k_h^2} = \gamma/\cos\psi$, that can be thought of as a properly scaled gravity body force (in the presence of vertical acceleration it would be $\overline{\gamma} = \gamma(1-k_v)/\cos\psi$).

With reference to Figure 8.43, zone 2 is the conventional passive zone in which the stress state is known, as represented by the small Mohr circle in Figure 8.43b, also

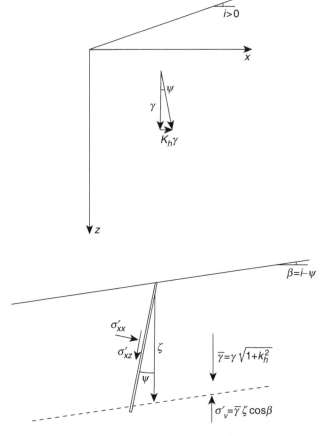

Figure 8.42 Initial and transformed geometry (Lancellotta, 2007).

shown for clarity in Figure 8.43c. If the segment OM represents the resultant stress $\sigma'_v = \overline{\gamma}\zeta\cos\beta$, acting at depth ζ on the plane parallel to the ground surface, it can be observed that the following relations hold (Figure 8.43b):

$$\overline{OM} = \overline{OH_2} - \overline{H_2M}; \quad \overline{OH_2} = s'_2 \cos\beta; \quad \overline{H_2M} = \sqrt{r^2 - \overline{C_2H_2}^2} =$$

$$s'_2\sqrt{\sin^2\varphi' - \sin^2\beta},$$

so that:

$$\overline{OM} = \overline{\gamma}\zeta\cos\beta = s'_2\left(\cos\beta - \sqrt{\sin^2\varphi' - \sin^2\beta}\right). \tag{8.61}$$

Moving through a fan of stress discontinuities to the zone 1 adjacent to the wall, the normal component σ'_{xx} of the passive earth resistance is obtained from the large

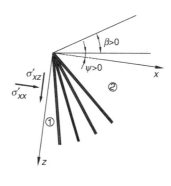

(a)

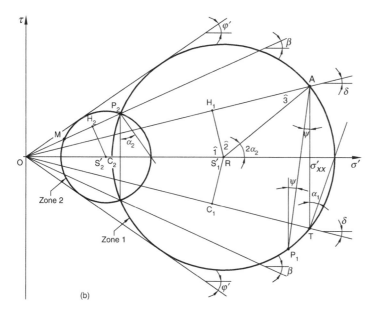

(b)

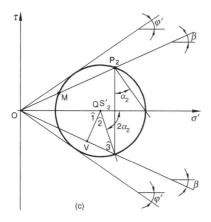

(c)

Figure 8.43 Stress discontinuity analysis: (a) Fan of stress discontinuities (b) Mohr circles relative to Zone 1 and Zone 2 (c) Mohr circle relative to conventional passive zone (Lancellotta, 2007).

Mohr circle in Figure 8.43b, by observing that:

$$\sigma'_{xx} = \overline{OT}\cos\delta; \quad \overline{OT} = \overline{OC_1} + \overline{C_1T} = s'_1(\cos\delta + \sqrt{\sin^2\varphi' - \sin^2\delta}), \text{ i.e.}$$

$$\sigma'_{xx} = s'_1\cos\delta(\cos\delta + \sqrt{\sin^2\varphi' - \sin^2\delta}). \tag{8.62}$$

By combining (8.61) and (8.62):

$$\sigma'_{xx} = \frac{s'_1}{s'_2}\frac{\cos\delta(\cos\delta + \sqrt{\sin^2\varphi' - \sin^2\delta})}{(\cos\beta - \sqrt{\sin^2\varphi' - \sin^2\beta})}\overline{\gamma}\zeta\cos\beta. \tag{8.63}$$

We now briefly recall that if, in order to smooth the overall stress field, we consider a fan of stress discontinuities (Figure 8.43a), across which the rotation of the principal direction assumes the finite value ϑ, then the shift between the two extreme Mohr circles is defined by (see Section 8.5, Atkinson, 1981; Bolton, 1979; Powrie, 1997; Lancellotta, 2002):

$$\frac{s'_1}{s'_2} = e^{2\vartheta\tan\varphi'}, \tag{8.64}$$

so that by inserting (8.64) into (8.63) the following solution is obtained:

$$\sigma'_{xx} = K_{PE}\cdot\overline{\gamma}\zeta\cos(i - \psi). \tag{8.65}$$

Finally, by taking into account the definition of $\overline{\gamma} = \gamma(1 - k_v)/\cos\psi$ and by observing that $z\cos i = \zeta\cos(i - \psi)$, the equation (8.65) in terms of the original variables writes:

$$\sigma'_{xx} = K_{PE}\cdot\gamma z(1 - k_v)\cos i/\cos\psi, \tag{8.66}$$

where:

$$K_{PE} = \left[\frac{\cos\delta}{\cos(i - \psi) - \sqrt{\sin^2\varphi' - \sin^2(i - \psi)}}\left(\cos\delta + \sqrt{\sin^2\varphi' - \sin^2\delta}\right)\right]e^{2\vartheta\tan\varphi'} \tag{8.67}$$

and, as proved in the worked Exercise 8.6:

$$2\vartheta = \sin^{-1}\left(\frac{\sin\delta}{\sin\varphi'}\right) + \sin^{-1}\left(\frac{\sin(i - \psi)}{\sin\varphi'}\right) + \delta + (i - \psi) + 2\psi. \tag{8.68}$$

The obtained solution is of value in engineering practice because it is a conservative estimate of the exact solution (see Kumar, 2001; Kumar and Chitikela, 2002, and the comparison in Figure 8.44 with the solution provided by Chang (1981), based on the limiting equilibrium approach with composite sliding surfaces). It allows the wall

Example 8.6
Proof of equation (8.68).

Solution. Consider in Figure 8.43b the larger Mohr circle, relating to zone 1. The following relations hold:

$$\hat{1} = \frac{\pi}{2} - \delta; \quad \hat{2} = \frac{\pi}{2} + \delta - 2\alpha_1; \quad \hat{3} = \frac{\pi}{2} - \hat{2} = 2\alpha_1 - \delta$$

$$\overline{H_1 R} = \overline{OR}\sin\delta = \frac{r}{\sin\varphi'}\sin\delta = r\sin\hat{3}$$

so that:

$$\sin(2\alpha_1 - \delta) = \frac{\sin\delta}{\sin\varphi'}.$$

By considering the other Mohr circle, relating to zone 2 (see details in Figure 8.43c), it follows:

$$\hat{1} = \frac{\pi}{2} - \beta; \quad \hat{2} = \frac{\pi}{2} + \beta - 2\alpha_2; \quad \hat{3} = 2\alpha_2 - \beta,$$

so that:

$$\sin(2\alpha_2 - \beta) = \frac{\sin\beta}{\sin\varphi'}.$$

Since:

$$2\vartheta = 2(\alpha_1 + \alpha_2 + \psi) = \sin^{-1}\left(\frac{\sin\delta}{\sin\varphi'}\right) + \sin^{-1}\left(\frac{\sin(i - \psi)}{\sin\varphi'}\right) + \delta + (i - \psi) + 2\psi,$$

the equation (8.68) is proved.

roughness to be taken into account, avoiding unjustified pessimistic assumptions (as an example, some codes suggest the use of the Mononobe-Okabe formula by neglecting the wall roughness), and in the absence of seismic action (i.e. $\psi = 0$) may be also used for the case of a sloping backfill.

8.12 Bearing capacity of shallow footings

The theory of bearing capacity deals with the *ultimate limit state* occurring when the applied load gives rise to a general collapse mechanism of the foundation. The analysis of such a mechanism, as well as the use of safety factors to guard against failure, is the subject of this and subsequent Sections 8.13 to 8.17.

However, the reader should be aware from the beginning that the definition of an allowable bearing load on the basis of the bearing capacity value does not represent

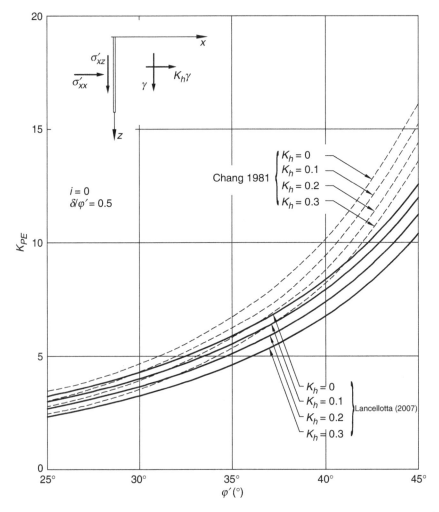

Figure 8.44 Coefficient of seismic passive earth resistance for a specific case ($i = 0, \delta/\varphi' = 0.5$) (Lancellotta, 2007).

the final solution, as in many cases the need to limit the settlements to tolerable values represents a more restrictive design requirement. This requirement, linked to a *serviceability limit state*, will be addressed in Chapter 9.

With reference to the ultimate bearing load, plasticity theorems help us to define bounds of collapse load, by using two approaches, already discussed in Section 1. For convenience we recall that the **lower bound theorem** states that if we find a stress field such that equilibrium equations and traction boundary conditions are satisfied and the stresses do not exceed the yield condition, the collapse cannot occur. Therefore, the stresses applied at the boundary are a lower bound of the collapse stresses and the stress field that satisfies these requirements is called a *stress admissible field*.

In order to construct a statically admissible stress field, the domain is first divided into a number of regions of uniform stress by introducing a convenient number of *stress discontinuities*. Then, moving from a region of known stress through regions of unknown stress, ensuring that regions are in equilibrium with each other and with the applied stresses, without violating the failure criterion, the boundary stresses causing failure can be obtained.

The **upper bound theorem** states that if for any compatible plastic mechanism (a mechanism for which no gaps, separations or overlaps occur) the rate of work of external loads equals the rate of internal dissipation then the collapse must occur. In practice, in the kinematic approach we try to envisage the worst possible mechanism, by subdividing the domain into a number of rigid blocks separated by kinematic discontinuities, and the collapse load is computed by an energy balance, without considering local states of stress at all.

We outline that a number of important idealizations are introduced, when we apply the above-mentioned theorems. The soil is assumed to be a rigid-perfect plastic material with an associated flow rule. Further, in most cases the soil is assumed to be isotropic, i.e. the strength is the same in all directions, and homogeneous. Aspects linked to the assumption of associated flow rule have been analysed in Section 8.1; the influence of inhomogeneity and anisotropy will be discussed in Section 8.14.

8.13 Undrained bearing capacity

When loads are applied to soft clays, a consolidation process takes place in time and the consequences are delayed deformations, a reduction of water content and an increase of effective stresses. A conclusion is reached that a higher value of shear resistance is available in the long-term conditions, therefore the critical conditions for the stability are attained during the loading process. The loading rate is usually higher than that required for excess pore pressure dissipation, so that a reasonable assumption can be made that the loading process takes place in **undrained conditions**, without any changes of water content or excess pore pressure dissipation.

This case is examined in terms of **total stress**, to avoid difficulties linked to the prediction of excess pore pressure and the failure criterion reduces to that of a purely coesive material (the Tresca criterion), with the shear strength being expressed by the **undrained shear strength** s_u:

$$\tau_f = s_u. \tag{8.69}$$

8.13.1 Upper bound solution

In order to discuss the relevant features of the upper bound method of plasticity, let us start by considering the simple two-block mechanism sketched in the upper part of Figure 8.45, where we assume for simplicity plane strain conditions. Blocks are labelled A and B and the surrounding material is labelled O and imagine that corners are relieved in order to have a kinematically admissible mechanism. Figure 8.45b shows the velocity vectors of the two blocks and a property of this diagram, called velocity diagram or *hodograph*, is that the vector joining two points represents the relative velocity of the corresponding regions. Therefore, the vector $\overline{OA}$ represents the velocity of region A relative to the surrounding material O; the same holds for the vector $\overline{OB}$, representing the velocity of region B with respect to the stationary material O; then $\overline{AB}$ represents the velocity of the region B relative to the region A.

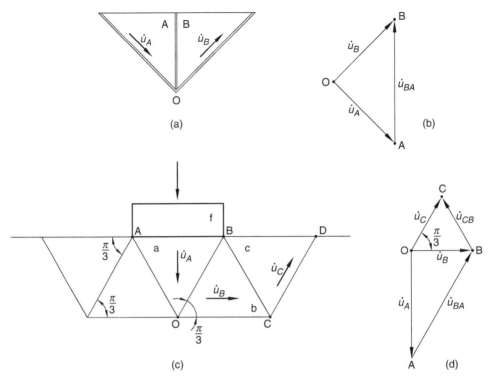

Figure 8.45 Velocity diagram or hodograph.

The Figure 8.45c shows a more sophisticated mechanism with an increased number of blocks and the associate hodograph is also shown. If the load P is given a unit relative velocity $\dot{u}_A$, by observing that the vector $\dot{u}_{BA}$, representing the velocity of block b relative to block a, must be parallel to the segment OB, one can deduce:

$$\dot{u}_B = \frac{\dot{u}_A}{\sqrt{3}} \quad ; \quad \dot{u}_{BA} = \frac{2}{\sqrt{3}}\dot{u}_A \, .$$

The power dissipated along a slip line is equal to the soil resistance times the relative velocity, and by equating the power done by the applied load to the dissipated power we obtain:

$$q_{\lim}\overline{AB}\cdot\dot{u}_A = 2s_u\left(\overline{OB}\cdot\dot{u}_{BA} + \overline{OC}\cdot\dot{u}_B + \overline{BC}\cdot\dot{u}_{CB} + \overline{CD}\cdot\dot{u}_C\right).$$

Taking into account the previous relations, the following solution is derived:

$$q_{\lim} = 5.77s_u \, . \tag{8.70}$$

As this solution is an upper bound of the true collapse load, it is of interest to explore other mechanisms, which could give a lower value than the one obtained. To this aim,

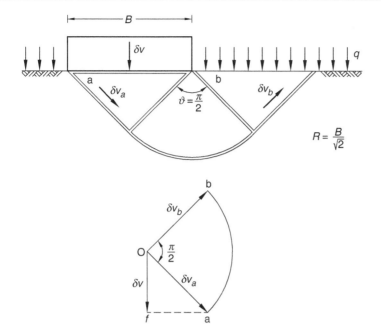

Figure 8.46 Undrained collapse mechanism.

let us imagine to insert between the two rigid blocks shown in Figure 8.46 a fan of slip planes. If the foundation is given a unit relative velocity δv, then:

$$\delta v_a = \delta v_b = \delta v \cdot \sqrt{2}.$$

The dissipated power in a fan is obtained by summing over the circular arc and the radial slip planes (Figure 8.47), i.e.:

$$\delta W = \int\limits_o^\theta s_u(Rd\theta) \cdot \delta w + \int\limits_o^\theta (s_u R) \cdot \delta w d\theta,$$

and, since $\theta = \dfrac{\pi}{2}$, we get:

$$\delta W = 2s_u R \delta v \frac{\pi}{2}\sqrt{2}.$$

Then, the total dissipated power is obtained by adding the power dissipated along the slip planes, equal to $2Rs_u\delta v\sqrt{2}$, so that:

$$\delta W = (2+\pi)s_u R \delta v \sqrt{2}.$$

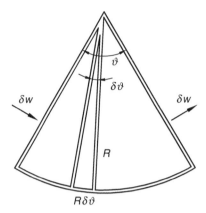

Figure 8.47 Dissipation within a fan of kinematic discontinuities.

The power done by the applied load and by the surface load q is equal to:

$$\delta L = F_u \delta v - q B \delta v$$

and, since $B = R\sqrt{2}$, the following solution is obtained:

$$q_{\lim} = \frac{F_u}{B} = q + (2 + \pi) s_u. \qquad (8.71)$$

This is a better approximation of the exact collapse load, being lower than the previous one. However, it still represents an upper bound solution, so that it is now of interest to move towards the application of the lower bound theorem.

8.13.2 Lower bound solution

In Figure 8.48 the footing and the soil above the founding depth D are represented by uniform stresses. The uniform stress $q_{\lim}$ is applied over the width of the footing, and the uniform stress q, supposed to act alongside, represents the overburden stress γD.

According to the lower bound theorem, we must find an admissible stress field, i.e. a stress field such that equilibrium equations and stress boundary conditions are satisfied and the stresses do not exceed the yield criterion.

A simple possibility is to introduce the vertical stress discontinuity sketched in Figure 8.48. Then, we move from the region of known state of stress, i.e. the surcharge q in region B, which is plotted as a point on the Mohr diagram. Since the region B is presumed to be at failure, assuming that the horizontal stress is greater than q, the Mohr's circle at failure can be constructed, the radius being equal to the undrained strength.

Moving to region A, it must be in equilibrium with region B across the stress discontinuity, so that the horizontal stresses must be equal. Assuming again that region A is at failure, we can construct the corresponding Mohr's circle, and we get:

$$q_{\lim} = q + 4 \cdot s_u. \qquad (8.72)$$

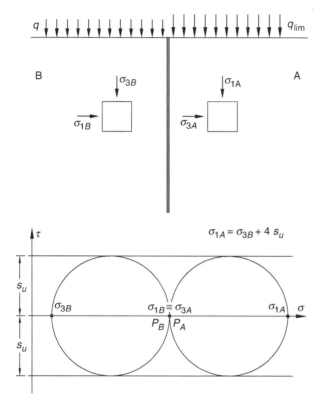

Figure 8.48 Lower bound solution for undrained loading.

The obtained solution represents a *lower bound* of the exact collapse load, so that it is of interest to envisage a better stress field to maximize this value.

Alternative possibilities can be explored by increasing the number of stress discontinuities, but this first requires to digress a little bit on some general theoretical aspects.

Figure 8.49 shows, as already discussed in Section 8.5, the jump ϑ in the direction of the major principal stress across a stress discontinuity, and in order to link this jump to the shift of Mohr's circles, we can proceed as follows. Suppose that the point M represents the stress state of the region 1. Then, the pole P_1 is obtained by drawing a line parallel to the stress discontinuity. Since the equilibrium requires that shear and normal stress must be the same on both sides of the discontinuity, the Mohr circles for the two zones must intersect at point M, and it follows that the pole of the Mohr'circle relative to the region 2 will be represented by point P_2.

Therefore, by considering the geometric relations shown in Figure 8.49, the change in the average total stress as a function of the rotation of the principal directions will be:

$$\Delta s = s_2 - s_1 = 2s_u \cdot \sin \vartheta . \tag{8.73}$$

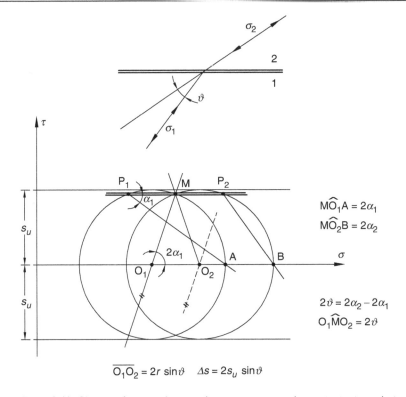

Figure 8.49 Change of principal stress direction across a discontinuity in undrained loading.

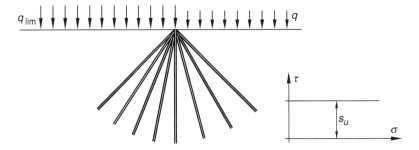

Figure 8.50 Lower bound solution: a fan of stress discontinuities.

For a fan zone, comprising n discontinuities (Figure 8.50), the change in the average total stress of the two extreme regions will be:

$$\Delta s = 2s_u \cdot n \sin\left(\frac{\pi}{2}\frac{1}{n}\right). \qquad (8.74)$$

The use of the expansion series $sin x = x - \dfrac{x^3}{3!} + \dfrac{x^5}{5!} - \dfrac{x^7}{7!}$, and the limit of (8.74) when $n \rightarrow \infty$ will reduce the previous expression to $2s_u \cdot \dfrac{\pi}{2}$, and, since:

$$q_{\text{lim}} = q + 2s_u + \Delta s$$

we obtain the solution:

$$q_{\text{lim}} = q + N_c \cdot s_u, \qquad\qquad\qquad (8.75)$$

where:

$$N_c = 2 + \pi \qquad\qquad\qquad (8.76)$$

is known as **bearing capacity coefficient**.

Note that, since the application of the lower bound theorem gives the same solution of the upper bound approach, this represents the exact value, first obtained by Prandtl (1921), by using the method of stress characteristics.

8.14 Non-homogeneity and anisotropy

8.14.1 The effect of non-homogeneity

The undrained shear strength of soft clays increases with depth, the ratio s_u/σ'_{vo} being constant for a given deposit. This gives the soil a character of inhomogeneity in the vertical direction and, in such a situation, a common practice is to rely on the use of a slip circle analysis, with the introduction of an averaged value of the undrained strength.

Davis and Booker (1973) provided a rigorous treatment of the problem, where the undrained strength increases linearly with depth:

$$s_u = s_o + \rho z, \qquad\qquad\qquad (8.77)$$

and by using the method of stress characteristics have shown that the limit load Q of a rigid strip footing is expressed in the form:

$$Q/b = F\left[(2+\pi)s_o + \rho b/4\right], \qquad\qquad\qquad (8.78)$$

where b is the breadth of the footing, s_o is the strength at the founding level and F is a dimensionless factor depending on the ratio $\rho b/s_o$ and on the roughness of the footing (see Table 8.1).

This solution proves that the usual practice of introducing an averaged value of the shear strength provides reliable results if the value at a depth of $b/4$ below the founding level is used in conjunction with the conventional solution (see Example 8.6). However, if this value is higher than twice the strength at the founding level, the design should be based on equation (8.78).

Table 8.1 Dimensionless factor F for rough footings

$\rho b/s_o$	1	2	3	4	6	8
F	1.20	1.36	1.45	1.50	1.57	1.62

Source: Deduced from Davis and Booker (1973).

Example 8.7
Figure 8.51 shows a strip footing resting on a soft clay with the undrained strength increasing linearly with depth, as expressed by the relation:

$$s_u = s_o + \rho z,$$

with value at the founding level $s_o = 10$ kPa and $\rho = 3.4$ kPa/m.
    Find the allowable bearing capacity.

Solution. The undrained strength at depth $b/4$ below the founding level is equal to 13.40 kPa, and this value is lower than $2s_o = 20$ kPa. Then, a sufficiently approximate estimate of the gross bearing capacity can be obtained by inserting this value into the solution obtained for the homogeneous case, i.e.:

$$q_{\lim} = s_u N_c + q = 68.88 + 28.50 = 97.38 \text{ kPa}.$$

    In order to obtain the allowable bearing capacity, we have to consider that the load factor (presumed to be equal 3) is applied to the net bearing capacity, defined as $q_{\lim}^N = q_{\lim} - q$ (see Section 8.16), so that:

$$q_{all} = \frac{q_{\lim}^N}{F} + q.$$

For the analysed case we have:

$$q_{all} = \frac{68.88}{3} + 28.50 = 51.46 \text{ kPa}.$$

8.14.2 The effect of anisotropy

For soft clays, particularly when dealing with low plasticity clays, the strength in simple shear and triaxial extension is usually significantly less than in triaxial compression. Therefore, when selecting the operational value of the undrained strength one needs to realize the dependence of this parameter on the imposed loading path (see Figure 5.42). For example, the soil element A will eventually reach failure in compression loading, but elements B and C will follow paths represented by simple shearing and extension loading respectively. As a consequence, prediction based on triaxial compression tests can be unsatisfactory, because they tend to overpredict the operational strength (Figure 5.43).

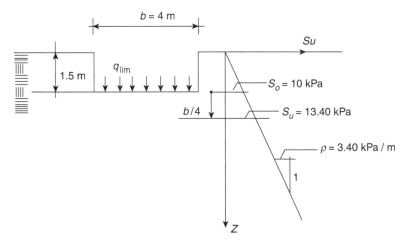

Figure 8.51 Illustrative example of undrained bearing capacity.

Furthermore, the development of strains along a failure surface forces a large number of soil elements well past the peak strength, leading to a *progressive failure* characterized by an operational strength lower than the peak.

In order to account for both these aspects, Koutsoftas and Ladd (1985) have suggested to evaluate the operational strength by means of the expression:

$$\frac{s_u}{\sigma'_{vo}} = (0.22 \pm 0.03) \cdot OCR^{0.8}. \tag{8.79}$$

Note that the above relation is supported by experimental evidences related to values of OCR up to 10, and a lot of care must be taken when analysing the behaviour of *stiff fissured clays*. Researches performed mostly on London clay (Marsland, 1974) suggest that the undrained strength of the soil mass is close to the lower limit of the scattered laboratoty test values. Moreover, it must be observed that it is at least doubtful if the behaviour of fissured clays can be defined fully undrained.

8.15 Bearing capacity of shallow footings: drained loading

When considering the case where full drainage occurs during loading, the analysis is carried out in terms of effective stresses. In order to highlight the factors contributing to the bearing capacity of a shallow footing, let us start the analysis by assuming (rather conservatively) that the failure zones can be made up of two Rankine wedges, as sketched in Figure 8.52 (Lambe and Whitman, 1969). This involves the introduction of a *frictionless plane ST*, which separates the Rankine active zone, below the footing, from the adjacent Rankine passive zone, and assumes that the footing and the soil above the founding depth are represented by uniform stresses.

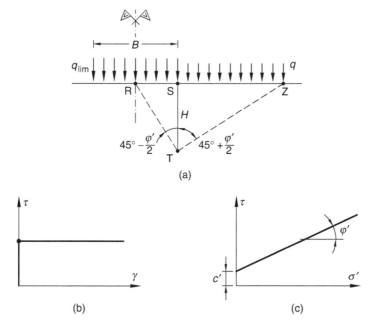

Figure 8.52 Drained bearing capacity of a foundation.
(a) active and passive wedges
(b) rigid perfectly-plastic material
(c) yield criterion

By computing:

$$P_a = \frac{1}{2}\gamma H^2 K_a + q_{\lim}HK_a - 2c'H\sqrt{K_a}$$
$$P_p = \frac{1}{2}\gamma H^2 K_p + qHK_p + 2c'H\sqrt{K_p} \tag{8.80}$$

and equating $P_a = P_p$, the following expression is found for the drained bearing capacity:

$$q_{\lim} = \frac{1}{2}\gamma BN_\gamma + c'N_c + qN_q . \tag{8.81}$$

The coefficients:

$$N_\gamma = \frac{\frac{1}{2}\left(\frac{K_p}{K_a}-1\right)}{\sqrt{K_a}} , \quad N_c = \frac{2\left(\sqrt{K_p}+\sqrt{K_a}\right)}{K_a} , \quad N_q = K_p^2 . \tag{8.82}$$

depend on the angle of shear resistance and are known as **bearing capacity coefficients**.

The obtained expression is overconservative, because it has been deliberately obtained by assuming a simple stress field, and for this reason it will be improved in the following points. However, this approach offers the advantage to show in a simple and intuitive way what are the main contributions to the bearing capacity: the first term $\gamma B N_\gamma / 2$ occurs because of the effect of gravity; the term $c' N_c$ arises due to the cohesive component of the strength of the soil; the term $q N_q$ results from the surcharge, acting alongside the foundation.

8.15.1 The effect of surcharge: lower bound solution

Consider a *weightless cohesionless soil* and, in order to build an admissible stress field, assume that the footing and the soil above the founding depth D are represented by uniform stresses. The uniform stress $q_{\lim}$ is applied over the width of the footing, and the uniform stress q, supposed to act alongside, represents the overburden stress γD.

We now introduce a fan of stress discontinuities between the active zone (below the footing) and the passive zones of either sides (see Figure 8.53).

As already discussed in Section 8.5, the ratio s'_1 / s'_2 between the abscissas of the Mohr circles of these two extreme zones is linked to the rotation ϑ of the direction of the major principal effective stress through the relation:

$$\frac{s'_1}{s'_2} = e^{2\vartheta} \tan \varphi'. \tag{8.83}$$

Being in the present case:

$$s'_2 = \frac{q}{1 - \sin \varphi'} \qquad q_{\lim} = s'_1 \cdot (1 + \sin \varphi') \qquad \vartheta = \frac{\pi}{2},$$

we obtain:

$$q_{\lim} = q \cdot N_q$$
$$N_q = \frac{1 + \sin \varphi'}{1 - \sin \varphi'} e^{\pi \tan \varphi'}. \tag{8.84}$$

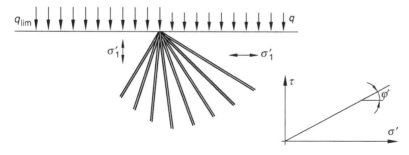

Figure 8.53 Stress field which includes a stress fan.

8.15.2 The effect of surcharge: upper bound solution

The result (8.84) is based on the lower bound approach, and it is certainly of interest to show how the same result can be obtained by using the upper bound approach. We stated in Section 8.1 that the slip lines in drained loading must be straight or logarithmic spirals. Therefore, in the mechanism sketched in Figure 8.54, the intermediate zone is bounded by a logarithmic spiral BC. Then, if r_o is the initial radius EB, the final radius EC is given by:

$$EC = r_1 = r_0 \cdot \exp\left(\frac{\pi}{2}\tan\varphi'\right). \tag{8.85}$$

By recalling that the upper bound approach requires a work balance, it is needed to construct a hodograph for computing the relative velocities. Because of the normality rule, the motion is at an angle φ' to the slip surface, and by taking into account this constraint the following relations can be derived (see Figure 8.54):

$$\delta v_2 = \delta v_1 \cdot \exp\left(\frac{\pi}{2}\tan\varphi'\right)$$

$$\delta v_3 = \delta v_0 \cdot \tan\left(\frac{\pi}{4}+\frac{\varphi'}{2}\right)\cdot \exp\left(\frac{\pi}{2}\tan\varphi'\right) \tag{8.86}$$

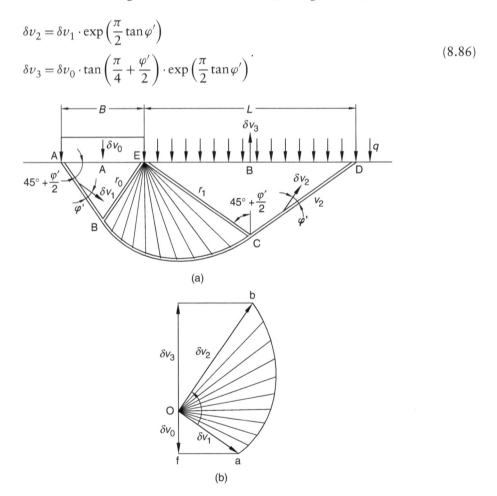

(a)

(b)

Figure 8.54 Upper bound mechanism which includes a slip fan.

Moreover, the intermediate zone cannot be a rigid block, because this would give rise to a kinematic inadmissibility, therefore it is composed by a fan of shearing planes. However, because of the normality condition, the relative displacement on these shearing planes is orthogonal to the resultant stress, so that no energy dissipation takes place. It follows that the rate of energy change of the foundation load must be equal to the rate of energy change of the surcharge:

$$q_{\lim} \cdot B \cdot \delta v_0 - q \cdot L \cdot \delta v_3 = 0,$$

and by using (8.86) the following solution is obtained:

$$q_{\lim} = q \frac{1 + \sin \varphi'}{1 - \sin \varphi'} e^{\pi \tan \varphi'}. \tag{8.87}$$

This solution is the same as the equation (8.84), i.e. the upper bound solution and the lower bound solution are the same, so that they are the correct solution (as derived by Prandtl, in 1921, by using the method of characteristics).

8.15.3 The contribution of cohesion

Let us now suppose that the failure envelope be represented as shown in Figure 8.52c. We can solve this case by following the approach suggested by Caquot (1934): the solution for the case in which both the cohesive and the frictional component are present can be obtained by assuming the soil as a purely frictional material, but subjected to a fictitious additional all around effective stress equal to $c' \cot \varphi'$.

Therefore, we can write:

$$q_{\lim} + c' \cot \varphi' = N_q \left(q + c' \cot \varphi' \right) \tag{8.88}$$

and it follows:

$$q_{\lim} = q N_q + c' N_c$$
$$N_c = \left(N_q - 1 \right) \cot \varphi' \tag{8.89}$$

8.15.4 The self-weight influence

In the case of a cohesionless soil, with unit weight γ but no surcharge at the founding level, by considering a soil element at the unloaded surface and near the edge of the footing the lower bound theorem would show that there is no possibility of applying a uniform load on the footing without violating the failure criterion. But clearly the soil effective weight also contributes to friction and must generate a contribution to the bearing capacity. An approximate but reasonable way to circumvent this difficulty is to consider, as suggested by Bolton (1979), the effect of gravity as an equivalent fictitious surcharge, corresponding to the effective overburden stress acting on a properly

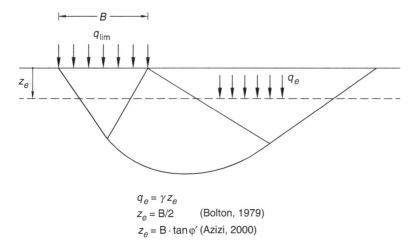

$$q_e = \gamma z_e$$
$$z_e = B/2 \quad \text{(Bolton, 1979)}$$
$$z_e = B \cdot \tan \varphi' \text{ (Azizi, 2000)}$$

Figure 8.55 Empirical determination of an equivalent depth of overburden to simulate the effect of gravity.

selected average depth. Assuming, as shown in Figure 8.55, this depth is equal to $B \tan \varphi'$ (Fethi Azizi, 2000), the equivalent surcharge will be:

$$q_e = \gamma B \tan \varphi', \tag{8.90}$$

and applying equation (8.84)

$$q_{\lim} + q_e = N_q q_e, \tag{8.91}$$

we obtain the solution:

$$q_{\lim} = \frac{1}{2} \gamma B N_\gamma, \tag{8.92}$$

where:

$$N_\gamma = 2(N_q - 1) \tan \varphi'. \tag{8.93}$$

The obtained expression of the bearing capacity coefficient N_γ is just a little more conservative than the solution obtained by Caquot and Kérisel (1953):

$$N_\gamma = 2 \left(N_q + 1 \right) \tan \varphi', \tag{8.94}$$

currently used in engineering practice (see Sloan and Yu (1996) for a rigorous solution).

8.15.5 The influence of groundwater table

The presence of groundwater table will have a significant influence on the bearing capacity, as can be deduced by examining the following cases.

If the water table rises at the founding level, the value of q in equation (8.81) will not be affected, but the unit weight below the founding level will be reduced to the buoyant unit weight, i.e. $\gamma' = \gamma - \gamma_w$. This implies that if a dry soil became saturated, the bearing capacity of a surface footing will be reduced by the ratio of the buoyant unit weight to the dry unit weight, typically 0.5 to 0.7.

If the water table rises above the founding level, both the unit weight and the surcharge will be reduced, this latter being $q' = \gamma D - u$, if u is the pore pressure acting at the founding level.

Note that in this case, because of the pore pressure acting at the founding level, equation (8.81) refers to the gross effective stress, as defined in the next section.

8.16 Load factor and strength mobilization

Before discussing the way safety factors are usually applied in geotechnical engineering, it is convenient to introduce some definitions.

The intensity of normal contact stress at the base of founding due to all loads above that level is called *gross bearing stress*, and the difference between the gross bearing stress and the total overburden stress at founding level is called *net bearing stress*.

The difference between the gross bearing stress and the pore pressure acting at the founding level is called *gross effective stress*, and an important result is that the net bearing stress and the net effective bearing stress are equal.

When dealing with shallow footings, it is a matter of tradition to express the margin of safety through the introduction of a **load factor**. Note that this load factor should always be *applied to the net bearing stress*, because it is the net stress which drives further shearing of the soil, and it is also intended to limit settlements.

An alternative approach, supported by Eurocodes, is to apply a factor of safety to the shear strength, so that a reduced value, called **mobilized shear strength** (or *design value*, in Eurocode terminology), is used for design. As an example, in drained conditions the design value φ'_d of the angle of shear resistance is obtained as $\tan\varphi'_d = \tan\varphi'/F_{\varphi'}$.

As the two definitions are totally different, the factor of safety (load factor) applied to the net bearing capacity (usually of the order of 3) is not equal to the safety factor applied to the shear strength (usually equal to 1.25).

There is also some debate about the choice of strength parameters, because this aspect requires both skill and experience, as well as some caution due to the idealizations introduced when deducing the theoretical solutions.

Soils on the dry or on the wet side of critical state have very different stiffness, so that using the same strength (namely the critical state strength) would require a different load factor to achieve the same tolerable settlement. Since regulation codes suggest the same load factor, the alternative is to make reference to the peak strength, provided that the stress level is properly taken into account. In dense sands, this can be done by referring to the relation introduced by Bolton (1986):

$$\varphi' = \varphi'_{cv} + 3\left[D_R(10 - \ln p'_f) - 1\right],$$

and the average mean stress p'_f at failure can be linked to the applied bearing stress, as suggested by De Beer (1970):

$$p'_f = \frac{1}{4}(q_{\lim} + 3q)(1 - \sin\varphi'),\tag{8.95}$$

through an iteration procedure, since the operational strength depends on the computed bearing stress.

In stiff fissured clays strength parameters along structural discontinuities are appropriate for drained loading, while the operational undrained strength is close to the lower limit of the scattered laboratory tests results, but it is at least doubtful if the behaviour of fissured clays can be considered undrained.

The above suggestions correspond to the common practice of selecting the *most probable* values of the strength parameters and of using the normal factors of safety in conjunction with these values.

But the analysis should be also carried out with the *lowest credible* strength parameters, in order to ensure that the safety factor is not lower than unity.

8.17 Bearing capacity: routine analysis

8.17.1 Drained bearing capacity

In the previous sections we used upper and lower bound theorems of plasticity to derive the contribution to bearing capacity of the surcharge stress, the gravity and cohesion. These contributions were obtained independently from each other and, since the mode of failure is not necessarily the same in each case, the superposition is uncorrect. However, Terzaghi (1943) first argued, as a reasonable approximation, that the total ultimate load could be obtained by using this superposition, and Davis and Booker (1971) and Sloan and Yu (1996) have shown, by using numerical analyses, that this superposition gives rise to errors which are on the safe side and do not exceed 25%.

Therefore, it is current practice to estimate the total bearing capacity by using the extended formula suggested by Hansen (1970):

$$q_{\lim} = \frac{1}{2}\gamma B N_\gamma s_\gamma i_\gamma b_\gamma g_\gamma + c' N_c s_c d_c i_c b_c g_c + q' N_q s_q d_q i_q b_q g_q.\tag{8.96}$$

The bearing capacity coefficients are a unique function of the angle of shear resistance (see Table 8.2); the additional coefficients depend on many other aspects, as is explained in the following.

(a) Previous theoretical analyses have been restricted to strip footings. Since it is reasonable to expect that the bearing capacity of square or circular footings will be greater than for a strip footing, because the failure mechanism now involves a three-dimensional geometry, the following **shape coefficients**, based on model

Table 8.2 Coefficients of bearing capacity

$\varphi'(°)$	0	22	24	26	28	30	32	34	36	38	40	42
N_c	5.14	16.88	19.32	22.25	25.80	30.14	35.49	42.16	50.59	61.35	75.31	93.71
N_q	1.00	7.82	9.60	11.85	14.72	18.40	23.18	29.44	37.75	48.93	64.20	85.38
N_γ	0.00	7.13	9.44	12.54	16.72	22.40	30.22	41.06	56.31	78.03	109.4	155.6

Source: Vesic (1975).

tests, have been suggested (Meyerhof, 1951):

$$s_q = s_\gamma = 1 + 0.1 \frac{1 + sin\,\varphi'}{1 - sin\,\varphi'} \frac{B}{L}; \qquad s_c = 1 + 0.2 \frac{1 + sin\,\varphi'}{1 - sin\,\varphi'} \frac{B}{L} \qquad (8.97)$$

being B and L (with $B < L$) the footing dimensions.

(b) When the founding level is at depth D, previous solutions consider the soil above the founding level simply as a surcharge without strength. This may be considered a reasonable assumption, since cracks and weathering can exist to relatively shallow depths (i.e. up to 2 m). For greater depth, the shear strength of the overlying soil contributes to the bearing capacity, and to account for this contribution the following **depth coefficients** have been suggested (Hansen, 1970; Vesic, 1973):

$$d_q = 1 + 2\tan\varphi'\left(1 - sin\varphi'\right)^2 \frac{D}{B} \qquad (D \leq B)$$

$$d_q = 1 + 2\tan\varphi'\left(1 - sin\varphi'\right)^2 \tan^{-1}\frac{D}{B} \qquad (D > B). \qquad (8.98)$$

$$d_c = d_q - \frac{1 - d_q}{N_c \tan\varphi'}$$

(c) High values of **horizontal load** H can produce failure by sliding. If this is not the case, the influence of this component must be taken into account. A simple way is to introduce the following coefficients, as suggested by Vesic (1973):

$$i_\gamma = \left(1 - \frac{H}{N + BLc'\cot\varphi'}\right)^{m+1}; \qquad i_q = \left(1 - \frac{H}{N + BLc'\cot\varphi'}\right)^{m}$$

$$i_c = i_q - \frac{1 - i_q}{N_c \tan\varphi'}; \qquad m = \frac{2 + \dfrac{B}{L}}{1 + \dfrac{B}{L}} \qquad (8.99)$$

(d) Sometimes designers adopt an **inclined founding base** in the presence of high horizontal components (see Figure 8.56). An exact solution has been obtained by Hansen (1970) for a weightless soil, and this corresponds to the introduction of the coefficient:

$$b_q = \left(1 - \alpha \tan\varphi'\right)^2. \qquad (8.100)$$

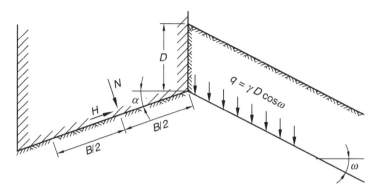

Figure 8.56 Inclined foundation and sloping ground surface (Vesic, 1975).

It is also suggested assuming $b_\gamma = b_q$ and the value of b_c can be obtained from the expression $b_c = b_q - \dfrac{1 - b_q}{N_c \tan \varphi'}$.

(e) Similarly, for a **sloping ground surface** (Figure 8.56), Hansen (1970) has obtained:

$$g_q = (1 - \tan \omega)^2 \qquad (8.101)$$

and the g_γ, g_c coefficients should be evaluated on the basis of the considerations made in the previous points.

(f) Finally, if the resultant has an **eccentricity** e, the value of the width B to be introduced into all previous formulas is that corresponding to the minimum surface for which the load is centred (Meyerhof, 1953). For rectangular footings this is given by:

$$B = B_R - 2e, \qquad (8.102)$$

and Figure 8.57 suggests how to manage the problem when dealing with circular footings.

8.17.2 Routine analyses for undrained loading

By following similar arguments, the complete bearing capacity formula that applies for undrained loading in terms of total stress is the following:

$$q_{\lim} = s_u N_c s_c^o d_c^o i_c^o b_c^o g_c^o + q. \qquad (8.103)$$

(a) The **shape coefficient**, on the basis of model tests, is given by:

$$s_c^o = 1 + 0.2 \frac{B}{L} \qquad (8.104)$$

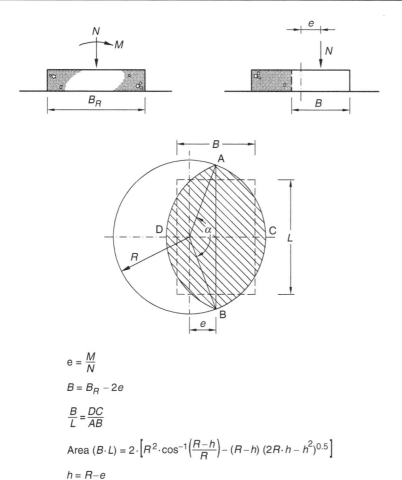

$e = \dfrac{M}{N}$

$B = B_R - 2e$

$\dfrac{B}{L} = \dfrac{DC}{AB}$

Area $(B \cdot L) = 2 \cdot \left[R^2 \cdot \cos^{-1}\left(\dfrac{R-h}{R}\right) - (R-h)\,(2R \cdot h - h^2)^{0.5} \right]$

$h = R - e$

Figure 8.57 Solution for eccentric load.

(b) An approximate solution for the **depth coefficient** has been provided by Meyerhof (1951), Skempton (1951) and Hansen (1961):

$$d_c^o = 1 + 0.4\dfrac{D}{B} \qquad (if:\ D \leq B)$$

$$\qquad\qquad\qquad\qquad\qquad\qquad\qquad\qquad\qquad\qquad\qquad (8.105)$$

$$d_c^o = 1 + 0.4\tan^{-1}\dfrac{D}{B} \qquad (if:\ D > B)$$

(c) Vesic (1975) suggests the introduction of the following coefficient for the presence of the **horizontal load component**:

$$i_c^o = 1 - \dfrac{mH}{BLs_uN_c}; \quad m = \dfrac{2 + \dfrac{B}{L}}{1 + \dfrac{B}{L}}. \qquad\qquad\qquad (8.106)$$

Example 8.8

Consider again the cantilever wall, already examined in the worked Example 8.4 and sketched in Figure 8.58. We require to compute the bearing capacity.

Solution. Figure 8.58 summarizes the actions transmitted to the foundation soil. By neglecting any surcharge acting on the founding level at the toe side, the bearing capacity is given by:

$$q_{\lim} = \frac{1}{2}\gamma B_R N_\gamma i_\gamma .$$

The dimension to be considered is the reduced value B_R, which takes into account the eccentricity of the resultant load, i.e. $B_R = b - 2e = 1.47\,\text{m}$; in addition we take into account the presence of the horizontal component through the coefficient $i_\gamma = (1 - H/N)^3 = 0.32$. Accordingly, the bearing capacity assumes the value $q_{\lim} = 183.49\,kPa$, to be compared with the actual bearing stress $q_s = \dfrac{N}{B_R \cdot 1} = 72.80\,kPa$.

Since the load factor $\left(F = \dfrac{q_{\lim}}{q_s}\right)$ is higher than 2.5, the bearing capacity safety requirement is fulfilled.

(d) The **inclination of the founding base** has an influence that can be evaluated by means of the following coefficient (Hansen, 1970):

$$b_c^o = 1 - \frac{2\alpha}{2 + \pi} . \tag{8.107}$$

(e) Finally, a **sloping ground surface** requires the introduction of the coefficient (see Vesic, 1975):

$$g_c^o = 1 - \frac{2\omega}{2 + \pi} \tag{8.108}$$

and the further term:

$$\frac{1}{2}\gamma B \left(1 - 0.4\frac{B}{L}\right)(-2\omega) . \tag{8.109}$$

should be added into (8.103).

8.18 Slope stability

Problems involving slope stability include natural slopes, cuttings and excavations, sides of embankments or earth dams and many others. When dealing with natural slopes, it must be recognized that landslides represent a natural hazard with important socio-economical impact in all countries, and this is sufficient to justify the interest on the subject.

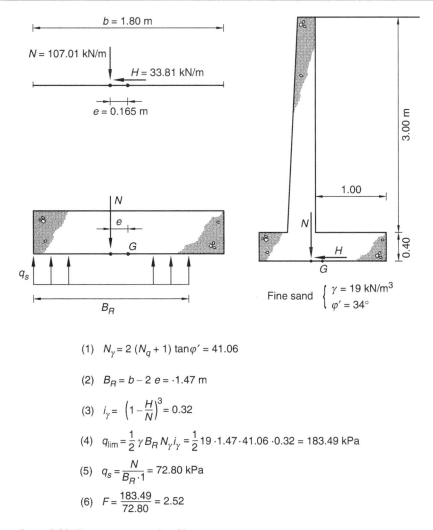

(1) $N_\gamma = 2\,(N_q + 1)\,\tan\varphi' = 41.06$

(2) $B_R = b - 2\,e = \cdot 1.47$ m

(3) $i_\gamma = \left(1 - \dfrac{H}{N}\right)^3 = 0.32$

(4) $q_{\lim} = \dfrac{1}{2}\,\gamma\,B_R\,N_\gamma\,i_\gamma = \dfrac{1}{2}\,19 \cdot 1.47 \cdot 41.06 \cdot 0.32 = 183.49$ kPa

(5) $q_s = \dfrac{N}{B_R \cdot 1} = 72.80$ kPa

(6) $F = \dfrac{183.49}{72.80} = 2.52$

Figure 8.58 Illustrative example of bearing capacity.

The stability of natural slopes may be threatened by rainfall, seismic actions, natural erosion or man-made excavations. Slow downhill movements can also be produced by soil creep.

Obviously, such causes can sometimes be concomitant; but even when this is not the case, the slope stability analysis is always a rather complex problem, because it is difficult to reduce to a design scheme and to define the available *in situ* soil strength. Therefore, before entering into any analysis it is necessary to collect a great deal of information.

First, the complementary geological and geotechnical nature of the problem must be considered. Even though the interest of the geotechnical engineer is limited to a specific site or slope, the mechanism in slope evolution must be clearly understood.

A second key element that must be introduced into the analysis is represented by the hydrogeological conditions and their change in time.

Finally, a careful site investigation, related to stratigraphical and structural features, is essential. In particular, when dealing with stiff clays, an understanding of geological processes is of paramount importance and a detailed description of the material as well as any evidence of principal displacement shears, induced by faults and bedding plane slips, as well as minor shears of limited extent is required (Skempton, 1966; Skempton and Petley, 1967).

As will be apparent in the sequel, natural slopes usually have very low factors of safety, and the material involved is characterized by variability and unusual complexity. Therefore, **monitoring** is frequently undertaken, in order to confirm design decisions or to corroborate safeguarding measures.

The first types of instrumentation to be installed are piezometers, because slope stability is sensitive to the pore pressure in the ground. Section 7.4 discusses the types of piezometers that have been developed, and details of these instruments can be found in Hanna (1973) and Dunnicliff (1988).

Slope movements are also of paramount importance, as they may indicate impending instability or the location of a plane of sliding. Because slope movements are predominantly horizontal, the most common instrument is the **inclinometer** (see Hanna, 1973; Dunnicliff, 1988; Lollino, 1992). This consists of a flexible tube installed in a borehole. A probe can be lowered into the tube, and changes in inclination of the tube can be made, by running the probe in guide channels. Usually the inclination is measured over a length of 0.5 m, so that by taking readings every 0.5 m the full profile of the tube can be computed.

Measurements are referred to a zero position at the bottom of the tube, if it is fixed into a hard stratum, or to the top of the tube, which must be located by surveying.

The reader can now appreciate, from these introductory notes, that there are specific courses devoted to this subject. Therefore, in this book the subject is limited to presenting only those methodologies currently used by engineers to analyse the stability of natural or man-made slopes. More insight can be gained from further reading, which includes Varnes (1978), Brunsden and Prior (1984), Anderson and Richards (1987), Hoek and Bray (1977), Simons *et al.* (2001) and the reference papers mentioned in the sequel.

8.18.1 Terminology

According to Skempton and Hutchinson (1969) the term *landslide* embraces all downslope movements, primarily occurring as the result of shear failures at the boundary of the moving mass. Within this context, *translational slides* usually result from the presence of weaker surfaces located at relatively shallow depth below the ground surface.

Rotational slides may occur in relatively uniform clay or shale, the curved surface promoting a back-tilt to the sliding mass, which sinks at the rear and heaves at the toe. Successive rotational slips may also occur on overconsolidated fissured clays, as has been observed in London clay.

Compound slides usually reflect the presence of pronounced heterogeneities at moderate depths.

Many examples of mass movements cannot be classed in the above-mentioned mechanisms. These take the characteristics of a *flow*, either fast or slow, wet or dry.

In particular, the term *debris flow* is usually associated to coarse materials, and *mud flow* refers to clays with silt and sand layers.

Common features to debris flows are a steep upper zone, where the the movement is triggered, usually as the result of high intensity rainfall; a downstream path along which transportation and erosion occur; a lower flatter zone where deposition in a fan-shaped area occurs. Because the velocity of flow may range from very slow rates to more than 20 m/s, consequences in this latter case may be catastrophic. Therefore, safeguarding measures are objects of intensive research efforts (Hungr *et al.*, 1984, Pirulli, 2005).

Catastrophic flowslides in loose unsaturated pyroclastic soils occurred in Italy on 5 May 1998 (as described by Cascini *et al.*, 1998; Olivares and Picarelli, 2003; Scotto di Santolo, 2002). These were triggered by heavy rainfall, which changed the stability conditions through an increase of saturation degree. Once the soil is saturated or nearly so, static liquefaction is likely to occur, promoting a flowslide.

A further example is the collapse of tailings dams near Stava, northern Italy, in July 1985. The tailings dams consisted of two adjacent basins, one up-slope of the other. The collapse of the upper basin released the stored tailings and the superincumbent water, which, in turn, overwhelmed the lower basin, also causing its collapse. These collapses resulted in a mud-flow, which flowed as a mass of sediment-charged water down the valley to Tesero, causing 268 deaths and the destruction of a great deal of property, making it the worst disaster of its type to have occurred in Europe, as described by Chandler and Tosatti (1995).

8.19 Limit equilibrium methods

All the procedures currently used are based on the limit equilibrium method and have the following assumptions in common:

- the failure occurs along a known or an assumed sliding surface;
- plane strain conditions are supposed;
- the actual strength of the soil is compared with the value required for the equilibrium of the soil mass and this ratio is a measure of the safety factor.

It must also be recalled that, in this method, soils are treated as rigid-plastic materials and due to this assumption the analysis does not consider deformations. Therefore, this method allows to only consider conditions at the onset of failure.

Note that, although the method is very similar to the kinematic approach, frequently the restrictions of a kinematically admissible mechanism are ignored. However, even if there is no formal proof that the limit equilibrium method can give a correct solution, this method is firmly established in practice for slope stability analysis, and the experience has shown that this approach gives answers which agree with observations of failures.

8.19.1 Infinite slope

A convenient introductory subject to the analysis of the stability of slope is related to slopes that can be considered as infinite in extent. As already discussed in Section 8.3, the definition of *infinite slope* requires that strata of different soil properties must be parallel to the surface and implies that the stress state is independent of the coordinate along the slope (see Example 8.1). Therefore, the state of stress acting on any vertical section, such as AD or BC (see Figure 8.59a) must be the same, so that the total forces on these sections have the same magnitude and the same line of action.

The shear force T on the base CD is given by:

$$T = W \sin i = N \tan i \qquad (8.110)$$

and the equilibrium is assured until this force is lower than that given by the soil strength $T_r = N \tan \varphi'$. Therefore, the safety factor is given by the ratio:

$$F = \frac{T_r}{T} = \frac{\tan \varphi'}{\tan i}. \qquad (8.111)$$

This result shows a dry slope is just stable when the slope angle i is equal to the angle of shear resistance φ', and for this reason this *slope angle* is usually named *angle of repose*. If the slope is completely *submerged* and water is under static conditions (Figure 8.59b), the equilibrium can be analysed by considering the forces acting at the boundaries of the soil element $ABCD$ and its weight. The resultant water pressure is

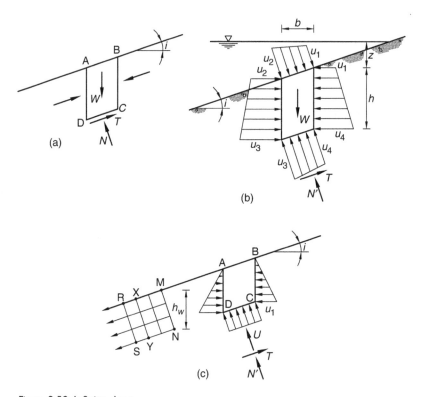

Figure 8.59 Infinite slope.

a vertical force (buoyancy) equal to:

$$U = bh\gamma_w, \tag{8.112}$$

and the result previously obtained (8.111) in terms of safety factor also applies to this case. This is equivalent to saying that the maximum slope angle for a given sand is the same independent of the fact that the slope is dry or submerged with water in hydrostatic conditions.

Figure 8.59c refers to the more complex case where the slope is subjected to seepage parallel to the ground surface. The problem could again be analysed by considering the soil element under the action of its weight and forces acting at its boundaries. An alternative procedure considers the equilibrium of the soil skeleton subjected to its buoyant weight, the boundary effective stresses and the seepage forces.

According to the analysis developed in Chapter 6, the seepage force is equal to:

$$F_i = bh\gamma_w i_w, \tag{8.113}$$

where i_w is the hydraulic gradient.

By considering that lines RS and XY in Figure 8.59c are equipotential, the hydraulic gradient is:

$$i_w = \frac{RX \sin i}{RX} = \sin i, \tag{8.114}$$

and the forces T and N' are:

$$\begin{aligned} T &= \gamma' bh \sin i + \gamma_w bh \sin i = \gamma bh \sin i \\ N' &= \gamma' bh \cos i \end{aligned}, \tag{8.115}$$

so that the safety factor is now:

$$F = \frac{\tan\varphi'}{\tan i} \frac{\gamma'}{\gamma}. \tag{8.116}$$

Compared with the previous case, this result shows that a seepage parallel to the slope has a significant effect on its stability, because the safety factor is about half the value relative to the case of no flow.

Also note that, in all previously considered cases, the obtained factor of safety is independent of the depth of failure surface. In practice, failure would occur on a surface of weakness within the soil mass, or close to the surface where usually changes to drive failure are likely to occur.

On the contrary, if we consider the case of flow parallel to the slope with water table below the surface, it can be proved that the factor of safety assumes the expression:

$$F = \frac{\tan\varphi'}{\tan i}\left(1 - \frac{\gamma_w}{\gamma} + \frac{\gamma_w z_w}{\gamma z}\right), \tag{8.117}$$

where z_w is the water table depth and z is the depth of the failure surface. It is now apparent that the safety factor decreases as the depth of the slipping surface increases, but in such a case it must also be observed that the assumption of infinite slope may be questionable.

8.19.2 Method of slices

The previous analysis is very effective in determining the safety factor when considering a homogeneous infinite slope and the failure mechanism is of the translational type. The method of slices has been developed to analyse more complex situations, and where more than one layer with different strength parameters is present, the expected groundwater regime differs from simple patterns and the failure surface may be composite.

In this method (see Figure 8.60) the soil mass is subdivided into a number of vertical slices and the equilibrium of each slice is then considered. If a number of n slices is considered, the unknowns are the following:

> n forces N_i' normal to the base of each slice;
> $(n-1)$ normal forces E_i' and $(n-1)$ shear forces X_i at the interface of slices;
> n coordinates a to locate the normal forces N_i';
> $(n-1)$ coordinates b to locate the interface forces E_i'.

If the further unknown of the safety factor is added (note that the safety factor gives the possibility of expressing the shear forces at the bottom in terms of N_i'), then the total number of unknowns is $(5n-2)$ to be compared with the number $3n$ of available equilibrium equations.

Presuming that the slices are so thin that the forces N_i' can be located on the centroid of each slice, then there are $(4n-2)$ unknowns, but the problem still remains statically undetermined. It is then necessary to introduce additional assumptions in order to remove the extra unknowns. These assumptions usually refer to the interface forces, and they explain the differences between various methods (see for example Morgenstern and Price, 1965; Sarma, 1973). In the sequel, we limit the presentation to some approximate methods of analysis, which has been proved to give satisfactory results.

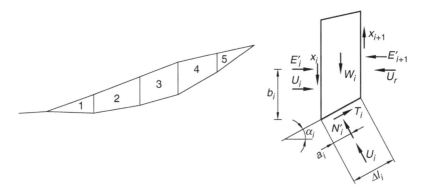

Figure 8.60 Method of slices.

8.19.3 Simplified Bishop method

In this method (Bishop, 1955), the failure surface is represented by a circular sliding surface (rotational failure). With reference to Figure 8.61, equating the moment about O of the weight of the soil with the moment of the forces acting on the sliding surface, the safety factor is given by:

$$F = \frac{R \cdot \sum \left[c' l_i + (N_i - U_i) \tan \varphi' \right]}{\sum W_i x_i}, \tag{8.118}$$

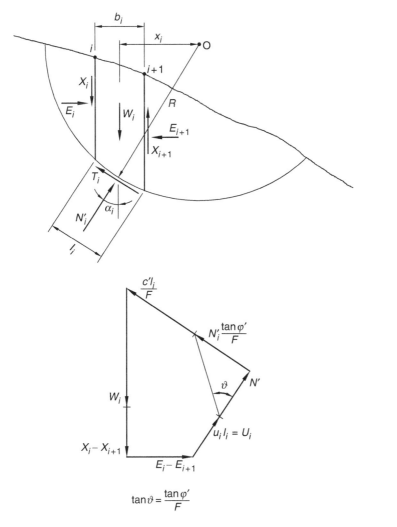

$$\tan \vartheta = \frac{\tan \varphi'}{F}$$

Figure 8.61 Simplified Bishop method.

the magnitude of the mobilized shear strength being:

$$T_i = \frac{c'l_i + (N_i - U_i)\tan\varphi'}{F}.$$ (8.119)

From the vertical equilibrium of the slice:

$$W_i - T_i\sin\alpha_i - N_i\cos\alpha_i - (X_{i+1} - X_i) = 0,$$ (8.120)

we can derive the value of N_i, and, by substituting this value into (8.118), the expression of the safety factor becomes:

$$F = \frac{\sum\left[(c'b_i + W_i(1 - r_u)\tan\varphi' + (X_i - X_{i+1})\tan\varphi')\dfrac{1}{M_\alpha}\right]}{\sum W_i\sin\alpha_i},$$ (8.121)

where:

$$M_\alpha = \cos\alpha_i\left(1 + \frac{\tan\varphi'\tan\alpha_i}{F}\right)$$ (8.122)

and r_u is the ratio which gives the pore pressure as a function of the total weight of the column of the soil above the considered point, i.e.:

$$r_u = \frac{u_i b_i}{W_i}.$$ (8.123)

Assuming that $X_i - X_{i+1} = 0$ throughout, the factor of safety can be computed by means of the approximate expression:

$$F = \frac{\sum\left[(c'b_i + W_i(1 - r_u)\tan\varphi')\dfrac{1}{M_\alpha}\right]}{\sum W_i\sin\alpha_i},$$ (8.124)

where M_α is given by (8.122).

Note that in equation (8.124) the safety factor appears within the summation on the r.h.s. as well on the l.h.s., so that an iterative procedure is needed. An initial value of the safety factor is guessed (closed to unity) and inserted in the r.h.s. and the value of F is computed. This value is the new input on the r.h.s. and the procedure is repeated until an almost constant value of F is attained (usually three to four iterations provide the required convergency).

8.19.4 Simplified Janbu method

The simplified method suggested by Janbu et al. (1956) assumes that the value of N_i can be computed from (8.120), by neglecting X_i. Then the safety factor for the horizontal equilibrium is given by:

$$F = \frac{\sum\left[(c'\Delta x_i + (\gamma z_i - u_i)\tan\varphi'\Delta x_i)\dfrac{1}{n_\alpha}\right]}{\sum \gamma z_i \Delta x_i \tan\alpha_i},$$ (8.125)

where:

$$n_\alpha = \cos^2\alpha_i \frac{1 + \tan\varphi'\tan\alpha_i}{F}.$$ (8.126)

The above solution has been successively improved by Janbu (1973), through a comparison with the rigorous method. This originated a correction factor f_o, which depends on the geometry of the problems and soil strength (see Figure 8.62), which should be applied to the simplified expression of the safety factor in order to forecast a more reliable value:

$$F_c = f_o \cdot F.$$ (8.127)

8.20 Landslides analysis and operational strength

Errors arising from the methods of analysis are small if compared with uncertainties associated to the definition of actual pore pressure acting in the ground and to the selection of soil strength parameters. The majority of authors who deal with slope stability analysis usually agree with this conclusion.

Pore pressure data are, in particular, of paramount importance. Maximum values are associated to exceptionally heavy rainfalls and quite often represent the triggering cause.

The selection of strength parameters has been the subject of considerable research, and some indications can be summarized, with reference to the following aspects:

* short and long-term conditions;
* first time slide and pre-existing slip surface.

(a) If a saturated clay is unloaded, as may occur in a cutting (Figure 8.63a), a reduction of mean total stress occurs. Low values of hydraulic conductivity prevent the soil from changing volume during the excavation, so that a negative excess pore pressure develops (Figure 8.63b). With time this pore pressure suction is dissipated, and migration of pore water takes place from the surrounding area of higher pore pressure unaffected by the excavation into the area of lower pore pressure. This will cause soil swelling and softening, with a reduction in strength, so that the minimum factor of safety will be achieved in long-term conditions (Bishop and Bjerrum, 1960). However, short-term failures in cuttings have also been referred (see, as an example, Skempton and La Rochelle 1965; Esu, 1966). The case documented by Skempton and La Rochelle (1965) is a failure which occurred at Bradwell in London clay, five days after completion of excavation. The back-figured average undrained strength was about 56% of the value obtained from

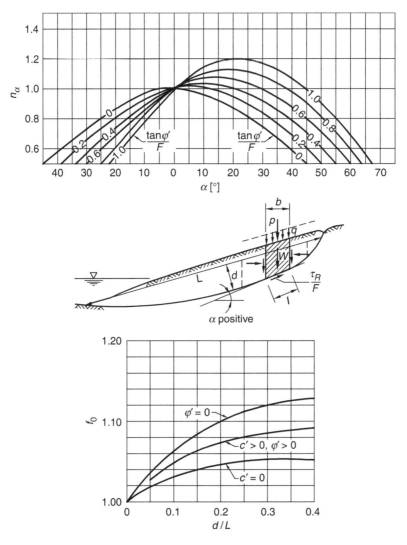

Figure 8.62 Simplified Janbu method.

conventional undrained tests on small specimens, this difference being explained by the presence of fissures and discontinuities that cannot be detected by small laboratory samples, and by the testing time interval (15 minutes) used in laboratory tests, which does not allow pore pressure equalization through the clay mass.

(b) Delayed failures of cutting slopes in stiff clays have been extensively studied and documented, and a recurrent conclusion is than the average operational strength at failure is significantly lower that the peak strength. As a possible explanation of this, progressive failure is generally postulated. The term **progressive failure** is used to indicate the *non-uniform mobilization of shear strength* along a potential

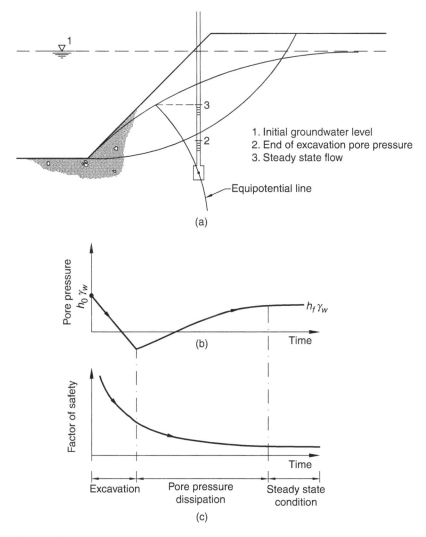

Figure 8.63 Changes in pore pressure and factor of safety during and after the excavation of a cut in clay (Bishop and Bjerrum, 1960).

failure surface. Conditions for progressive failure to take place are the following (Bjerrum, 1967): the soil exhibits a *brittle behaviour*; stress concentration takes place; boundary conditions allow differential strains. If a brittle soil is loaded in a non-uniform manner, some soil elements will reach peak strength and a failure surface will start to develop. With further straining, the strength will reduce from peak to post-peak and towards residual values, so that when the failure surface has fully developed the average strength of the soil mass will be less than the peak strength. The limit equilibrium method of analysis cannot capture such complex phenomenon, whereas Potts *et al.* (1997) have shown the capability of the finite element method in accurately reproducing the progressive failure.

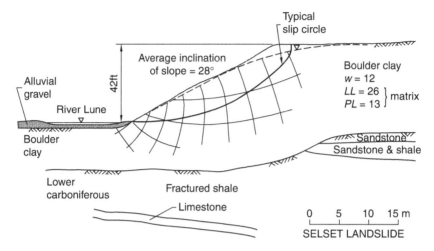

Figure 8.64 An example of first slide at Selset (Skempton and Brown, 1961).

(c) An example of *first slide in intact stiff clays* occurred at Selset and has been reported by Skempton and Brown (1961). The River Lune is eroding its valley through a thick deposit of clay till, and, during flooding, the river is cutting into the toe of the slope. Clear evidence of a rotational slip was detected in 1955 and the piezometers also indicated a small upward component of groundwater flow from the underlaying bedrock. The till consists of stones and boulders in a massive intact sandy clay matrix, characterized as follows:

$$w_P = 13\%; \quad w_N = 12\%; \quad w_L = 26\%; \quad CF = 25\%$$
$$\varphi'_{peak} = 32^0; \quad c'_{peak} = 8.6 \text{ kPa};$$

   The retrospective analysis (see Figure 8.64) was performed by using the Bishop method and the conclusion was reached that the introduction of strength parameters corresponding to the critical state gives an underestimation of the safety factor of about 30%, whereas, when using peak strength values and two limiting conditions concerning the pore pressure distribution the authors obtained safety factors ranging from 0.99 to 1.14.
   A similar conclusion was reached during the analysis of the Lodalen slope, a railway cutting originally made in 1925 and further widened in 1949, in a post-glacial intact clay, lightly overconsolidated. A slide occurred five years later, in October 1954 (Sevaldson, 1956). The safety factor obtained using the Bishop method and the peak strength parameters provided values close to unity.
   Both these examples, and many others, support the conclusion that first-time slides in non-fissured clays correspond to shear strength only slightly less than the peak strength.

(d) As already discussed in Section 5.9, clays are described as stiff if their undrained strength exceeds 75 kN/m^2. Therefore, if reference is to depths relevant to the analysis of slopes, stiff clays are heavily overconsolidated clays. In addition, we note that they are often fissured, a feature that influences both the laboratory and field behaviour of such materials, and that they experience a brittle behaviour. This implies that as the sample is being sheared in drained conditions, displacements can thought to be the result of relatively uniform strains only until the peak strength is reached. After the peak, displacements are the result of localized movements on a newly formed failure surface.

Skempton (1970) and Burland *et al.* (1996) have shown that the shear strength on the newly formed discontinuity is comparable to that of a pre-existing joint, as given by a zero value of c' and a friction angle, which, at low to moderate stress level, is similar to that of the reconstituted material (Figure 5.35). This is an important conclusion, by considering that back analyses of first sliding failure in fissured clays (Chandler and Skempton, 1974; Skempton, 1977) suggest that the **operational effective strength** envelope lies between the lower bound envelope for the strength on fissures and the post-peak strength for initially intact samples, and it is almost the same as for the reconstituted sample (i.e. the critical state value in the context of critical state theory).

(e) When dealing with the long-term stability of cuttings and natural slopes, back-analyses of *failures along pre-existing surfaces* strongly support the conclusion that the involved shear strength is the residual value, as extensively proved by Skempton (1964). Residual conditions are characterized by zero cohesion and by a friction angle dependent on the percentage of clay particles that can be reoriented during shearing. In soils mainly composed of bulky particles ($CF < 20\%$), shearing does not induce any preferred orientation and φ'_r is equal or slightly smaller than φ'_{cv}, whereas in soils composed of platy particles ($CF > 50\%$), shearing induces particles orientation and φ'_r is significantly smaller than the critical state value (Figure 5.37) (Lupini *et al.*, 1981; Skempton, 1985).

However, it must be observed that the development of a smooth shear surface can be inhibited in clay soils containing lithorelicts, as clay shales, and in this case the residual strength of the natural soil may result higher than the residual envelope of the reconstituted material (Leroueil *et al.*, 1997).

(f) We stated that most of the stiff clays are characterized by discontinuities such as bedding planes, fissures, joints and faults. But, in addition, stiff clays may have also been sheared by tectonic events. In particular, differential shear displacements may have occurred in a bedded sequence of stiff soils or rocks, as a consequence of tilting or folding. The term *flexural slip* is used to indicate such differential displacements. Therefore, in slope stability problems a crucial aspect is related to the presence or not of pre-existing discontinuities. Generally, such discontinuities are anticipated through site investigation if they are produced by previous landslides. On the contrary, they can be missed if they are produced by tectonic activity, because usually their effects on the surface are absent. Skempton (1966) introduced the *terminology* related to tectonic shear zone, as follows. Slip *surfaces* and *shears* are synonymous for surfaces of discontinuity along which slip takes place, the term slip indicating any finite displacement. Slip is not always confined to a single surface but it is observed to occur along many slip surfaces distributed through a zone, called *shear zone*. *Displacement shears* is the term used to indicate those particular slip surfaces which follow the direction of general displacement and have undergone large relative displacements. Slip surfaces inclined with respect to the displacement shears are the *Riedel shears* (so called because they were first demonstrated by Riedel in 1929 and Cloos, 1928, as referred to by Skempton, 1966) and the *thrust shears*.

All these slip surfaces divide the shear zone into lenses, that, as the shearing process advances, are distorted into a *sigmoidal section*. Finally, the term *slickenside* denotes a polished, grooved and striated surface produced by displacement along a fault or at the bottom of a landslide.

Chandler (2000) argues that the flexural slip is an important widespread feature of depositional basins, since it can promote shear zone having a strength close to residual, and, therefore, the presence of tectonic shear zones is of particular significance in engineering works (also see Skempton and Petley, 1967; Hutchinson, 1995; Mesri and Shahien, 2003; Morgenstern and Tchalenko, 1967).

8.21   Summary

When analysing limit analysis problems, we abandon the quest for the actual state of a soil structure and, instead, we concentrate on the way the structure might collapse. Obviously, it is not envisaged that the structure will actually collapse; rather the analysis is based on loads increased by a hypothetical factor, so that if the structure is loaded by smaller working loads it will not collapse.

In this view, plasticity theorems help us to define bounds of the true collapse load, by using two approaches: the *static approach* and the *kinematic approach*. The lower bound theorem (static approach) states that if we find a stress field such that equilibrium equations and traction boundary conditions are satisfied and the stresses do not exceed the yield condition, the collapse cannot occur. Therefore, the stresses applied at the boundary are a lower bound of the collapse stresses and the stress field that satisfies these requirements is called a *stress admissible field*.

The upper bound theorem (kinematic approach) states that if for any compatible plastic mechanism (a mechanism for which no gaps, separations or overlaps occur) the rate of work of external loads equals the rate of internal dissipation then collapse must occur. The answers provided by the two approaches are in general different from each other, but usually the upper and the lower bounds are fairly close to one another, so that the true collapse load is then closely bracketed.

When we apply these approaches, soil may be considered as a purely cohesive material ($\tau_f = s_u$), an assumption appropriate for total stress analyses under undrained conditions, or purely frictional ($\tau_f = \sigma_f' \tan \varphi'$), an assumption appropriate for effective stress analyses under drained conditions, applicable to coarse-grained soils in almost all cases and to fine-grained soils in the long term.

The problem of a soil mass which is in equilibrium and has everywhere attained the failure condition was first considered by Rankine (1857), and for this reason is referred as the **Rankine limiting stress field**.

If the obtained limiting values of active and passive stresses are used for the calculation of earth thrust against vertical retaining walls, it is important to realize that the condition of conjugate planes should be fulfilled.

Coulomb solved in 1773 the problem of soil thrust against a retaining wall, by using a procedure that nowadays is referred as the *limit equilibrium method*. This procedure basically consists of devising an arbitrary mechanism of collapse and the collapse load is found from the equilibrium of forces that act at boundaries.

Recall that this method assumes a planar sliding surface, which is no longer consistent when friction forces act at the interface between the wall and the soil. For this reason, while the Coulomb theory gives reasonable values of the active thrust, for the passive case this theory gives unsafe results, especially when $\delta > \varphi'/3$. As in practice a realistic value of δ is close to φ', a more refined analysis is required to estimate the passive earth resistance, by using procedures based on the lower bound theorem of plasticity theory or solutions based on curved (log-spiral) failure surface.

Even if there is no formal proof that the limit equilibrium method can give a correct solution, this method is firmly established in practice for slope stability analysis, and the experience has shown that this approach gives answers which agree with observations of failures.

Errors arising from the methods of analysis are small if compared with uncertainties associated to the definition of actual pore pressure acting in the ground and to the selection of soil strength parameters.

The selection of strength parameters has been the subject of considerable research, and some indications have been summarized, with reference to the following aspects: short and long-term conditions and first time slide and pre-existing slip surface.

8.22 Further reading

An introduction to metal plasticity, well suited for beginning readers, is the book by C.R. Calladine, *Plasticity for Engineers*, Ellis Horwood, 1985.

At intermediate level there are two books that may be suggested as a link between soil mechanics and more specialized books on plasticity: J.H. Atkinson (1981) *Foundations and Slopes*, McGraw-Hill; R.O. Davis and A.P.S. Selvadurai (2002) *Plasticity and Geomechanics*, Cambridge University Press.

More specialized reading on general aspects of limit analysis and soil plasticity are found in the following books: W.F. Chen (1974) *Limit analysis and soil plasticity*, Elsevier; J. Salençon (1974) *Théorie de la plasticité pour les applications à la Mécanique des Sols*, Eyrolles; Hai-Sui Yu. (2006), *Plasticity and Geotechnics*, Springer.

The slip lines theory was beyond the scope of this book. Readers wishing to learn about it should refer to the book by V.V. Sokolowskii (1960) *Static of Soil Media*, Butterworths.

In addition, an introduction to the method of characteristics is given by M.B. Abbot (1966) *An Introduction to the Method of Characteristics*, Thames and Hudson.

Chapter 9

Serviceability limit state

Civil engineers traditionally feel comfortable that the use of safety factors (namely load factor), based on past experience, against the attainment of a collapse limit state may also be seen as guaranteeing the serviceability and performance of similar structures under working loads.

Modern codes introduce some cultural shifts, requiring to estimate explicitly the performance of soil structures and to predict settlements, which is the aspect that in many cases dictates the choice of a foundation solution or the design of a foundation.

In this view, this chapter deals with the performance and serviceability of structures, and, in particular, attention will mainly be focused on current methods for predicting the settlement of shallow footings: the one-dimensional (oedometer) method when dealing with footings on clay, or empirical methods based on field tests for footings on sand.

Simple calculations based on closed form solutions are attractive, because they offer the advantage of focusing on the most important parameters to be considered, as well as on their relative influence, and for this reason the use of the theory of elasticity will also be discussed.

Solutions from the theory of elasticity, which can be applied to determine the stresses, strains and displacements induced by surface loads, will be provided for the most recurrent cases. Furthermore, one of the most challenging aspects is represented by the problem of relating settlements to damage. This requires us to appreciate how many factors may have an influence on the relation between settlements and damage, so that it is appropriate to discuss the main steps of a building process, since the probability of damage diminishes the larger the ratio of immediate to consolidation settlement and the smaller the ratio between the imposed and the dead loads.

In this perspective, a final section provides guidelines for limiting values of settlement, as concerned to visual appearance, serviceability and structural damage.

9.1 Description of ground and foundation movements

When designing a structure, engineers are interested in forecasting both the total amount and differential settlements the structure will suffer, and how fast settlements will occur. For these reasons it is useful to introduce some preliminary considerations.

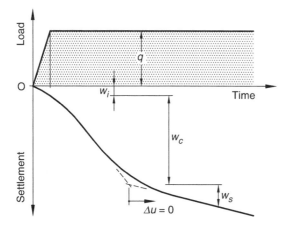

Figure 9.1 Evolution of settlement with time.

9.1.1 Evolution of settlement with time

When a load is applied to a foundation of limited extent, resting on fine-grained soils, the following phenomena can be observed (Figure 9.1).

(a) Excess pore pressures develop in the ground during loading and due to the low hydraulic conductivity of the soil a working hypothesis is assumed that in this initial stage the soil behaves in *virtually undrained conditions*.

As a consequence, deformations should occur without volume change and the settlement at the surface is the result of pure distortion. This component of settlement is usually named **immediate settlement** and even though it is not actually elastic, it is customary to estimate its amount by using the theory of elasticity or, alternatively, by using empirical rules that predict its amount as a fraction of the overall settlement (see Section 9.4).

(b) The subsequent time-dependent process of pore pressure dissipation and effective stress change causes the **consolidation settlement**. We stated in Chapter 4 that the term **compression** is commonly used when describing the change in void ratio due to the change in effective stresses, without any reference to the time interval over which this compression occurs. The time dependent phenomenon which describes the evolution of pore pressure and the deformation of the porous medium is defined **consolidation**, and this process has been described in detail in Chapter 6.

(c) Finally, after the consolidation process has been completed, the **secondary settlement** takes place at *constant effective stresses*. We have already discussed that due to the viscous nature of the soil, both primary and secondary compressions takes place from the beginning. Therefore, the above assumption is only a working engineering hypothesis, which, in many cases, helps to solve the problem in an easy manner.

The importance of each component depends on the geometry of the problem, the nature of the soil and the loading programme. In general, it can be observed that the immediate settlement is relevant in soft clays of medium to high plasticity, where significant plastic strains can take place, as well as undrained creep phenomena. The consolidation settlement is by far the predominant component and is of major concern. The secondary settlement may usually be relevant in organic soils.

9.1.2 Spatial distribution of settlement

As far as the spatial distribution of settlement is concerned, a wide terminology has been used in literature to identify ground and foundation movements, so that it is appropriate to specify the meaning of the most recurrent terms. This is done in the following with reference to definitions put forward by Burland and Wroth (1975).

Figure 9.2 sketches the settlement w suffered by different points of a structure, and the maximum differential settlement is highlighted as Δw_{max}.

The term **tilt** (denoted by ω in Figure 9.2) is used to describe the *rigid body rotation*. The **relative rotation** β describes the rotation of the straight line joining two reference points relative to the tilt. Note that this is identical to the *angular distortion* defined by Skempton and MacDonald (1956).

The **angular strain** is given by:

$$\alpha = \frac{w_B - w_A}{L_{AB}} + \frac{w_B - w_C}{L_{BC}}, \tag{9.1}$$

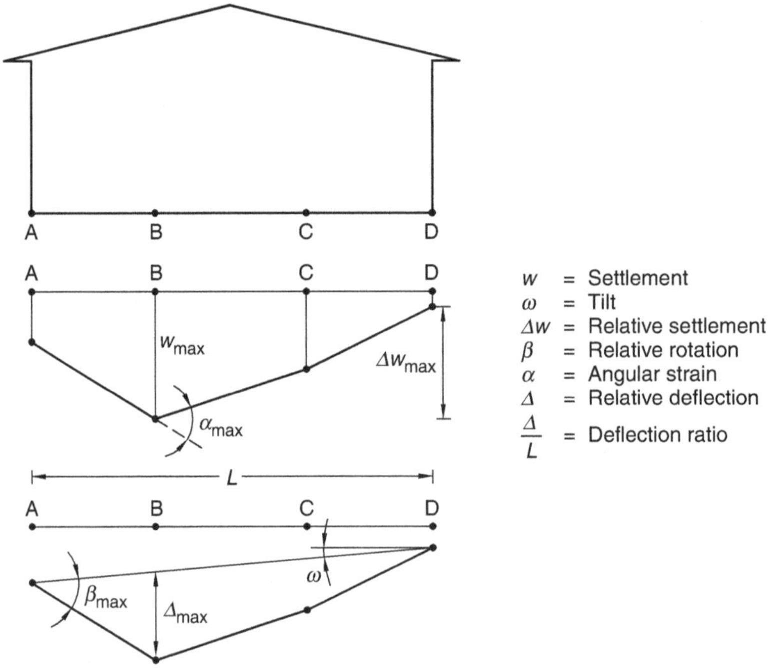

w	= Settlement
ω	= Tilt
Δw	= Relative settlement
β	= Relative rotation
α	= Angular strain
Δ	= Relative deflection
$\frac{\Delta}{L}$	= Deflection ratio

Figure 9.2 Definition of foundation movements (adapted from Burland and Wroth, 1975).

and is a reference parameter for predicting crack width in buildings in which movements occur at existing weakness lines. In general, the angular strain is assumed positive if it corresponds to *sagging* or upward concavity, and it is assumed negative for *hogging* or downward concavity.

The **relative deflection** Δ is the maximum displacement relative to the straight line connecting two reference points at a distance L apart, and the ratio Δ/L is defined as **deflection ratio** or *sagging ratio* or *hogging ratio*.

The above referred definitions will be used in Section 9.6 in order to establish some rational criteria for limiting movements and related damage.

9.2 Methods of computing settlements: introductory notes

A recurrent dilemma in soil mechanics is the choice of a soil constitutive model and related mechanical parameters. The desire to capture basic features such as geological history, recent stress history and coupling between volumetric and shearing behaviour claims for advanced soil models, which offer the advantage that mechanical parameters do not depend on the actual boundary problem under consideration. Shortcomings are represented by the increased complexity of the numerical analysis, that may require experts to properly handle the problem.

Therefore, necessity may dictate a more relaxed alternative of relying on simple models. However, since this alternative condenses all difficulties on the selection of a representative soil stiffness, we must be able to take into account, at least, the relevant loading path, the current state and the strain level.

This chapter follows this relaxed alternative, so that the attention will be focused on the currently in use methods for predicting the settlement of shallow footings: the one-dimensional (oedometer) method when dealing with footings on clay, or empirical methods based on field tests for footings on sand.

Simple calculations based on closed form solutions are also attractive, because they offer the advantage of focusing on the most important parameters to be considered, as well as on their relative influence, and for this reason the use of the theory of elasticity will be addressed in the next section.

However, the reader must be aware of the fact that many other problems of engineering interest require more sophisticated approaches. One such example is represented by settlements associated to ground excavations: as proved by Simpson *et al.* (1979), the use of a simple model, such as the elastic one, would give unrealistic and misleading results.

9.3 Prediction using the theory of elasticity

Theory of elasticity is attractive because it provides closed form solutions for the analysis of some boundary problems of geotechnical interest. Examples include the stress distribution induced within the soil mass by load applied at the surface, as well as the displacement of a loaded area of finite dimension. There are certainly limitations as far as the use of elasticity based predictions is concerned, and this aspect will be addressed at the end of this section (see Section 9.3.5). However, simple calculations based on closed form solutions still offer the advantage of focusing on the most important parameters to be considered, as well as on their relative influence. Furthermore, these solutions represent benchmark problems against which solutions obtained by means of sophisticated numerical methods can be checked.

In general, solving a boundary value problem of linearized and isotropic elasticity involves 15 unknowns, represented by the six independent components of the stress tensor σ_{ij}, the six independent components of the strain tensor ε_{ij}, and the three components of displacement u_i.

Field equations involve the equilibrium equations:

$$\sigma_{ji,j} = 0, \quad \forall \mathbf{x} \in D, \tag{9.2}$$

the expressions of strain in terms of displacement field:

$$\varepsilon_{ij} = \frac{1}{2}\left(u_{i,j} + u_{j,i}\right), \quad \forall \mathbf{x} \in D, \tag{9.3}$$

and the constitutive relationships:

$$\sigma_{ij} = \lambda \varepsilon_{kk}\delta_{ij} + 2\mu\varepsilon_{ij}. \tag{9.4}$$

The **boundary conditions** for the field equations may be expressed as *traction boundary conditions*, with the traction components t_i prescribed at the stressed boundary ∂D_f:

$$\sigma_{ji}n_j = t_i(\mathbf{x}) \quad \forall \mathbf{x} \in \partial D_f, \tag{9.5}$$

as *displacement boundary conditions*, with the displacement components u_i prescribed at the constrained boundary ∂D_u:

$$u_i = b_i(\mathbf{x}) \quad \forall \mathbf{x} \in \partial D_u, \tag{9.6}$$

as well as *mixed boundary conditions*. Note that, since the aim of this analysis is to compute the stresses and the deformations induced by the applied loads, the quantities in equations (9.2) and (9.3) are intended to be incremental quantities. In particular, the incremental stresses must be summed to the initial state of stress in order to obtain the actual state of stress. The initial state of stress accounts for body forces that do not appear in equations (9.2).

9.3.1 The Boussinesq's problem

A problem of particular interest for geotechnical engineering is that relative to a finite force applied at a point on the surface of an elastic half-space. Since this force acts on a surface of zero area, it is referred to as a **vertical point load**, and can be considered as a mathematical idealization if we consider the actual distribution of forces encountered in reality. This problem was solved by the French mathematician Joseph Boussinesq in 1878 and it is sketched in Figure 9.3.

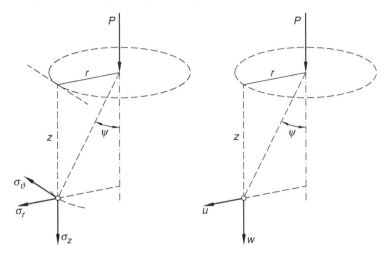

Figure 9.3 Boussinesq's problem.

We first note that we are faced with conditions of axial symmetry, so that, referring to cylindrical coordinates (r, ϑ, z) and considering the displacement components
(u_r, u_ϑ, u_z), we deduce that the component u_ϑ must vanish and the components u_r and
u_z must be independent of ϑ.

Stress components must also be independent of ϑ and, in particular, the component
$\tau_{\vartheta r}, \tau_{\vartheta z}$ must be zero, so that the stress tensor assumes the form:

$$\sigma_{ij} = \begin{bmatrix} \sigma_r & 0 & \tau_{rz} \\ 0 & \sigma_\vartheta & 0 \\ \tau_{zr} & 0 & \sigma_z \end{bmatrix}.$$

The equilibrium equations (in absence of body forces) write:

$$\frac{\partial \sigma_r}{\partial r} + \frac{\partial \tau_{zr}}{\partial z} + \frac{\sigma_r - \sigma_\vartheta}{r} = 0$$
$$\frac{\partial \tau_{rz}}{\partial r} + \frac{\partial \sigma_z}{\partial z} + \frac{\tau_{rz}}{r} = 0$$

$$\tag{9.7}$$

and strain components in terms of displacement field write:

$$\varepsilon_r = \frac{\partial u_r}{\partial r}; \quad \varepsilon_\vartheta = \frac{u_r}{r}; \quad \varepsilon_z = \frac{\partial u_z}{\partial z}; \quad \varepsilon_{zr} = \frac{1}{2}\left(\frac{\partial u_r}{\partial z} + \frac{\partial u_z}{\partial r}\right). \tag{9.8}$$

By combining (9.7) and (9.8) with the elastic constitutive law provides a set of
partial differential equations. These can be integrated for the prescribed boundary
conditions, specified as follows: on the surface $(z = 0)$ the stresses are zero, except at
the origin $(r = 0)$, where the stresses must equilibrate the applied load; at any point
in the half-space at infinite distance from the surface the displacements must vanish.
Boussinesq solved this problem and obtained the stress distribution (see Section 138

of Timoshenko and Goodier, 1971):

$$\sigma_z = \frac{3P}{2\pi} \frac{z^3}{R^5}$$

$$\tau_{rz} = \frac{3P}{2\pi} \frac{z^2 r}{R^5}$$

$$\sigma_r = \frac{P}{2\pi} \left[\frac{3zr^2}{R^5} - \frac{1-2\upsilon}{R(R+z)} \right] \qquad (9.9)$$

$$\sigma_\vartheta = \frac{P}{2\pi} (1-2\upsilon) \left[\frac{1}{R(R+z)} - \frac{z}{R^3} \right]$$

where $R^2 = z^2 + r^2$ and υ is the Poisson's ratio.

In order to prove that this stress field satisfies the prescribed boundary conditions, we first observe that at the origin ($z = 0$; $r = 0$) the stresses become infinite. Therefore, to prove that the stress field near the origin equilibrate, the applied force, we need to imagine around the origin a hemispherical surface of small radius centred on the point of application of the force. The stress vector acting on this hemispherical surface has a non-vanishing horizontal component t_r and a vertical component t_z, while the component t_ϑ is zero. The horizontal components have zero resultant, because there is an equal and opposite component on the opposite sides of the surface; on the contrary, if we integrate the vertical component over the hemispherical surface we obtain the applied force.

The related displacement field is the following:

$$u_z = \frac{P}{4\pi RG} \left[2(1-\upsilon) + \frac{z^2}{R^2} \right]$$

$$u_r = \frac{P}{4\pi RG} \left[\frac{rz}{R^2} - \frac{(1-2\upsilon)r}{R+z} \right] \qquad (9.10)$$

In particular, the displacement components at the surface are:

$$u_z = \frac{P(1-\upsilon)}{2\pi Gr}$$

$$u_r = -\frac{(1-2\upsilon)P}{4\pi Gr} \qquad (9.11)$$

It can be verified that both components u_z and u_r approach zero as the distance R becomes larger and larger; for $r \to 0$, i.e. at the point of application of the load, the displacement is infinitely large.

This again is a consequence of the singularity in the surface load, since in the origin the stress is infinitely large.

There are other point load problems that may be of interest, although not considered in this book. These include the *point load acting at the interior of an infinite elastic medium* (solved in 1848 by William Thompson, who became Lord Kelvin); *the horizontal point load acting at the surface of an elastic half-space* (solved by the Italian

V. Cerruti, in 1882) and the *point load (either vertical or horizontal) at the interior of an elastic half-space* (solved by Raymond Mindlin, in 1936). These solutions are discussed in detail by Davis and Selvadurai (1996).

9.3.2 Loaded area

The differential equations and the boundary conditions related to the point load problem are linear, and the small displacements do not affect the action of external force. Therefore, we can use the principle of superposition to find the stress and the displacement field due to a load acting on a finite area. In the following we discuss only a limited number of solutions of practical interest and we suggest to the reader the book by Poulos and Davis (1974) for a complete compilation.

(a) Many foundations may be represented as a **circular** pad of radius a, loaded by a uniform load q (Figure 9.4). From an engineering point of view, assuming that the contact stresses are equal to the applied uniform load means that the stiffness of the foundation can be neglected, and the response of the soil is uncoupled from that of the structure. In these conditions, if we integrate the Boussinesq solution over the loaded area and we consider the points under the *centreline*, the stress components are given by the following expressions:

$$\sigma_z(r=0)=q\left\{1-\frac{1}{\left[(a/z)^2+1\right]^{3/2}}\right\}$$
$$\sigma_r=\sigma_\vartheta=\frac{q}{2}\left\{(1+2\upsilon)-2(1+\upsilon)\left[1+\left(\frac{a}{z}\right)^2\right]^{-1/2}+\left[1+\left(\frac{a}{z}\right)^2\right]^{-3/2}\right\}$$

(9.12)

Note that the soil stiffness plays no role in the stress distribution for a homogeneous linear elastic material, and, in addition, the Poisson's ratio does not influence the vertical stress.

Since we can give to the vertical stress σ_z at any location (r,z) the expression:

$$\sigma_z=q\cdot f(a,r,z),$$

(9.13)

we can explore the locus of points with equal vertical stress, i.e the locus characterized by the condition:

$$\frac{\sigma_z}{q}=\text{const}.$$

(9.14)

This will produce a family of surfaces $f(a,r,z)=\text{const}$, which resemble the layers of a bulb, and originates the definition of **bulb of stresses** (Figure 9.5) as the space bounded by a surface, which is the locus of vertical stresses corresponding to a given fraction of the applied uniform load.

The expression obtained for the vertical stress tells us that the induced vertical stress reduces progressively with depth and reaches a value of about 10% of the the applied

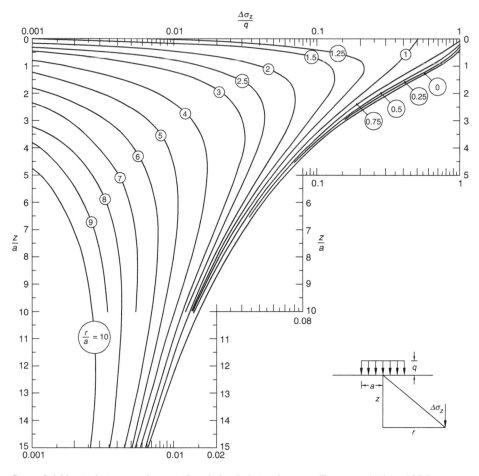

Figure 9.4 Vertical stress under a uniformly loaded circular area (Foster and Alvin, 1954).

uniform load at a depth twice the diameter of the footing. This is a consequence of the spreading of the load with depth.

We consider now the Boussinesq solution relative to the vertical displacement of the surface produced by the vertical point load:

$$u_z = \frac{P(1-\upsilon)}{2\pi \, Gr}.$$

If we replace the point load by the applied stress q multiplied by an element of area $dA = rd\vartheta \, dr$ and we integrate over the loaded area:

$$\int_0^{2\pi} \int_0^a \frac{(qrd\vartheta \, dr)(1-\upsilon)}{2\pi \, Gr},$$

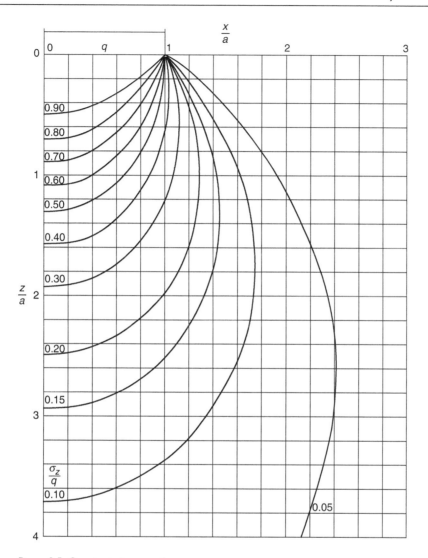

Figure 9.5 Constant stress surfaces.

we obtain the following expression for the vertical displacement beneath the centre of the loaded area:

$$w_c = \frac{qa(1-v)}{G}.$$ (9.15)

By using the same procedure, beneath the rim the displacement assumes the expression:

$$(w)_{r=a} = \frac{2}{\pi}\frac{qa(1-v)}{G},$$ (9.16)

and the average displacement is equal to:

$$\bar{w} = 0.85w_c. \tag{9.17}$$

(b) If we consider a **rectangular** loaded area, the vertical stress σ_z beneath the corner of this area loaded by a uniform load q is given by (see Figure 9.6 for symbols):

$$\sigma_z = \frac{q}{2\pi}\left[\frac{2mn\cdot(m^2+n^2+1)^{0.5}}{m^2+n^2+1+m^2n^2}\cdot\frac{(m^2+n^2+2)}{(m^2+n^2+1)} +\tan^{-1}\frac{2mn\cdot(m^2+n^2+1)^{0.5}}{m^2+n^2+1-m^2n^2}\right]. \tag{9.18}$$

The vertical stress under points other than corners may be obtained by using the principle of superposition, as shown in Exercise 9.1.

In particular, when considering the case of a **square loaded area**, whose dimensions are $(2b \times 2b)$, the maximum vertical displacement beneath the centre line is given by:

$$w_c = 1.12qb\frac{1-\upsilon}{G}, \tag{9.19}$$

and the settlement beneath the corner is equal to $w_b = \frac{1}{2}w_c$, with an average value $\bar{w} = 0.85w_c$.

(c) The French engineer Alfred Flamant derived in 1892 the solution for a *vertical line load*, and, by applying the principle of superposition, this solution has been used to derive the stresses due to a load q per unit area on a **strip of infinite length** (this is a plane strain problem). Expressed in terms of auxiliary angles, as defined in Figure 9.7a, the stresses are:

$$\sigma_z = \frac{q}{\pi}(\alpha + \sin\alpha\cos(\alpha + 2\beta'))$$

$$\sigma_y = \frac{q}{\pi}(\alpha - \sin\alpha\cos(\alpha + 2\beta')). \tag{9.20}$$

$$\tau_{zy} = \frac{q}{\pi}(\sin\alpha\sin(\alpha + 2\beta'))$$

It can be proved that the principal stresses and the maximum shear stresses at any point are given by:

$$\sigma_I = \frac{q}{\pi}(\alpha + \sin\alpha)$$

$$\sigma_{III} = \frac{q}{\pi}(\alpha - \sin\alpha), \tag{9.21}$$

$$\tau_{max} = \frac{q}{\pi}\sin\alpha$$

and the loci of constant maximum shear stress are represented by circles passing through the edges of the loaded area. In particular, the maximum value of shear stresses

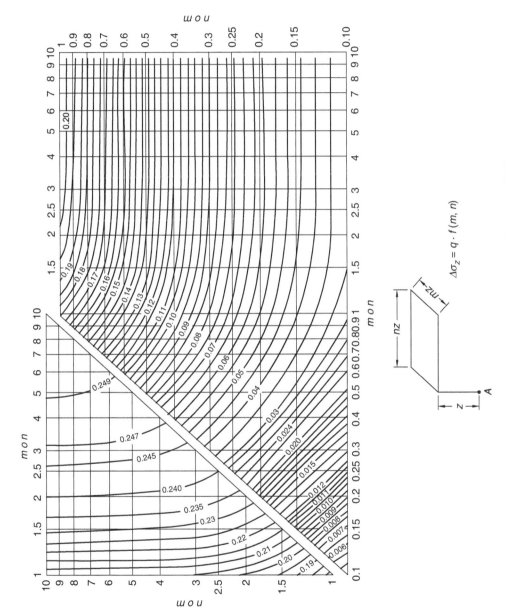

Figure 9.6 Chart for determining vertical stresses underneath corners of rectangular areas (Newmark, 1942).

Example 9.1

With reference to Figure 9.8, *find* the vertical stress at depth of 5 m below point P, external to the loaded area BNRD.

Solution. The effect of the loaded area BNRD can be obtained by considering the area ABCP, by subtracting the effect of the two areas MNCP and ADEP and summing up the effect of the area MREP.

If the intensity of the applied uniform load is equal to 100 kPa, by using the solution in Figure 9.6 we obtain:

Area	m	n	$f(m,n)$	$\Delta\sigma_z$
ABPC	3	2.6	0.242	+ 24.2
ADEP	2.6	1	0.203	− 20.3
MNCP	3	1	0.205	− 20.5
MRPE	1	1	0.177	+ 17.7

and therefore: $\Delta\sigma_z = 1.1$ kPa.

---

is attained when $\alpha = \pi/2$ (Figure 9.7c) and occurs at points lying on a semicircle with the strip footing as a diameter. Since the magnitude of the maximum shear stresses is q/π, it can be predicted that **local yield** of the soil (in undrained condition) occurs when:

$$q_{crit} = \pi s_u,$$ \hfill (9.22)

whereas the complete plastic failure will occur when $q = (2 + \pi)s_u$, as discussed in Chapter 8.

There are two aspects that deserve attention when dealing with strip loads. First, if we evaluate the vertical stress under the edge of the strip load we find that $\sigma_z = q/2$. This could render us suspicious about the solution, but we can explain this apparently unexpected result if we consider two equal strips in contact to each other: by using the superposition we must obtain the applied load q, and this tells us that each contribution must be $q/2$.

The second aspect is related to the fact that, if the load line is infinite (plane strain problem), the half-space surface will experience an infinite settlement. This depends on the mathematical abstract concept of infinite strip, and in order to compute settlements we need to take into account the finite length of the actual loaded area.

9.3.3 Elastic layer underlain by a rigid base

When a rigid base is found at a given depth, the stress distribution changes due to the presence of this rigid boundary, as illustrated in Figure 9.9. The curve labelled *a* is the vertical stress distribution at depth *h* for the case of a half-space. The curve *b* is the solution (Biot, 1935) for the layer of finite thickness, with no friction at the interface with the rigid base. The curve *c* is the same as *b*, except that perfect adhesion has been assumed at the interface between the elastic layer and the rigid base. This analysis

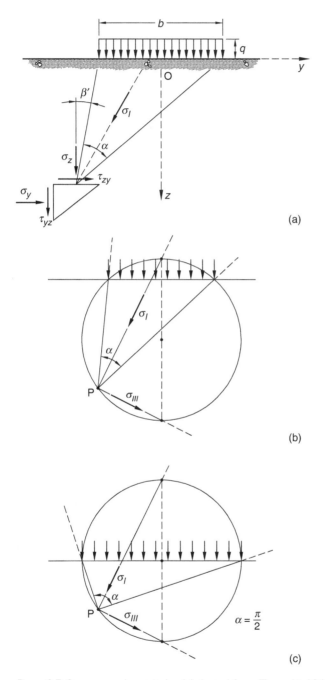

Figure 9.7 Stresses under strip load (adapted from Terzaghi, 1943; Tsytovich, 1976).

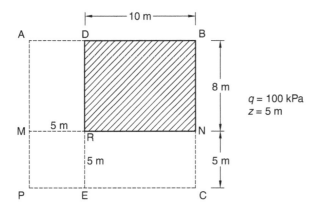

Figure 9.8 Illustrative example of determining the vertical stress under corners of rectangular areas.

proves that, in presence of a rigid base, the vertical stress predicted by the the half-space assumption can be significantly in error, and for this reason, it is of practical interest to complete the compilation of the available elastic solutions by adding those obtained by Egorov (1958) and summarized in Figures 9.10 and 9.11.

The settlement of the loaded area is expressed in the form:

$$w = 2aq\frac{1 - v^2}{E}I,$$
(9.23)

where I is a factor depending on the shape of the loaded area, the thickness of the elastic layer relative to the dimension of the foundation ($2a$ is the width of the rectangular area or the diameter of a circular area), and depends on whether the loaded area is supposed to be flexible (Figure 9.10) or rigid (Figure 9.11).

9.3.4 Rigid foundations

The stress distribution on the surface of an elastic half-space can be specified in advance only for flexible foundations. When we deal with rigid foundations, the settlement is uniform under the loaded area, but this is no longer the case for the vertical contact stress. In fact, the stress distribution is now unknown and the only constraint is that the resultant of the contact stress must equilibrate the applied load. Since a uniform contact stress induces a sagging shaped pattern of settlement, in order to produce a uniform settlement we can guess that the vertical contact stress must increase moving from the centreline (Figure 9.12).

Boussinesq (1885) solved this problem for the case of a rigid circular foundation and obtained the following distribution of the contact stress (Figure 9.12)

$$q(r) = \frac{P}{2\pi a\sqrt{a^2 - r^2}}.$$
(9.24)

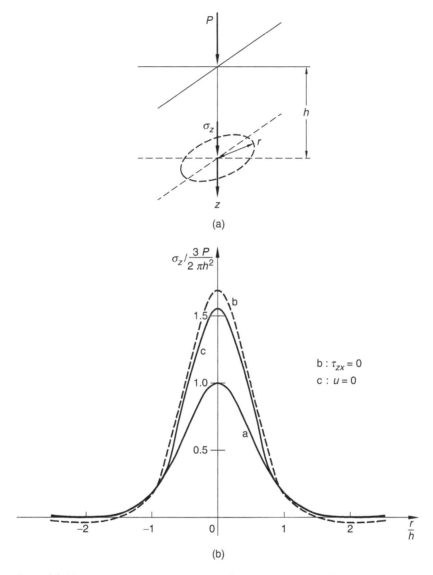

Figure 9.9 Vertical stress distribution at interface between an elastic layer and a rigid base (Biot, 1935).

The related settlement is given by:

$$w = \frac{P(1-\upsilon)}{4Ga} I_R .$$ (9.25)

According to this solution, the vertical contact stress varies with radius from a value of about $q/2$ at the centreline to a value theoretically infinite at the edge, and this infinite value is related to the discontinuity of the slope of the displacement pattern that we observe, when we move from some point outside the loaded area towards the

$$w = \frac{2aq}{E}(1 - v^2) I_F$$

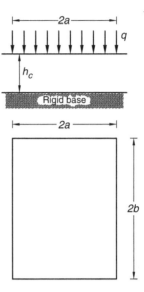

$\frac{h_c}{a}$	Coefficient of influence I_F				
	Circle	$b/a = 1$	2	5	10
0.20	0.08	0.08	0.08	0.08	0.08
0.50	0.22	0.21	0.21	0.21	0.21
1.00	0.45	0.44	0.43	0.43	0.43
2.00	0.68	0.72	0.78	0.78	0.78
3.00	0.78	0.84	0.99	1.02	1.02
5.00	0.87	0.95	0.19	1.34	1.34
10.00	0.93	1.04	1.36	1.69	1.77
∞	1.00	1.12	1.52	2.10	2.53

Figure 9.10 Vertical displacement under the centre of a flexible loaded area (adapted from Egorov (1958) and Harr (1966)).

$$w = \frac{2aq}{E}(1 - v^2) I_R$$

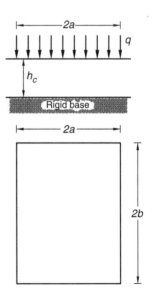

$\frac{h_c}{a}$	Coefficient of influence I_R				
	Circle	$b/a = 1$	2	5	10
0.20	0.096	0.096	0.098	0.099	0.099
0.50	0.225	0.226	0.231	0.236	0.238
1.00	0.396	0.403	0.427	0.441	0.446
2.00	0.578	0.609	0.698	0.748	0.764
3.00	0.661	0.711	0.856	0.952	0.982
5.00	0.740	0.800	1.010	1.201	1.256
10.00	0.818	0.873	1.155	1.475	1.619
∞	0.849	0.946	1.300	1.826	2.246

Figure 9.11 Vertical displacement of a rigid foundation (adapted from Egorov (1958) and Harr (1966)).

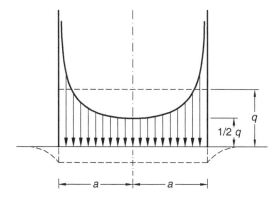

Figure 9.12 Contact stress distribution for rigid circular foundation (Boussinesq, 1885).

footing edge. In reality, soils have a finite strength and plastic zones develop, thereby reducing this value to a finite stress.

9.3.5 Shortcomings of the elastic approach

The reliability of linear elastic analyses has been investigated by comparison with more sophisticated computations (see Burland *et al.*, 1977).

In particular, two major aspects have received a lot of attention: the prediction of stress distribution and the selection of soil stiffness.

Complex computations for anisotropic, inhomogeneous and non-linear materials have proved that errors in the vertical stress, as predicted by the linear isotropic elasticity, are very small. On the contrary, other stress components, as well as the displacements, are very sensitive to the adopted soil model. However, the vertical stress may significantly differ from that predicted by the Boussinesq solution in the case of a stiffer stratum overlying a more compressible layer (Figure 9.13), because the upper layer tends to operate a more pronounced spreading of the load.

The use of closed form solutions, provided by the elasticity theory, requires the selection of an appropriate, unique value of the soil stiffness, and this represents by no means the most difficult task.

In principle, these solutions can be used to predict both the immediate (undrained) settlement and the final long-term settlement. When dealing with immediate settlement, although the soil behaviour is controlled by changes in effective stresses, it is still usual practice to describe the undrained behaviour of soil in terms of changes in *total stresses*, and in this case the undrained Young modulus E_u (with the subscript u) and the undrained Poisson's ratio $v = 0.5$ (corresponding to a constant volume constraint) are the appropriate parameters.

The long-term (drained) settlement requires to estimate the Young modulus and Poisson's ratio from drained tests. However, in both cases soil stiffness depends on the stress history, the current state and the loading path, so that it is really difficult to single out a unique value to be introduced into a closed form solution of a boundary problem.

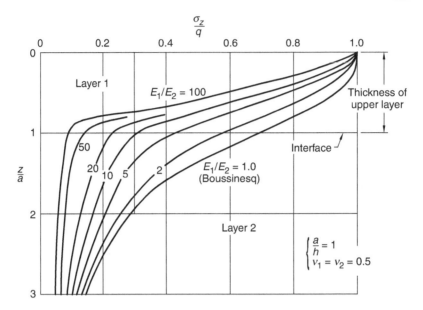

Figure 9.13 Vertical stress distribution in presence of a stiff layer overlying a more compressible stratum (Burmister, 1956).

9.4 Settlement of foundations on clay

Predicting the settlement of a structure on clay involves three major steps:

1 Definition of a **design profile**, usually a layered soil profile, properly characterized in terms of stress history $(\sigma'_p, \sigma'_{vo})$ and compressibility $(C_c$ and $C_r)$.
2 For the layered soil profile the **vertical stresses** are computed assuming that the soil is isotropic, linearly elastic and homogeneous, ignoring the differences in soil properties of the different layers.
3 **The vertical strains** of each layer are computed using the one-dimensional (oedometric) approach, and the reductions in thickness of the layers are added to obtain the settlement of the soil surface.

Each of the mentioned steps is underpinned by some arbitrary assumptions, discussed at the end of this section, but experimental evidences support the reliability of this rather simple approach, and this justifies its popularity. In the sequel, this approach will be illustrated in detail with reference to a well-documented case history: the leaning Tower of Pisa (further reading on this case includes Burland *et al.*, 2003; Commissione Polvani, 1971; Calabresi *et al.*, 1993; Callisto, 1996; Costanzo, 1994; Lancellotta, 1993; Lancellotta *et al.*, 1994; Pepe, 1995; Rampello and Callisto, 1998).

(a) The construction of a **design profile** requires performing a convenient number of borings, such that a reliable stratigraphical section can be defined, taking into account the inherent spatial variability of natural soil deposits. As already discussed in Chapter 4, in the subsoil profile of the Piazza dei Miracoli (see Figure 4.11), the 'horizon C', found between the depths of 40 m and at least 120 m, is composed of marine sands deposited during the Flandrian transgression. Above these dense sands is the 'horizon B', a 30 m thick marine clay deposit, formed at a time of rapid eustatic rise. Between the so called 'lower clay' and the 'upper clay', known locally as 'Pancone Clay', there is a 2 m thick layer of intermediate sand and a 4 m thick layer of stiff 'intermediate clay'. The rate of eustatic rise decreased during the last 10,000 years, so that the upper sediments (the 'horizon A' in Figure 4.11) became increasingly estuarine in character. These sediments are in fact characterized by pronounced spatial variability, inclined layering and they differ over short horizontal distances. By considering Cone Penetration Test profiles (see Figure 9.14), the material to the south of the Pisa Tower appears to be finer-grained than that to the north, and the sand layer is locally thinner. This soil variability was probably responsible for the initial tilting of the Pisa Tower.

A lot of care is required when investigating the groundwater regime. Pore pressure values may be different from those corresponding to the hydrostatic regime (see Figure 9.15a), as it can be the case when the clay stratum is still experiencing consolidation induced by the application of recent fill (Figure 9.15b). Deviations can also occur in the presence of a steady state flow, driven by boundary conditions corresponding to different hydraulic heads (Figura 9.15c). In the case of the subsoil profile of Pisa, pumping from the horizon C has resulted in downward seepage from horizon A, with a pore pressure distribution with depth slightly below the hydrostatic distribution (Figure 9.16).

(b) The second step consists of computing the **vertical stresses** induced by the *net contact stress* $q_N = q - \gamma D$, assuming that the soil is isotropic, linearly elastic and homogeneous, and ignoring the differences in soil properties of different layers.

(c) Finally, in order to compute the settlement, the soil profile is subdivided into a convenient number of layers and the **vertical strain** of each layer is computed using the one-dimensional (oedometric) approach (see Figure 9.17). The reduction in thickness of the layers is added to obtain the settlement of the soil surface.

In particular, if H_o and H_s are the initial thickness of a thin layer and the height of the solid fraction, the one-dimensional assumption allows to write (Figure 9.17c):

$$H_o = H_S + e_o H_S = H_S(1 + e_o).$$

Then the reduction in thickness of the layer can be expressed in terms of reduction of void ratio, i.e:

$$\Delta H = -(H_f - H_o) = -H_S(e_f - e_o) = -H_S \Delta e, \tag{9.26}$$

which in turn may be computed by taking into account the stress history and soil compressibility:

$$-\Delta e = C_r \log \frac{\sigma'_p}{\sigma'_{vo}} + C_c \log \frac{\sigma'_{vo} + \Delta \sigma'_z}{\sigma'_p}. \tag{9.27}$$

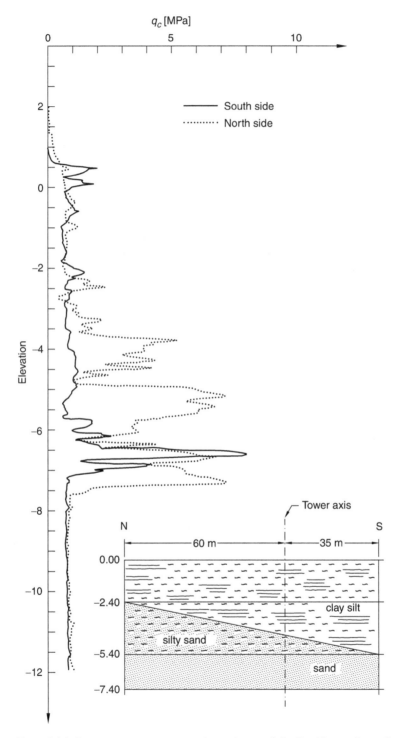

Figure 9.14 Cone penetration tests performed around the Pisa Tower (Lancellotta et al., 1994).

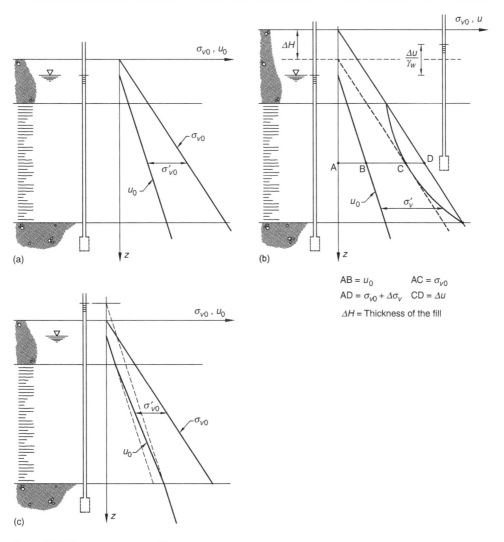

$AB = u_0$ $AC = \sigma_{v0}$

$AD = \sigma_{v0} + \Delta\sigma_v$ $CD = \Delta u$

ΔH = Thickness of the fill

Figure 9.15 Pore pressure profiles.

This is the one-dimensional approach, discussed in Chapter 4, C_r being the **recompression index** and C_c the **compression index** (see Figure 9.17 and Example 9.2).

9.4.1 Reliability of the conventional one-dimensional method

There are a number of assumptions in the conventional one-dimensional method. The starting point is that the consolidation settlement results from the dissipation of pore pressure, generated by the applied load. Therefore, in order to compute the foundation displacement we need to estimate the induced pore pressure and to compute the vertical strains, by selecting the appropriate soil stiffness. The conventional

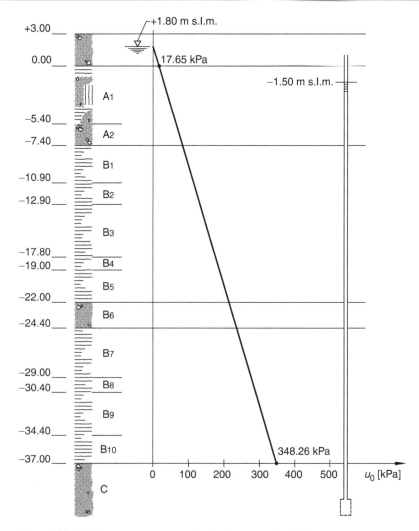

Figure 9.16 Pisa Tower: piezometric level in horizon A and C.

one-dimensional approach implicitly assumes that $\Delta u = \Delta\sigma_z$ and the soil stiffness to be used is the constrained modulus E_{oed} .

Skempton and Bjerrum (1957) have discussed how to remove the first assumption, and Simons and Som (1969, 1970) have given some suggestions to relax the simplification related to the use of the constrained modulus.

However, although this approach is underpinned by the above assumptions, it has been successfully used, so that its accuracy must be assessed mainly on the basis of observed performance.

Designers are interested in making a distiction between the immediate and the long-term settlement, because it is only the components of deformation acting on a structure

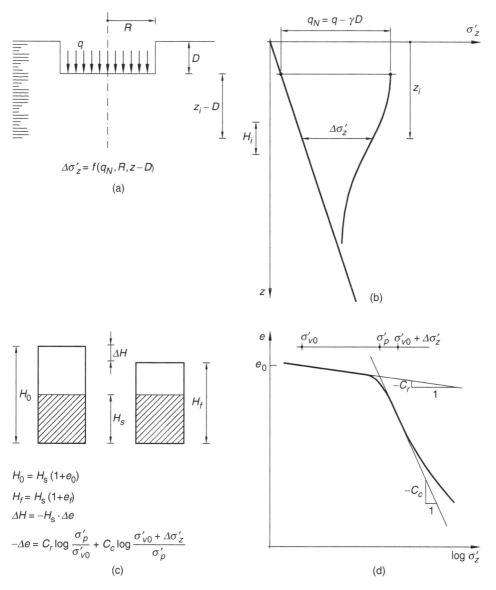

Figure 9.17 Synoptic sketch of one-dimensional approach.

that have reached a certain degree of rigidity which may cause some damage. In this perspective, the immediate settlement could be evaluated by referring to elastic solutions complemented by an appropriate choice of the undrained Young modulus and Poisson's ratio. An alternative is to rely on empirical rules, which give the immediate settlement as a fraction of the overall settlement, computed by means of the one-dimensional approach.

Example 9.2
Referring to soil profile depicted in Figure 4.11 (Chapter 4), it is required to estimate the consolidation settlement of the Pisa Tower.

Let us start by subdividing the horizons A and B into a convenient numbers of layers, as shown in the first column of Table 9.1. The elevation values in the second column refer to the bottom of each layer, so that the elevation corresponding to the top of the first layer is 0.00 and the bottom is –2.40, the thickness H_i being 2.40 m. In doing so, it is presumed that the founding level of the Pisa Tower was originally at 0.00 m, as argued by P. Sampaolesi in the book *Il Campanile di Pisa*, edited by Opera della Primaziale di Pisa, 1956 (see page 20).

Since the ground level is +3.00 m, the total overburden stress at elevation 0.00 m is equal to $\sigma_{vo} = 18 \cdot 3 = 54 \, \text{kN/m}^2$ and, by using the unit weight values given in the third column, the overburden stress σ_{vo} can also be computed at other elevations.

At this point we note that computing the vertical effective stress requires to take into account the pore pressure profile shown in Figure 9.16, and for ease of computation the relevant values of u_o have been reported in Table 9.1 (fifth column).

Moving to columns 8, 9 and 10 we read in the order: the average depth of each layer, as computed from the founding level; the effective vertical stress at this depth and the increment of the vertical stress produced by the foundation load. This increment has been evaluated through the elastic solution:

$$\Delta\sigma_v = q_N \left\{ 1 - \frac{1}{\left[1 + (a/z)^2 \right]^{3/2}} \right\},$$

where the *net bearing stress* q_N has been derived by considering the Tower weight (141.75 MN) as uniformly acting on a circular section of radius $a = 9.78$ m, so that the *gross bearing stress* is $q = \dfrac{141.75}{\pi R^2} \cdot 1000 = 472 \, \text{kPa}$, and therefore $q_N = q - \sigma_{vo} = 472 - 54 = 418 \, \text{kPa}$.

Soil behaviour in one-dimensional condition is characterized by the following parameters, as deduced from oedometer tests: overconsolidation ratio OCR and yield stress σ'_p, compression and recompression index. The relevant values of these parameters are given in columns 11 to 15. Therefore, we can now evaluate the change of thickness of each layer, by using equations (9.26) and (9.27). A detailed example is given for the layer B2, that spans from elevation −10.90 to −12.90:

$$\Delta H = \frac{H_o}{1+e_o} \left(C_r \log \frac{\sigma'_p}{\sigma'_z} + C_c \log \frac{\sigma'_z + \Delta\sigma'_z}{\sigma'_p} \right)$$

$$= \frac{200}{1+1.4} \left(0.13 \log \frac{184.07}{141.59} + 0.80 \log \frac{141.59 + 225.25}{184.07} \right) = 21.20 \, \text{cm}.$$

As a further example, consider the layer B5, ranging from elevation −19.00 to −22.00. In this case the final vertical stress (220.07 + 110.68 = 330.75 kPa) does not exceed the yield stress (550.18 kPa), so that the change of thickness is due to the reduction of void ratio occurring in the overconsolidated domain:

$$\Delta H = \frac{H_o}{1+e_o} C_r \log \frac{\sigma_z' + \Delta \sigma_z'}{\sigma_z'}$$

$$= \frac{300}{1+0.7} 0.06 \log \frac{220.07 + 110.68}{220.07} = 1.87 \, cm.$$

Finally it is relevant to note that, when dealing with sand layer A2, reference has been made to the oedometer modulus E_{oed}, evaluated through the empirical correlation given in Figure 9.18. This required the value of the tip cone resistance $q_c = 2$ MPa, as well as an estimate of the relative density through the correlation (Lancellotta, 1983; Figure 7.39)

$$D_R = 68 \left(\log \frac{q_c}{\sqrt{\sigma_{vo}' \cdot p_a}} - 1 \right).$$

Assuming conservatively the layer as being normally consolidated, we deduced $E_{oed} = 8q_c = 16$ MPa.

By summing up the contribution of all layers, the settlement of the surface equal to 165 cm can be predicted, that compares quite well with the expected value based on the actual elevation -1.90 m the founding level has reached.

Following this route, data collected by Simons and Som (1970), Morton and Au (1975), Burland *et al.* (1977) and Somerville and Shelton (1972) support the following conclusion:

(a) When considering structures on *soft clay*, the settlement computed by using the one-dimensional approach can be considered to be representative of the consolidation settlement, and the ratio between the immediate settlement and the overall settlement may be assumed to be less than about 0.10;

(b) In *stiff clays*, the immediate settlement can be 0.32 to 0.74 of the overall settlement and the settlement predicted by the one-dimensional approach is representative of the overall settlement.

9.5 The use of field tests to predict the settlement of footings on sand

Despite the knowledge of the main factors affecting the soil behaviour, due to the difficulties of sampling coarse-grained soils the prediction of settlements of foundations on sand still relies on empirical approaches. These are based on correlations between the observed behaviour of footings and the results of *in situ* tests, such as the Marchetti dilatometer (Schmertmann, 1986; Leonards and Frost, 1988; Marchetti *et al.*, 2001),

Table 9.1 Computing the settlement of the Pisa Tower

Layer	Elevation (m s.l.m.)	γ (kN/m³)	σ_{vo} (kPa)	u_o (kPa)	σ'_{vo} (kPa)	H_i (m)	$z_i - D$ (m)
	0.00	18.00	54	17.65	36.34	–	
A1	−2.40	18.90	99.36	41.20	58.16	2.40	1.20
A1	−5.40	18.90	156.06	70.63	85.43	3.00	3.90
A2	−7.40	18.20	192.46	90.25	102.21	2.00	6.40
B1	−9.00	17.40	220.30	104.20	116.10	1.60	8.20
B1	−10.90	17.40	253.36	120.76	132.60	1.90	9.95
B2	−12.90	17.40	288.76	138.19	150.57	2.00	11.90
B3	−14.90	16.70	322.16	155.62	166.54	2.00	13.90
B3	−17.80	16.70	370.59	180.90	189.69	2.90	16.35
B4	−19.00	19.80	394.35	191.36	202.99	1.20	18.40
B5	−22.00	20.10	454.65	217.51	237.14	3.00	20.50
B6	−24.40	19.10	500.49	238.43	262.06	2.40	23.20
B7	−26.40	18.70	537.89	255.87	282.03	2.00	25.40
B7	−29.00	18.70	586.51	278.53	307.98	2.60	27.70
B8	−30.40	19.20	613.39	290.73	322.66	1.40	29.70
B9	−32.40	19.20	651.79	308.16	343.63	2.00	31.40
B9	−34.40	19.20	690.19	325.60	364.19	2.00	33.40
B10	−37.00	19.20	740.11	348.26	391.85	2.60	35.70

the pressuremeter test (Baguelin *et al.*, 1978), the plate loading test (Terzaghi and Peck, 1967; Parry, 1978) and penetration tests (Burland and Burbidge, 1985; Schmertmann, 1970). In the following we concentrate on procedures based on penetration tests, but it is relevant to outline how the use of one procedure rather than others can be a matter of tradition and of locally collected experience. In this respect, our choice does not mean that we deserve major credit to sounding procedures. It may certainly be considered subjective and, to relax this subjectivity, to some extent we suggest that the interested reader refer to the above listed references to learn about other approaches.

9.5.1 Burland and Burbidge method

The key aspect of the method suggested by Burland and Burbidge (1985) is that we can give the settlement w of the footing the following form:

$$\frac{w}{Z_I} = q' I_c, \tag{9.28}$$

where Z_I is a depth of influence, q' is the applied uniform stress and I_c may be thought of as a compressibility index. Then, by collecting over 200 cases of measurements of settlement, Burland and Burbidge have suggested correlating the compressibility index to the *SPT* blow count N:

$$I_c = \frac{1.7}{N^{1.4}}. \tag{9.29}$$

$\sigma_z'(z)$ (kPa)	$\Delta\sigma_z$ (kPa)	OCR	σ_p' (kPa)	e_o	C_c	C_r	E_{oed} (MPa)	ΔH_i (cm)
–	–	–	–	–	–	–	–	–
47.25	417.24	4	189.00	0.8	0.35	0.035	–	21.03
71.80	396.76	2.5	179.50	1.0	0.35	0.035	–	23.97
93.82	349.37	2	187.64	–	–	–	16	3.19
97.56	307.14	2.5	243.90	1.6	0.95	0.16	–	16.78
124.35	266.38	2	248.70	1.6	0.95	0.16	–	17.14
141.59	225.25	1.3	184.07	1.4	0.80	0.13	–	21.20
158.56	189.34	1.2	190.27	1.5	0.80	0.18	–	17.91
178.16	153.81	1.2	213.79	1.5	0.70	0.13	–	16.71
196.34	130.21	2	392.68	0.6	0.25	0.065	–	1.08
220.07	110.68	2.5	550.18	0.7	0.30	0.060	–	1.87
249.60	90.94	1	249.60	–	–	–	60	0.36
272.05	78.28	1	272.05	1.2	0.70	0.10	–	7.69
295.01	67.53	1	295.01	1.2	0.70	0.10	–	7.41
315.32	59.81	1	315.32	0.9	0.35	0.08	–	1.95
333.15	54.20	1	333.15	0.9	0.35	0.08	–	2.41
353.91	48.52	1	353.91	0.7	0.35	0.08	–	2.30
378.02	43.00	1	378.02	0.8	0.35	0.08	–	2.37

The depth of influence Z_i is assumed to be the depth at which the settlement is 25% of the surface settlement. There are not too many field measurements for assessing this depth. However, data collected by Burland and Burbidge show that, although the scatter is large, this depth can be correlated to foundation breadth B as follows:

$$Z_I = B^{0.7}. \tag{9.30}$$

Certainly it can be argued that, for a given value of the breadth B, the depth of influence is not unique, because it depends on the profile of stiffness with depth, but nevertheless the suggested trend is in agreement with that predicted for the elastic layer with a linearly increasing stiffness with depth.

Substituting (9.30) and (9.29) into (9.28), the settlement of a square footing, resting on a *virtually normally consolidated sand*, can be predicted as:

$$w = q'B^{0.7}I_c \text{ (mm)}, \tag{9.31}$$

where the applied average effective stress q' is expressed in kPa and B in m.

For loading at the *base of an excavation* for which the effective overburden stress is σ_{vo}', it is assumed that where $q' \leq \sigma_{vo}'$ the compressibility index is 1/3 of the value given by (9.29), so that the settlement now assumes the expression:

$$w = \sigma_{vo}'B^{0.7}\frac{I_c}{3} + (q' - \sigma_{vo}')B^{0.7}I_c \text{ (mm)}. \tag{9.32}$$

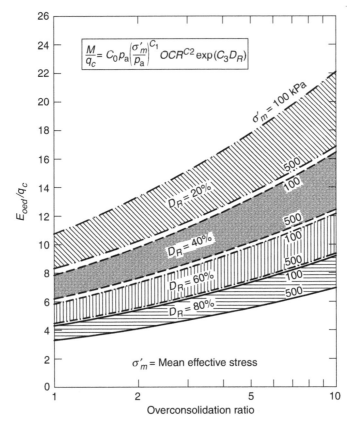

Figure 9.18 Correlation between the constrained modulus E_{oed} and the cone resistance q_c (Jamiolkowski et al., 1988).

The same expression applies to *overconsolidated sand*: where $q' \leq \sigma'_p$, the relevant contribution is only given by $q' B^{0.7} I_c / 3$; while, if $q' > \sigma'_p$ the expression (9.32) must be used, in which σ'_{vo} is substituted by σ'_p.

The value of N in equation (9.29) is the arithmetic mean of the measured values over the depth of influence, provided that N increases or is constant with depth. Instead, if N decreases with depth, then N is the average over a depth of influence taken as $2B$ or the bottom of the softer layer, whichever is the lesser.

Although the N values are not corrected for overburden stress, for very fine and silty sand beneath the water table it is suggested to apply the correction proposed by Terzaghi and Peck (1948), where N is greater than 15:

$$N(corrected) = 15 + 0.5(N - 15).$$
(9.33)

Further corrections have also been suggested to take into account factors like the *shape of the foundation*, the *thickness of the compressible layer* and the *time dependent fraction* of the overall settlement.

(a) The correction factor to be applied when the ratio of length to breadth of the foundation is greater than unity, i.e. $L/B > 1$, is the following:

$$f_s = \left[\frac{\frac{1.25L}{B}}{\frac{L}{B} + 0.25} \right]^2 > 1,$$

(9.34)

and it can be observed that this factor tends to 1.56 as L/B tends to infinity.

(b) When the thickness H of the sand or gravel layer is less than the depth of influence, then a correction f_H should be applied such that:

$$f_H = \frac{H}{Z_I} \left(2 - \frac{H}{Z_I} \right) < 1.$$

(9.35)

(c) Finally, the case records collected by Burland and Burbidge prove that footings on sand and gravels exhibit time dependent settlements. Therefore, the settlement at any time t, greater than three years, will be increased by a factor equal to:

$$f_t = \left(1 + R_3 + R \log \frac{t}{3} \right)$$

(9.36)

where R_3 is the fraction that takes place during the first three years.

Values of R_3 and R depend on the nature of the imposed loads, so that values of 0.3 and 0.2 are suggested when dealing with static loads, while values of 0.7 and 0.8 are suggested when dealing with cyclic loading. As an example, by considering the expect settlement the foundation can experience in a time interval of 30 years, the value of f_t is equal to 1.5 for static loads, and to 2.5 in presence of cyclic loading (this could be the case of special structures such as chimneys).

Burland and Burbidge (1985) also raised the question of *reliability* of the forecasted settlement value, and for this reason they made an analysis of the frequency distribution of the collected data (Figure 9.19). According to this analysis, they give to the value computed by means of equation (9.32) the meaning of *mean value*, but the structure could suffer a settlement as large as 1.5 times the mean value, due to spatial soil variability.

9.5.2 Schmertmann method

This method, originally suggested by Schmertmann (1970) and further improved by Schmertmann *et al.* (1978) is based on the results of *cone penetration tests*.

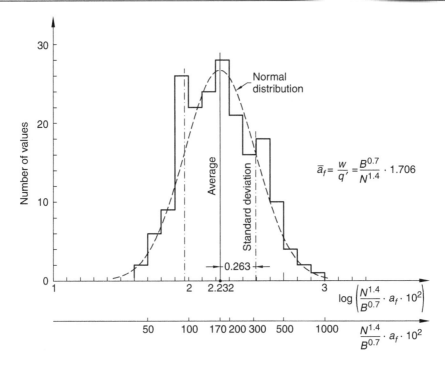

Figure 9.19 Frequency distribution of settlement observations (Burland and Burbidge, 1985).

Example 9.3
Given a square footing resting on the the soil profile shown in Figure 9.20, compute the settlement produced by an uniform load of 200 kPa.

A conservative approach would consider the sand soil as being normally consolidated. Accordingly, by using the Burland and Burbidge (1985) approach, the settlement is predicted through the following equation:

$$w = f_s f_t f_H \left[\left(q' - \frac{2}{3} \sigma'_{vo} \right) B^{0.7} I_c \right],$$

where $f_s = f_H = 1$, for the case under consideration. The effective overburden stress at the founding level is equal to 27.96 kPa and the arithmetic mean of N_{SPT} over the depth of 1.90 m is equal to 30 blows/30 cm. Therefore, the compressibility index is:

$$I_c = \frac{1.71}{30^{1.4}} = 0.015$$

and the immediate settlement is predicted as equal to:

$$w = (200 - 18.64) \cdot 2.5^{0.7} \cdot 0.015 = 5.17 \,(\text{mm}).$$

By considering the time dependent fraction, the total settlement in 30 years will reach the value:

$$w = f_t \cdot 5.17 = 7.76 \, (\text{mm}).$$

This can be considered a *mean* estimate, but spatial soil variability could cause footings within the same structure to reach larger values of settlement, of the order of 12 mm.

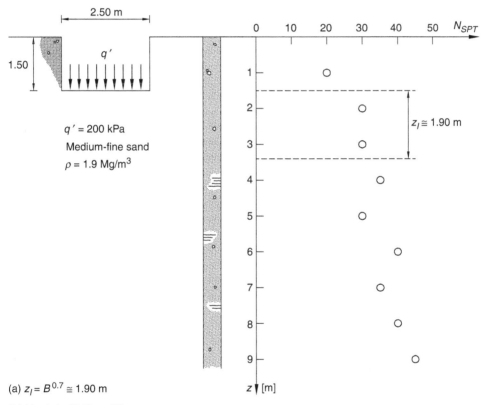

(a) $z_I = B^{0.7} \cong 1.90$ m

(b) $N_{AV}(z_I) = 30$ blows/30 cm

(c) $\sigma'_{v0} = \gamma'D = 1.9 \cdot 9.81 \cdot 1.50 = 27.96$ kPa

(d) $I_c = \dfrac{1.71}{30^{1.4}} = 0.015$; $f_S = f_H = 1$; $f_t = 1.5$

(e) $\boxed{w = f_S f_H \, f_t \left[\left(q' - \dfrac{2}{3}\sigma'_{v0}\right) B^{0.7} I_c\right] = 7.76 \text{ mm}}$

Figure 9.20 Illustrative example of application of the Burland and Burbidge method for computing settlement.

In order to highlight the central idea of this method, consider a circular footing on elastic half-space. The settlement w is the integration of the vertical strain, i.e.:

$$w = \int_o^\infty \varepsilon_z dz \, ,$$

and the vertical strain at any depth is given by:

$$\varepsilon_z = \frac{1}{E}\left(\Delta\sigma_z - 2\upsilon\Delta\sigma_r\right) .$$

If we explore the strain distribution with depth, we reach the conclusion that its maximum does not occur at the surface (because of the contribution of shearing), but at a depth of about half the radius. Schmertmann (1970) suggested to account for this distribution by simply assuming:

$$\varepsilon_z = \frac{\Delta q}{E}I_z \tag{9.37}$$

where I_z is a dimensionless influence coefficient, varying with depth according to the vertical strain distribution. Figure 9.21 reproduces the suggested simplified profile of I_z (as modified by Schmertmann et al., 1978): it assumes that the peak occurs at a depth equal to half the width of a square footing, and to the footing width for strip loadings.

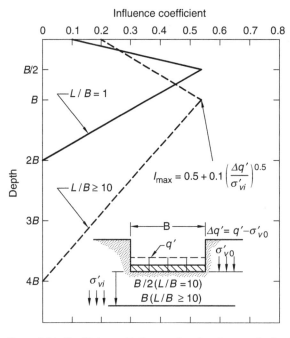

Figure 9.21 Coefficient of influence I_z related to strain distribution (Schmertmann et al., 1978).

The simplified profile also assumes that this factor is zero at depths greater than $2B$ (square footing) or $4B$ (strip footing), to account for the fact that the contribution of vertical strains becomes insignificant at greater depth (as a combined effect that stresses reduce because of spreading and the soil stiffness increases).

In order to account for the depth of embedment and creep effects, some empirical coefficients have also been introduced, and by evaluating the settlement as a sum over a convenient number of layers will give the possibility to account for changes of soil stiffness with depth. Therefore the settlement is evaluated as:

$$w = C_1 C_2 \Delta q \sum_{1}^{n} \left(\frac{I_z}{E}\right)_i \Delta z_i . \tag{9.38}$$

In this expression, Δz_i is the thickness of each layer (see Example 9.4), Δq is the *net contact stress* and C_1 and C_2 are two empirical correction factors, expressing the influence of the depth of embedment and of the soil creep:

$$\begin{aligned} C_1 &= 1 - 0.5\left(\frac{\sigma'_{vo}}{\Delta q}\right) \geq 0.5 \\ C_2 &= 1 + 0.2 \cdot \log\left(\frac{t}{0.1}\right) \quad (t \text{ expressed in years}). \end{aligned} \tag{9.39}$$

The value of the soil modulus is correlated to the cone penetration tip resistance q_c. In particular it is suggested to assume $E = 2.5q_c$ under conditions of axial symmetry, and $E = 3.5q_c$ for plane strain conditions. This assumption strictly holds for NC sands and may result in an overestimation of the settlement when dealing with overconsolidated sands (as expected by considering the correlation shown in Figure 9.22).

9.5.3 Non-linear soil behaviour

The empirical relation (9.31) suggested by Burland and Burbidge (1985) implicitly assumes that the compressibility index I_c does not depend on stress and strain level.

Example 9.4
Given the square footing resting on the soil profile shown in Figure 9.23, compute the settlement induced by the applied load of 200 kPa.

By using the *Schmertmann method*, the depth of influence to be considered is equal to $2B = 5$ m. The net bearing stress is equal to 172.04 kPa and the overburden effective stress at depth $B/2$ is equal to 51.26 kPa. Therefore, the coefficient I_z reaches a maximum value of 0.68, and its change with depth has to be considered in parallel with that of the tip resistance of the cone penetration test (see Figure 9.23).

By subdividing the soil profile into a convenient number of layers, an immediate settlement of 9.36 mm is predicted. By also considering the time dependent fraction, the overall settlement will reach in 30 years ($C_2 = 1.495$) the value of 14 mm.

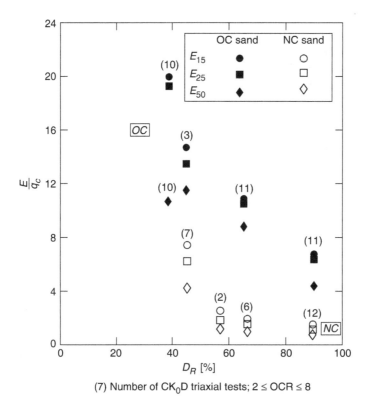

Figure 9.22 Dependence of the ratio E/q_c on OCR (Jamiolkowski *et al.*, 1988).

This can be argued by observing that, for a given value of the foundation breadth B and the blow-count N, the ratio w/q' is constant. Since one expects that the soil stiffness must depend on stress and strain level, Berardi and Lancellotta (1991) have reviewed the case histories collected by Burland and Burbidge (1985) and have suggested the following approach in order to account for soil non linearity. Consider the settlement as predicted by the theory of elasticity:

$$w = \frac{q}{E'} B(1 - v^2) \cdot I,\tag{9.40}$$

and give the soil stiffness the following expression, in order to account for the current stress level:

$$E' = K_E \cdot p_a \left(\frac{\sigma'_{vo} + 0.5\Delta\sigma'_{vo}}{p_a} \right)^{0.5},\tag{9.41}$$

where the overburden effective stress and the increment of the vertical stress are supposed to be evaluated at a depth corresponding to half of the depth of influence H_i.

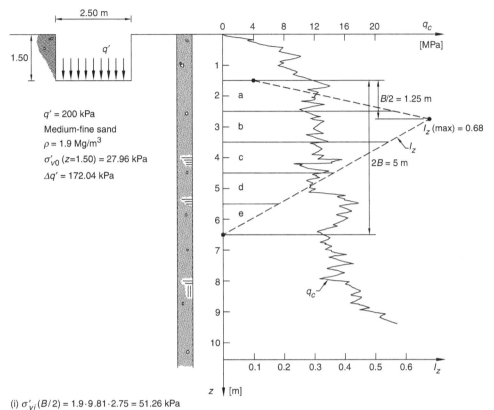

$q' = 200$ kPa

Medium-fine sand

$\rho = 1.9$ Mg/m³

$\sigma'_{v0}\,(z=1.50) = 27.96$ kPa

$\Delta q' = 172.04$ kPa

(i) $\sigma'_{vi}(B/2) = 1.9 \cdot 9.81 \cdot 2.75 = 51.26$ kPa

(ii) $I_z\,(\text{max}) = 0.5 + 0.1\left(\dfrac{200 - 27.96}{51.26}\right)^{0.5} = 0.68$

(iii) $C_1 = 1 - 0.5\,\dfrac{27.96}{172.04} = 0.919 \qquad C_2 = 1 + 0.2\,\log\left(\dfrac{30}{0.1}\right) = 1.495$

Layer (-)	Δ_z (m)	I_z (-)	q_c (MPa)	E' (MPa)	$\dfrac{\Delta z \cdot \Delta q'}{E'} \cdot I_z$ (mm)
a	1.00	0.335	12	30	1.92
b	1.00	0.618	12	30	3.54
c	1.00	0.463	12	30	2.66
d	1.00	0.280	12	30	1.61
e	1.00	0.093	14	35	0.46
					$\Sigma = 10.19$ mm
$w = C_1 \cdot C_2 \cdot \Sigma \dfrac{\Delta z \cdot \Delta q'}{E'} \cdot I_z = 0.919 \cdot 1.495 \cdot 10.19 = 14$ mm					

Figure 9.23 Prediction of settlement of a footing on sand by using the Schmertmann (1970) method.

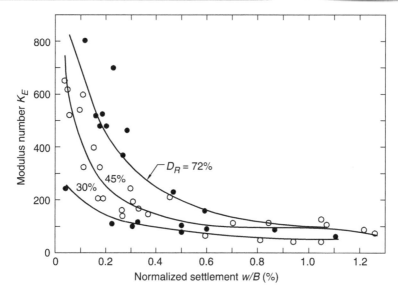

Figure 9.24 Dependence of the modulus number K_E on the normalized settlement (Berardi and Lancellotta, 1991).

The modulus number K_E, which appears in equation (9.41), may be thought as the value the soil modulus assumes when the current vertical stress is equal to the reference stress p_a (i.e. $K_E = E'$ if $\sigma'_{vo} + 0.5\Delta\sigma'_z = p_a$).

Performing a back analysis of the collected case records by using (9.40) and (9.41) gives the modulus number K_E, which can be plotted as a function of the normalized settlement w/B (Figure 9.24).

Therefore, the modulus number appears to be dependent on the relative density and on the amount of the normalized settlement. In particular, the greater the settlement experienced by the foundation, the lower the modulus number, as expected by considering the decay of soil stiffness with strain level.

In order to highlight the influence of the relative density, it is convenient to select a reference value of soil modulus, such as corresponding to a normalized settlement equal to 0.1%. Then, the relationship shown in Figure 9.25 proves how this reference values changes with the degree of packing of soil particles.

Finally, it has been proved (Berardi, 1999) that data of Figure 9.24 can also be conveniently normalized:

$$\frac{E'}{E'_{0.1}} = 0.008\left(\frac{w}{B}\right)^{-0.7}, \tag{9.42}$$

and by substituting (9.42) into (9.40) gives:

$$\frac{q}{E'_{0.1}} = \frac{1}{125 \cdot I \cdot (1 - v^2)} \cdot \left(\frac{w}{B}\right)^{0.3}. \tag{9.43}$$

The obtained relationship, jointly used with Figure 9.25, allows to predict the settlement of the footing by taking into account all the previously mentioned factors.

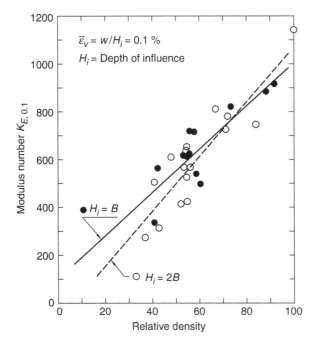

Figure 9.25 Dependence of the modulus number K_E on the relative density (Berardi and Lancellotta, 1991).

We summarize for sake of clarity the main steps to be followed in practice (see Example 9.5):

(a) make an estimate of the relative density (through *CPT* or *SPT* tests) and use Figure 9.25 to evaluate the soil number K_E (corresponding to $w/B = 0.1\%$);

(b) given the applied load q', compute the induced vertical stress $\Delta\sigma'_z$ at depth $H_i/2$ so that equation (9.41) can be used to evaluate the soil modulus $\bar{E}'_{0.1}$;

(c) use equation (9.43) to forecast the settlement w of the foundation.

9.6 Damage criteria and limiting values of settlement

9.6.1 Limiting movements

When dealing with the problem of relating settlements to damage, it must be stressed that serviceability and damage are rather relative concepts as they depend on the function of the structure and also on the reaction of the users.

In addition, it is not a simple task to give general guidelines as structures differ one from another, both in their structural conception and in details.

In order to appreciate how many factors may have an influence on the relation between settlements and damage, it is instructive to briefly recall the main steps of a building process (Burland *et al.*, 1977).

Example 9.5
Given the footing in Figure 9.20 (see Example 9.3), compute the immediate settlement by using the approach suggested by Berardi and Lancellotta (1991).

The N_{SPT} blow-counts within a layer of 2.5 m beneath the founding level allow to estimate a relative density of 80 to 85%, when reference is made to the empirical correlation suggested by Skempton (1986):

$$D_R^2 = \frac{N_1}{60},$$

where:

$$N_1 = C_N N_{SPT}$$

$$C_N = \begin{cases} \dfrac{2}{1+\sigma_{vo}'/100} \text{ fine sand} \\[4mm] \dfrac{3}{2+\sigma_{vo}'/100} \text{ coarse sand} \end{cases}.$$

Therefore, we can obtain a modulus number $K_{E(0.1)} = 800$ by using the correlation in Figure 9.25.

The induced vertical stress at depth $B/2$ is equal to $\Delta\sigma_z' = 0.70 \cdot 172 = 120$ kPa and by applying the relation (9.41) we can estimate the soil modulus as: $E_{0.1}' = 800 \cdot 100 \sqrt{\dfrac{51.26+120/2}{100}} = 84.000$ kPa.

Finally, by using equation (9.43), we obtain $\left(\dfrac{w}{B}\right)^{0.3} = \dfrac{172}{84.000} \cdot 125 \cdot 0.609 \cdot (1 - 0.15^2) = 0.152$ and we can predict an immediate settlement $w = 4.7$ mm.

When structural loads are applied to a foundation raft, the immediate settlement takes place and part of the existing structure may suffer some distortion. The overall stiffness of the structure increases as the construction evolves, and when cladding is added the stiffness is substantially increased. Then loads start to be imposed. As expected, during the construction process not all components of the structure are subjected to the same relative deflections: those suffered by structural members depend on their location and elevation, and a portion of these deflections will affect cladding and finishes, giving rise (eventually) to architectural damage.

The probability of damage is diminishing the larger is the ratio of immediate to consolidation settlement and the smaller is the ratio between the imposed and the dead loads. The probability of aesthetic damage is also reduced the later the stage at which finishes are imposed. Furthermore, it must be recalled that movements depend on settlements as well as on factors such as creep, shrinkage and temperature.

Because of all these reasons, the indications given in the following, for the planning and the designing of a structure, are not indended to substitute the evaluation of specific details that can affect its performance.

The question of limiting settlements concerns three basic considerations: visual appareance, serviceability, and structural damage.

(a) Deviation from the vertical in excess of about 1/250 can be distinguishable to the casual eye and may produce an unpleasant and sometimes alarming feeling. A deflection ratio of more than 1/250 will also be visible for horizontal members.

(b) The most comprehensive analysis of limiting settlements of structures is that of Skempton and MacDonald (1956), based on the performance of 98 buildings. According to this analysis, limiting values of the relative rotation β (see Section 9.1) should be of the order of 1/500 to avoid cracking in walls and partitions and of the order of 1/150 to avoid structural damage.

Polshin and Tokar (1957) quoted values of 1/500 and 1/250 and similar conclusions have also been reached by Meyerhof (1956).

(c) When dealing with *load bearing walls*, expecially when subjected to hogging rather than sagging, the most significant parameter is the deflection ratio, and according to Burland and Wroth (1975), the limiting values are:

$$\frac{\Delta}{L} = 2 \cdot 10^{-4} \quad \text{for} \quad \frac{L}{H} = 1$$
$$\frac{\Delta}{L} = 4 \cdot 10^{-4} \quad \text{for} \quad \frac{L}{H} = 5$$

where H is the building height.

9.6.2 Predicting differential settlements

Visual and structural damage can be avoided by using the above indications provided that differential settlements can be predicted. Accuracy and reliability of the presently used methods for forecasting the amount of settlement have been discussed in previous sections. The main source of differential settlements is the inherent soil variability and to overcome difficulties linked to this aspect reference can be made to the empirical relations suggested by Grant *et al.* (1974) between the maximum settlement and the relative rotation:

(a) foundations on sands:

$$w_{max}(\text{mm}) = 15.000\beta_{max} \text{ (isolated footings)}$$
$$w_{max}(\text{mm}) = 18.000\beta_{max} \text{ (mat foundations)}$$

(b) foundations on clays:

$$w_{max}(\text{mm}) = 30.000\beta_{max} \text{ (isolated footings)}$$
$$w_{max}(\text{mm}) = 35.000\beta_{max} \text{ (mat foundations)}.$$

Terzaghi and Peck (1948) suggested that differential settlements of footings on sands is lower than 75% of the maximum settlement and a limiting value of 25 mm was recommended for the latter.

Skempton and MacDonald (1956) reached the conclusion, based on the observation of 11 buildings, that the limiting differential settlement is about 25 mm and the maximum allowable settlements are about 40 mm for isolated footings and 40 to 65 mm for raft foundations on sand. They also concluded that the maximum differential settlement for foundations on clays is 40 mm and the limiting values of total settlement are about 65 mm for isolated footings and 65 to 100 mm for raft foundations.

Data collected by Bjerrum (1963) showed, in addition, that no damage to building on rafts on clay has been recorded for differential settlements of less than 125 mm and for total settlements of less than 250 mm. Damage has instead been reported for buildings on isolated footings on clay for differential settlements in excess of 50 mm and total settlements higher than 150 mm.

9.7 Summary

When designing a structure, engineers are interested in forecasting both the total and differential settlements the structure will suffer, and how fast they will occur. For these reasons it is important to distinguish between immediate, consolidation and secondary settlement and to account for the main phases of a building process.

Theory of elasticity is attractive because it provides closed form solutions for the analysis of some boundary problems of geotechnical interest. Examples include the stress distribution induced within the soil mass by load applied at the surface, as well as the displacement of loaded area of finite dimension. There are certainly limitations as far as the reliability of elasticity based predictions is concerned, but simple calculations based on closed form solutions still offer the advantage of focusing on the most important parameters to be considered, as well as on their relative influence.

Settlement of a structure on clay is currently predicted by using the *one-dimensional (oedometric) approach*. Despite the knowledge of the main factors affecting soil behaviour, due to the difficulties of sampling coarse-grained soils, the prediction of settlements of foundations on sand still relies on empirical approaches. These are based on correlations between the observed behaviour of footings and the results of *in situ* tests.

However, the reader must be aware of the fact that many other problems of engineering interest require more sophisticated approaches. One such example is represented by settlements associated with ground excavations. In this case, the use of a simple model, such as the elastic one, would give unrealistic and misleading results.

9.8 Further reading

A suggested reading for the use of elasticity in geotechnical engineering is the book by R.O. Davis and A.P.S. Selvadurai (1996) *Elasticity and Geomechanics*, Cambridge University Press.

The book by H.G. Poulos and E.H. Davis (1974) *Elastic Solutions for Soil and Rock Mechanics*, Wiley, presents a complete compilation of elastic solutions.

Methods of solution for elastic problems are discussed in detail in the book by S.P. Timoshenko and J.N. Goodier (1970): *Theory of Elasticity (3rd edn)*, McGraw-Hill.

The book by A.P. Selvadurai (1979) *Elastic Analysis of Soil-Foundation Interaction. Developments in Geotechnical Engineering, Vol. 17*, Elsevier, offers a complete description of the soil-structure interaction problem and presents solutions of many engineering problems.

The state of the art paper by B.B., Burland, B.B. Broms and V.F.B. De Mello (1977) *Behaviour of foundations and structures*, IX ICSMFE, Tokyo, 2, 495–546, is a further basic reference for soil-structure interaction problems.

Bibliography

Abbreviations:
AGI: Associazione Geotecnica Italiana
ASCE: American Society of Civil Engineering
ASTM: American Society for Testing and Materials
CIRIA: Construction Industry Research and Information Association, London
JGE: Journal of Geotechnical Engineering
JGED: Journal of the Geotechnical Engineering Division
JSMFD: Journal of the Soil Mechanics and Foundation Division
ECSMFE: European Conference of Soil Mechanics and Foundation Engineering
ICE: Institution of Civil Engineering, London
ICSMFE: International Conference of Soil Mechanics and Foundation Engineering
ICSMGE: International Conference of Soil Mechanics and Geotechnical Engineering
RIG: Rivista Italiana di Geotecnica
STP: Special Technical Publication
TRP: Transportation Research Board

Abbott M.B. (1966) *An Introduction to the Method of Characteristics*. Thames and Hudson, London.
Achenbach J.D. (1975) *Wave Propagation in Elastic Solids*. Elsevier, 425 pp.
Alonso E.E., Gens A. and Josa A. (1990) The constitutive model for partially saturated soils. *Géotechnique*, 3, 405–430.
Alpan I. (1967) The empirical evaluation of the coefficients K_o and K_{or}. *Soils and Foundations*, 7, 31–40.
Al Tabbaa A. and Wood D.M. (1989) An experimental based 'bubble' model for clay. *Proc. 3rd Int. Conf. Numerical Models in Geomechanics*, Niagara Falls.
Anderson M.G. and Richards K.S. (1987) *Slope Stability*. Wiley.
Arroyo M., Greening P.D. and Muir Wood D. (2003a) An estimate of uncertainty in current laboratory pulse test practice. *Rivista Italiana di Geotecnica*, 1, 38–56.
Arroyo M., Muir Wood D. and Greening P.D. (2003b) Source near-field effects and pulse tests in soil samples. *Géotechnique*, 53, 3, 337–345.
Arroyo M., Muir Wood D., Greening P.D., Medina L. and Rio J. (2006) Effects of sample size on bender-based axial G_o measurements. *Géotechnique*, 56, 1, 39–52.
Arthur J.R.F., Bekenstein S., Germaine J.T. and Ladd C.C. (1981) Stress path tests with controlled rotation of principal stress direction. *ASTM, STP740, Laboratory Shear Strength of Soils*, 516–540.
Arthur J.R.F., Chua K.S. and Dunstan T. (1977) Induced anisotropy in a sand. *Géotechnique*, 27, 1, 13–30.

Arthur J.F.R., Dunstan T., Al-Ani Q.A.J. and Assadi A. (1977) Plastic deformation and failure in granular media. *Géotechnique*, 27, 53–74.

Arulnathan R., Boulanger R.W., Riemer M.F. (1998) Analysis of bender elements tests. *Geotechnical Testing Journal*, 21, 120–131.

Atkinson J.H. (1981) *Foundations and Slopes. An introduction to applications of critical state soil mechanics*. McGraw-Hill, London.

Atkinson J.H. (1993) *The Mechanics of Soils and Foundations*. McGraw-Hill, London, 337 pp.

Atkinson J.H. (2007) Peak strength of overconsolidated clays. *Géotechnique*, 57, 2, 127–135.

Atkinson J.H. and Bransby P.L. (1978) *The Mechanics of Soils. An Introduction to Critical State Soil Mechanics*. McGraw-Hill, London, 375 pp.

Atkinson J.H., Richardson D. and Stallebrass S.E. (1990) Effect of recent stress history on the stiffness of overconsolidated soil. *Géotechnique*, 40, 3, 531–540.

Atkinson J.H. and Sallfors G. (1991) Experimental determination of stress-strain-time characteristics in laboratory and in situ tests. *General Report, Proc. X ECSMFE*, Firenze, Vol. III, 915–956.

Atterberg A. (1911) Die Plastizitat der Tone. *International Miteilungen fur Bodenkunde*, 1, 10–43.

Atterberg A. (1911) Lerornas forhallande till vatten, deras plasticitetsgranser och plasticitesgrader. Kungl. *Lantbruks akademiens Hadlingar och Tidskrift*, 2, 132–158.

Azzouz A.S., Baligh M.M. and Ladd C.C. (1983) Corrected field vane strength for embankment design. *JGE, ASCE*, May, 730–734.

Baguelin F., Jezequel J.F., Le Mee H. and Le Mehaute A. (1972) Expansion of cylindrical probes in cohesive soils. *JSMFD, ASCE*, SM11, 1129–1142.

Baguelin F., Jézéquel J.F. and Shields D.H. (1978) *The Pressuremeter and Foundation Engineering*. Trans Tech Publications, 617 pp.

Baligh M.M. (1976) Cavity expansion in sands with curved envelopes. *JGED, ASCE*, GT11, 1131–1147.

Baligh M.M. (1985) The strain path method. *JGED, ASCE*, 9, 1108–1136.

Baligh M.M. and Levadoux J.N. (1980) *Pore Pressure Dissipation After Cone Penetration*, Research Report R80-11, M.I.T., Cambridge, Mass.

Baligh M.M., Vivatrat V. and Ladd C.C. (1980) Cone penetration in soil profiling. *JGED, ASCE*, 4, 447–461.

Barron R.A. (1948) Consolidation of fine grained soils by drain wells. *Transaction, ASCE*, 113, 718–742.

Barros P.L.A. (2006) A Coulomb-type solution for active thrust with seepage. *Géotechnique*, 56, 3, 159–164.

Barton M.E. and Palmer S.N. (1989) The relative density of geologically aged, British fine and fine-medium sands. *Quarterly J. of Engineering Geology*, Vol. 22.

Battaglio M., Bruzzi D., Jamiolkowski M. and Lancellotta R. (1986) Interpretation of CPT's and CPTU's, 1st part: Undrained penetration of saturated clays. *4th Int. Geotechnical Seminar*, Singapore, 129–143.

Bazaraa A.R.S.S. (1967) Use of standard penetration test for estimating settlement of shallow foundations on sand. Ph.D. Thesis, Un. Illinois, Urbana.

Bear J. (1972) *Dynamics of Fluids in Porous Media*, Elsevier, Edizione Dover, 1988, 764 pp.

Bear J. and Verruijt A. (1987) *Modeling Groundwater Flow and Pollution*. D. Reidel Publishing Company, 414.

Been K. and Jefferies M.G. (1985) A state parameter for sands. *Géotechnique*, 35, 2, 99–112 (also see Discussion, *Géotechnique*, 1986, 123–132).

Been K. and Jefferies M. (2004) Stress-dilatancy in very loose sand. *Can. Geotech. J.*, 41, 972–989.

Been K., Jefferies M.G. and Hachey J. (1991) The critical state of sands. *Géotechnique*, 41, 3, 365–381.

Begemann H.K.S. (1953) Improved methods fo determining resistance to adhesion by sounding through a loose sleeve placed behind the cone. *Proc. 3rd ICSMFE*, Zurich, 1, 213–217.

Bellotti R., Bizzi G. and Ghionna V. (1982) Design, construction and use of a calibration chamber. *Proc. 2nd Symp. on Penetration Testing*, Amsterdam, 2, 439–446.

Bellotti *et al.* (1985) Laboratory validation of in-situ tests. Italian Geotechnical Society Jubilee Volume, *XI ICSMFE*, San Francisco.

Bellotti R., Ghionna V.N., Jamiolkowski M., Lancellotta R. and Manfredini G. (1986) Deformation characteristics of cohesionless soils from in situ tests. *Proc. Symp. In Situ 86*, ASCE, Blacksburg.

Berardi R. (1992) Cedimenti di fondazioni superficiali su sabbia. Ph.D. Thesis, Politecnico di Torino.

Berardi R. (1999) Non linear elastic approaches in foundation design. *2nd Int. Symp. On Pre-Failure Deformation Characteristics of Geomaterials*, IS Torino 99. Balkema (Eds Jamiolkowski, Lancellotta, Lo Presti).

Berardi R. and Lancellotta R. (1991) Stiffness of granular soils from field performance. Technical note. *Géotechnique*, 1, 149–157.

Berardi R. and Lancellotta R. (1994) Prediction of settlements of footings on sand: accuracy and reliability. *Proceedings of Settlement '94, Geotechnical Eng. Div.*, ASCE, 1, 640–651.

Berardi R. and Lancellotta R. (2002) Yielding from field behavior and its influence on oil tank settlements. *Journal of Geotechnical and Geoenvironmental Engineering*, ASCE, 128, 5, 404–415.

Binger W.V. (1948) Analytical studies of the Panama Canal slides. *2nd ICSMFE*, Rotterdam, 2, 54–60.

Biot M.A. (1935) Effect of certain discontinuities on the pressure distribution in a loaded soil. *Physics*, 6, 12.

Biot M.A. (1941) General theory of three-dimensional consolidation. *J. Appl. Phy.*, 12, 155–165.

Biot M.A. (1956) Theory of elastic waves in fluid-saturated porous solid. *J. Acoust. Soc. Am.*, 28, 168–191.

Bishop A.W. (1955) The use of the slip circle in the stability analysis of slopes. *Géotechnique*, 5, 7–17.

Bishop A.W. (1966) The strength of soils as engineering materials. 6th Rankine Lecture. *Géotechnique*, 2, 91–130.

Bishop A.W. and Bjerrum L. (1960) The relevance of the triaxial test to the solution of stability problems. *Research Conf. on Shear Strength of Cohesive Soils*, ASCE, Boulder, Colorado, 437–501.

Bishop A.W. and Green G.E. (1965) The influence of end restraint on the compression strength of a cohesionless soil. *Géotechnique*, 15, 243–266.

Bishop A.W., Green G.E., Garga V.K., Andresen A. and Brown J.D. (1971) A new ring shear apparatus and its application to the measurement of residual strength. *Géotechnique*, 21, 4, 273–328.

Bishop A.W. and Henkel D.J. (1962) The measurement of soil properties in the triaxial test. Arnold, London, 227 pp.

Bishop A.W., Kennard M. and Penman A.D.M. (1961) Pore pressure observations at Selset dam. *Proc. Conf. Pore Pressure and Suction in Soils*, Butterworths, London, 91–102.

Bishop A.W. and Wesley L.D. (1975) A hydraulic triaxial apparatus for controlled stress path testing. *Géotechnique*, 25, 4, 657–670.

Bjerrum L. (1955) Stability of natural slopes in quick clay. *Géotechnique*, 5, 1, 101.

Bjerrum L. (1963) Discussion Sect. *6th ECSMFE*, Wiesbaden, 2, 135–137.

Bjerrum L. (1967) Progressive failure in slopes of overconsolidated plastic clay and clay shales. 3rd Terzaghi Lecture. *JSMFD, ASCE*, 93, 5, 3–49.

Bjerrum L. (1967) Engineering geology of Norwegian normally consolidated marine clays as related to settlements of buildings. 7th Rankine Lecture. *Géotechnique*, 17, 2, 81–118.

Bjerrum L. (1972) Embankments on soft ground. *ASCE Spec. Conf. Performance of Earth and Earth-Supported Structures*, Purdue, 2, 1–54.

Bjerrum L. (1973) Problems of soil mechanics and construction of soft clays. *Proc. 8th ICSMFE*, Moscow, 3, 111–159.

Bjerrum L. and Rosenqvist I. (1956) Some experiments with artificially sedimented clays. *Géotechnique*, 6, 3, 124–136.

Bolton M.D. (1979) *A Guide to Soil Mechanics*. Macmillan Press, 439 pp.

Bolton M.D. (1986) The strength and dilatancy of sands. *Géotechnique*, 1, 65–78.

Bolton M.D. and Powrie W. (1987) Collapse of diaphragm walls retaining clay. *Géotechnique* 37, 3, 335–353.

Bolton M.D. and Powrie W. (1988) Behaviour of diaphragm walls in clay prior to collapse. *Géotechnique* 38, 2, 167–189.

Boussinesq J. (1878) Equilibre d'elasticité d'un solid isotrope sans pesanteur, supportant different poids. *C. Rendus Ac. Sc. Paris*, vol. 86, 1260–1263.

Boussinesq J. (1885) *Application des potentiels a l'etude de l'equilibre e du mouvement des solides elastiques*. Gauthier-Villar, Paris.

Brignoli E.G.M., Gotti M. and Stokoe K.H. (1996) Measurements of shear waves in laboratory specimens by means of piezoelectric transducers, *Geotechnical Testing Journal*, 19, 384–397.

British Standards Institution (1994) *Code of Practice for earth retaining structures, BS 2002*. London: BSI, London.

Bromhead E.N. (1979) A simple ring shear apparatus. *Ground Engineering*, 12, 40–44.

Broms B.B. (1971) Lateral earth pressure due to compaction of cohesionless soils. *Proc. 4th Budapest Conference on SMFE*, 373–384, Akademiai Kiado, Budapest.

Brunsden D. and Prior D.B. (1984) *Slope Instability*. Wiley.

Burghignoli A. and Calabresi G. (1975) Determinazione del coefficiente di consolidazione di argille tenere su campioni di grandi dimensioni. *XII Italian National Conference*, vol. 3, Cosenza.

Burghignoli A., Pane V., Cavalera L., Sagaseta C., Cuellar V. and Pastor M. (1991) Modelling stress-strain-time behaviour of natural soils. *Proc. X ECSMFE*, Firenze, General Report, vol. 3, 959–979, Balkema.

Burland J.B. (1989) Small is beautiful: the stiffness of soil at small strains. 9th Laurits Bjerrum Memorial Lecture. *Canadian Geot. Journal*, 26, 499–516.

Burland J.B. (1990) On the compressibility and shear strength of natural clays. Rankine Lecture. *Géotechnique*, 40, 3, 329–378.

Burland J.B., Broms B.B. and de Mello V.F. (1977) Behaviour of foundations and structures. *S.O.A. Report, IX ICSMFE*, Tokyo, 2, 495–546.

Burland J.B. and Burbidge M.C. (1985) Settlement of foundations on sand and gravel. *Proc. I.C.E.*, 78, 1, 1325.

Burland J.B., Jamiolkowski M. and Viggiani C. (2003) The stabilization of the leaning Tower of Pisa. *Soils and Foundations*, 43, 5, 63–80.

Burland J.B., Longworth T.I. and Moore J.F.A. (1997) A study of ground movement and progressive failure caused by a deep excavation in Oxford Clay. *Géotechnique*, 27, 4, 557–591.

Burland J.B., Potts D.M. and Walsh N.M. (1981) The overall stability of free and propped embedded cantilever retaining walls. *Ground Engineering*, 14, 5, 28–38.

Burland J.B., Rampello S., Georgiannou V.N. and Calabresi G. (1996) A laboratory study of the strength of four stiff clays. *Géotechnique*, 46, 3, 491–514.

Burland J.B. and Wroth C.P. (1975) Settlement of buildings and associated damage. Review Paper. *Proc. Conf. Settlement of Structures*, Cambridge, Pentech Press, 611–654.

Burmister D.M. (1951) The application of controlled test methods in consolidation testing, Symp. on Consolidation Testing of Soils. *ASTM, STP 126*, 89–98.

Burmister D.M. (1956) Stress and displacement characteristics of a two-layer rigid base soil system: influence diagrams and practical applications. *Proc. Highway Research Board*, vol. 35, 773–814.

Butterfield R. (1979) A natural compression law for soils (an advance on e-logp'). Technical Note. *Géotechnique*, 469–480.

Butterfield R. and Gottardi G. (1994) A complete three-dimensional failure envelope for shallow footings on sand. *Géotechnique*, 44, 1, 181–184.

Cadling L. and Odenstad S. (1950) *The Vane Borer*. Swedish Geotechnical Institute, 2, Stockholm.

Calabresi G. (1980) : L'influenza delle dimensioni dei campioni sui parametri di resistenza delle argille sovraconsolidate, intatte e fessurate. *XIV Convegno Italiano di Geotecnica*, Firenze, 2, 393–402.

Calabresi G. (2004) Terreni argillosi consistenti: esperienze italiane. *IV Conferenza 'Arrigo Croce'*, RIG, 1, 14–57.

Calabresi G. and Manfredini G. (1973) Shear strength characteristics of the jointed clay of S. Barbara. *Géotechnique*, 23, 2, 233–244.

Calabresi G., Rampello S. and Callisto L. (1993) *The Leaning Tower of Pisa: Geotechnical characterization of the Tower's subsoil within the framework of the Critical State Theory*, Studi e Ricerche, 1/93, Università di Roma 'La Sapienza'.

Calabresi G., Rampello S. and Viggiani G. (1990) Il comportamento meccanico delle argille consistenti. *Conf. Rock Mechanics and Rock Engineering*, Politecnico di Torino.

Callisto L. (1996) Studio sperimentale su un'argilla naturale: il comportamento meccanico dell'argilla di Pisa. Ph.D. Thesis, Un. La Sapienza, Roma.

Callisto L. and Calabresi G. (1998) Mechanical behaviour of a natural soft clay. *Géotechnique*, 48, 4, 495–513.

Callisto L. and Rampello S. (2004) An interpretation of structural degradation of three natural clays. *Can. Geotech. J.*, 41, 392–407.

Caquot A. (1934) *Equilibre des massifs a frottement interne*. Gauthier-Villars, Paris.

Caquot A. and Kerisel J. (1948) *Tables for The Calculation of Passive Pressure, Active Pressure and Bearing Capacity of Foundations*. Gauthier-Villars, Paris.

Caquot A. and Kerisel J. (1953) Sur le terme de surface dans le calcul des fondations en milieu pulverulent. *3rd ICSMFE*, Zurich, 1, 336–337.

Carder D.R. and Symons I.F. (1989) Long-term performance of an embedded cantilever retaining wall in stiff clay. *Géotechnique*, 39, 1, 55–75.

Carlson L. (1948) Determination in situ of the shear strength of undisturbed clay by means of a rotating auger. *2nd ICSMFE*, Rotterdam, 1, 265–270.

Carter J.P., Randolph M.F. and Wroth C.P. (1979) Stress and pore pressure changes in clay during and after the expansion of a cylindrical cavity. *Int. Journal for Numerical and Analytical Methods in Geomechanics*, 3, 305–323.

Casagrande A. (1932) Research on the Atterberg Limits of soils. *Public Roads*, 13, 8, 121–136.

Casagrande A. (1936) Characteristics of cohesionless soils affecting the stability of slopes and earth fills. *Journal of Boston Society of Civil Eng.*, Jan.; reprinted in *Contributions to Soil Mechanics*, BSCE, 257–276.

Casagrande A. (1936) The determination of the pre-consolidation load and its practical significance. *Proc. ICSMFE*, Cambridge, 3, 60–64.

Casagrande A. (1937) Seepage through dams. *J. New England Water Works Ass.* (reprint in *Contributions to Soil Mechanics*, Boston Society of Civil Engineers, 295–336).

Casagrande A. (1938) *Notes on Soil Mechanics. First Semester*. Harvard University (as cited by Holtz and Kovacs, 1981).

Casagrande A. (1948) Classification and identification of soils. *Transaction, ASCE*, vol. 113, 901–930.

Casagrande A. (1949) Soil mechanics in the design and construction of the Logan airport. In *Contribution to Soil Mechanics*, Boston Society of Civil Eng., 176–205.

Casagrande A. (1958) Notes on the design of the liquid limit device. *Géotechnique*, 8, 84–91.

Cascini L., Guida D., Romanzi G., Nocera N. and Sorbino G. (1998) A preliminary model for the landslides of May 1998 in Campania Region. *Proc. 2nd Int. Symp. on Geotechnics of Hard Soils-Soft Rocks*, Napoli, 3, 1623–1649.

Cerruti V. (1884–1885) Sulla deformazione di uno strato isotropo indefinito limitato da due piani paralleli. *Atti Accademia Nazionale dei Lincei*, Serie 4, vol. 1, 521–522.

Chandler R.J. (1970) A shallow slab slide in the Lias clay near Uppingham, Rutland. *Géotechnique*, 20, 253–260.

Chandler R.J. (1974) Lias clay: the long-term stability of cutting slopes. *Géotechnique*, 24, 1, 21–38.

Chandler R.J. (1977) Back analysis techniques for slope stabilisation works: a case record. *Géotechnique*, 27, 479–495.

Chandler R.J. (1988) The in-situ measurement of the undrained shear strength of clays using the field vane. Vane Shear Strength Testing in Soils. *ASTM*, Philadelphia, 13–44, STP 1014.

Chandler R.J. (2000) Clay sediments in depositional basins: the geotechnical cycle. The Third Glossop Lecture. *Quarterly Journal of Engineering Geology and Hydrogeology*, 33, 7–39.

Chandler R.J. and Skempton A.W. (1974) The design of permanent cutting slopes in stiff fissured clays. *Géotechnique*, 24, 4, 457–466.

Chandler R.J. and Tosatti G. (1995) The Stava tailings dams failure, Italy, July 1985. *Proc. Instn. Civ. Engrs. Geotech. Engng.*, 113, Apr., 67–79.

Chang M.F. (1981) Static and seismic lateral earth pressures on rigid retaining walls. Ph.D. Thesis, Purdue University.

Charles J.A. (1982) An appraisal of the influence of a curved failure envelope on slope stability. *Géotechnique*, 32, 4, 389–392.

Chen W.F. (1975) *Limit Analysis and Soil Plasticity*. Elsevier.

Chen Y.C., Ishibashi I. and Jenkins J.T. (1988) Dynamic shear modulus and fabric: Part I, depositional and induced anisotropy. *Géotechnique*, 38, 1, 25–32.

Clayton C.R.I., Milititsky J. and Woods R.I. (1993) *Earth Pressure and Earth-Retaining Structures*. Blackie Academic & Professional.

Clayton C.R.I., Simons N.E. and Matthews M.C. (1982) *Site Investigation*. Granada, London.

Clarke B.G., Carter J.P. and Wroth C.P. (1979) In situ determination of the consolidation characteristics of saturated clays. *Proc. 7th ECSMFE*, 2, 207–213.

Cloos H. (1928) Experiment zur inneren Tektonik. *Centralbl. Mineral. Geol. U. Pal.*, B, 609–621. l

Collins I.F. (1969) The upper bound theorem for rigid/plastic solids generalized to include Coulomb friction. *J. Mech. Phys. Solids*, 17, 323–338.

Collins I.F. (1973) A note on the interpretation of Coulomb's analysis of the thrust on a rough retaining wall in terms of the limit theorems of plasticity theory. *Géotechnique*, 442–447.

Collins K. and McGown A. (1974) The form and function of microfabric features in a variety of natural soils. *Géotechnique*, 24, 2, 223–254.

Comina C. (2005) Imaging heterogeneities and diffusion in sand samples. Ph.D. Thesis, Politecnico di Torino.

Commissione Polvani (1971) *Ricerche e studi su la Torre pendente di Pisa ed i fenomeni connessi alle condizioni di ambiente*, Ministero dei Lavori Pubblici, 3 vol., I.G.M., Firenze.

Conte E. (1998) Consolidation of anisotropic soil deposits. *Soils and Foundations*, v.38, n.4, 227–237.

Cooper M.R., Bromhead E.N., Petley D.J. and Grant D.I. (1998) The Selbourne cutting experiment. *Géotechnique*, 48, 83–102.

Cooper H.H. Jr. and Jacob C.E. (1946) A generalized graphical method for evaluating formation constants and summarizing well-field history. *Trans. Amer. Geophysical Union*, 27, 526–534.

Costanzo D. (1994) Prove meccaniche sull'argilla di Pisa. Ph.D. Thesis, Politecnico di Torino.

Cotecchia F. (1996) The effects of structure on the properties of an Italian pleistocene clay. Ph.D. Thesis, Imperial College, London.

Cotecchia F. and Chandler R.J. (1997) The influence of structure on the pre-failure behaviour of a natural clay. *Géotechnique*, 47, 3, 623–644.

Cotecchia F. and Chandler R.J. (2000) A general framework for the mechanical behaviour of clays. *Géotechnique*, 50, 4, 431–447.

Coulomb C.A. (1773) Essai sur une application des regles de maximis et minimis a quelques problemes de statique relatifs a l'architecture, *Mem. Div. Sav. Acad.*, vol. 7.

Cryer C.W. (1963) A comparison of the 3-dimensional theories of Biot and Terzaghi. *J. Mec. Appl. Math.*, 16.

Dafalias Y.F. and Herrmann L.R. (1982) *Bounding surface formulation of soil plasticity, Soil Mechanics, Transient and Cyclic Loads*. Pande G.N. and Zienkiewicz O.C. eds. Wiley, 253–282.

Dakoulas P. and Gazetas G. (2008) Insight into seismic earth and water pressure against caisson quay walls. *Géotechnique*, 58, 2, 95–111.

Darcy H. (1856) *Les fontaines publiques de la ville de Dijon*. Dalmont, Paris.

Davis E.H. and Booker J.R. (1971) The bearing capacity of strip footings from the standpoint of plasticity theory. *Proc. 1st Aust. N.Z. Conf. on Geomechanics*, 276–282.

Davis E.H. and Booker J.R. (1973) The effect of increasing strength with depth on the bearing capacity of clays. *Géotechnique* 23, 4, 551–563.

Davis E.H. and Poulos H.G. (1972) Rate of settlement under three-dimensional conditions. *Géotechnique* 1, 95–114.

Davis R.O. and Selvadurai A.P.S. (1996) *Elasticity and Geomechanics*. Cambridge University Press.

Davis R.O. and Selvadurai A.P.S. (2002) *Plasticity and Geomechanics*. Cambridge University Press.

De Beer E. (1970) Experimental determination of the shape factors and the bearing capacity factors of sand. *Géotechnique*, 20, 4, 387–411.

D'Elia B. (1991) Deformation problems in the Italian structurally complex clay soils. *Invited Lecture, X ECSMFE*, Firenze, v.4, 1159–1170.

De Freitas M.H. and Watters R.J. (1973) Some field examples of toppling failure. *Géotechnique* 4, 495–514.

De Josselin de Jong G. (1957) Application of stress functions to consolidation problems. *4th ICSMFE*, London, 1, 320–323.

De Mello V.F.B. (1977) Reflections on design decisions of practical significance to embankment dams. *Géotechnique*, 27, 3, 279–355.

De Ruiter (1971) Electrical penetrometer for site investigation. *JSMFD, ASCE*, SM2.

De Ruiter J. (1982) The static cone penetration test: state of the art report. *2nd Eur. Symp. Penetration Testing*, Amsterdam, 389–405.

Desrue J., Chambon R., Mokni M. and Mazerolle F. (1996) Void ratio evolution inside shear bands in triaxial sand specimens studied by computed tomography. *Géotechnique*, 46, 3, 1–18.

Drescher A. and Vardoulakis, I. (1982) Geometric softening in triaxial tests on granular material. *Géotechnique*, 32, 291–303.

Di Maio C. (1996) Exposure of bentonite to salt solution: osmotic and mechanical effects. *Géotechnique*, 46, 4, 696–707.

Di Maio C. and Fenelli G.B. (1994) Residual strength of kaolin and bentonite: the influence of their constituent pore fluid. *Géotechnique*, 44, 2, 217–226.

Di Prisco C. (1993) Studio sperimentale e modellazione matematica del comportamento anisotropo delle sabbie. Ph.D. Thesis, Politecnico di Torino.

Di Prisco C., Matiotti R. and Nova R. (1995) Theoretical investigation of the undrained stability of shallow submerged slopes. *Géotechnique*, 45, 3, 479–496.

D'Onofrio A., Silvestri F. and Vinale F. (1999) New torsional shear device. *Geotechnical Testing Journal, ASTM*, 22, 2, 107–117.

Drnevich V.P. (1978) Resonant column testing. Problems and solutions. Dynamic geotechnical testing. *ASTM STP 654, ASTM*, 384–398.

Drnevich V.P., Hardin B.O. and Shippy D.J. (1978) Modulus and damping of soils by the resonant column method. Dynamic geotechnical testing. *ASTM STP 654, ASTM*, 91–125.

Drucker D.C. (1956) On uniqueness in the theory of plasticity. *Quart. Appl. Math.*, 14, 35.

Drucker D.C. (1959) A definition of stable inelastic material. *J. Applied Mechanics*. Trans. ASME, 26, 101–106.

Drucker D.C. and Prager W. (1952) Soil mechanics and plastic analysis or limit design. *Quart. Appl. Math.*, 10, 2, 157–165.

Duncan J.M. and Chang C. (1970) Nonlinear analysis of stress and strain in soils. *JSMFE, ASCE*, 96, 5, 1629–1653.

Dunnicliff J. (1988) *Geotechnical Instrumentation for Monitoring Filed Performance*. Wiley, New York, 577 pp.

Dunnicliff J. (1995) Monitoring and instrumentation of landslides. Keynote paper. *Landslides*, Bell ed., Balkema, 1881–1895.

Dupuit J. (1863) *Etudes theoriques et pratiques sur le mouvement des eaux dans les canaux decouverts et a travers les terrains permeables*. Dunod, Paris.

Ebelin R.M. and Morison E.E. (1992) The seismic design of waterfront retaining structures. *Technical Report ITL-92-11*. Washington DC: US Army Corps of Engineers.

Egorov K.E. (1958) Concerning the question of the deformation of bases of finite thickness. *Mekhanika Gruntov*, Sb. Tr., 34 (as cited by Harr (1966)).

Esu F. (1966) Short term stability of slopes in unweathered jointed clays. *Géotechnique*, 16, 4, 321–328.

Esu F. and D'Elia B. (1976) Time-dependent behaviour of deep excavations in overconsolidated jointed clays. *Proc. 6th ECSMFE*, 1.1, 41–44.

Evangelista A. (1973) Analisi della prova di consolidazione a gradiente idraulico controllato. *RIG*, 1, 3–11.

Fahey M. and Carter J.P. (1986) Some effects of rate of loading and drainage on pressuremeter tests in clays. *Proceedings of Specialty Geomechanics Symposium*, Adelaide, 50–55.

Fang Y. and Ishibashi I. (1986) Static earth pressures with various wall movements. *JGE, ASCE*, 112, 3, 317–333.

Fethi A. (2000) *Applied Analyses in Geotechnics*. E & FN Spon, 753 pp.

Finno R.J., Harris W.W., Mooney M.A. and Viggiani G. (1997) Shear bands in plane strain compression of loose sand. *Géotechnique*, 47, 1, 149–165.

Fioravante V., Jamiolkowski M. and Lancellotta R. (1994) An analysis of pressuremeter holding tests. *Géotechnique*, 44, n.2, 227–238.

Flaate K. and Peck R.B. (1973) Braced cuts in sand and clay. *Norwegian Geotechnical Institute*, Publ. 96.

Flamant A.A. (1892) Sur la repatition des pressions dans un solide rectangulaire chargé transversalement. *Compte Rendus Académie des Sciences*, vol. 114, 1465–1468.

Foster C.R. and Ahlvin R.G. (1954) Stresses and deflections induced by a uniform circular load. *Proc. Highway Research Board*, v. 33, 467–470.

Foti S. (2000) Multistation methods for geotechnical characterization using surface waves. Ph.D. Thesis, Politecnico di Torino.

Foti S. (2003) Small-strain stiffness and damping ratio of Pisa clay from surface wave tests. *Géotechnique*, 53, 5, 455–461.

Foti S., Lai C.G. and Lancellotta R. (2002) Porosity of fluid-saturated porous media from measured seismic wave velocities. *Géotechnique*, 5, 359–373.

Foti S. and Lancellotta R. (2004) Soil porosity from seismic velocities. *Géotechnique*, Tech. Note, 54, 8, 551–554.

Foti S., Lancellotta R., Marchetti D., Monaco P. and Totani G. (2006) Interpretation of SDMT tests in a transversely isotropic medium. *Int. Conf. On Flat Dilatometer*, Washington DC, 275–280.

Fox E.N. (1948) The mean elastic settlement of uniformly loaded area at a depth below ground surface. *2nd ICSMFE*, Rotterdam, 1, 129.

Freeze R.A. and Cherry J.A. (1979) *Groundwater*. Prentice-Hall, 604 pp.

Giani G.P. (1992) Rock slope stability analysis. Balkema, 361 pp.

Ghionna V.N., Jamiolkowski M., Lacasse S., Ladd C.C., Lancellotta R. and Lunne T. (1983) Evaluation of self-boring pressuremeter. *Int. Symp. on Soil and Rock Investigation by in Situ Testing*, Paris, 2.

Ghionna V.N., Jamiolkowski M., Lancellotta R., Tordella M.L. and Ladd C.C. (1981) Performance of self-boring pressuremeter tests in cohesive deposits. *M.I.T. Report*, FHWA/RD-81/173, 182 pp.

Gibbs H.J. and Holtz W.G. (1957) Research on determining the density of sands by spoon penetration testing, *Proc. IV ICSMFE*, 1, 35.

Gibson R.E. (1958) The progress of consolidation in a clay layer increasing in thickess with time. *Géotechnique*, 8, 171–182.

Gibson R.E. (1963) An analysis of system flexibility and its effects on time lag in pore water pressure measurements. *Géotechnique*, 13, 1–11.

Gibson R.E. (1967) Some results concerning displacements and stresses in a non-homogeneous elastic half-space. *Géotechnique*, 17, 58–67.

Gibson R.E. (1974) The analytical method in soil mechanics. *Géotechnique*, 2, 115–139.

Gibson R.E. and Anderson W.F. (1961) In situ measurement of soil properties with the pressuremeter. *Civil Eng. Pubbl. Works Rev.*, 56, 615–618.

Gibson R.E., England G.L. and Hussey M.J.L. (1967) The theory of one dimensional consolidation of saturated clays. *Géotechnique*, 17, 261–273.

Gibson R.E. and McNamee J. (1957) The consolidation settlement of a load uniformly distributed over a rectangular area. *4th ICSMFE*, London, 1, 297–299.

Giunta G. (1993) Comportamento dei terreni sabbiosi a bassi livelli di deformazione. Ph.D. Thesis, Politecnico di Torino.

Grant R., Christian J.T. and Vanmarcke E.R. (1974) Differential settlement of buildings. *ASCE*, GT9, 973.

Green G. (1839) *Trans. Cambridge Phil. Soc.*, 7, 1.

Hanna T.H. (1973) *Foundation Instrumentation*. Trans Tech Publications.

Hansbo S. (1979) Consolidation of clay by band-shaped prefabricated drains. *Ground Engineering*, 12, 5, 16–25.

Hansbo S. (1981) Consolidation of fine-grained soils by prefabricated drains. *Proc. X ICSMFE*, Stockholm, 677–682.

Hansen J.B. (1961) A general formula for bearing capacity. *Bulletin N.11*, Danish Geotechnical Institute.

Hansen J.B. (1970) A revised and extended formula for bearing capacity. *Bulletin N.28*, Danish Geotechnical Institute.

Hardin B.O. (1970) Suggested methods of test for shear modulus and damping of soils by the resonant column. Special procedures for testing soil and rock for engineering purposes. *ASTM STP 479, ASTM*, 516–529.

Harr M.E. (1962) *Groundwater and Seepage*. McGraw-Hill.

Harr M.E. (1966) *Foundations of Theoretical Soil Mechanics*. McGraw-Hill.

Harris W.W., Viggiani G., Mooney M.A. and Finno R.J. (1995) Use of stereophotogrammetry to analyse the development of shear bands in sand. *Geotechnical Testing Journal*, 18, 4, 405–420.

Head K.H. (1980) *Manual of Soil Laboratory Testing*. Pentech Press, London.

Hegg U., Jamiolkowski M., Lancellotta R. and Parvis E. (1983) Behaviour of oil tanks on soft cohesive ground improved by vertical drains. *Proc. VIII ECSMFE*, Helsinki, 2, 627–632.

Hegg U., Jamiolkowski M., Lancellotta R. and Parvis E. (1983) Performance of large oil tanks on soft ground. *Proc. Conf. on Piling and Ground Treatment, ICE*, 125.

Henkel D.J. (1960) The shear strength of saturated remolded clays. *ASCE Spec. Conf. on Shear Strength of Cohesive Soils*, Boulder, 533–554.

Hepton P. (1988) Shear wave velocity measurements during penetration testing. *Proc. Penetration Testing in UK, ICE*, 275–278.

Heyman J. (1972) *Coulomb's Memoir on Statics. An Essay in the History of Civil Engineering*. Cambridge University Press.

Hoek E. and Bray J.W. (1977) *Rock Slope Engineering* (2nd edn), The Institution of Mining and Metalurgy, London, 358 pp.

Holden J.C. (1974) Penetration testing in Australia. *ESOPT*, 1, 155–162.

Holtz R.D. and Kovaks W.D. (1981) *An Introduction to Geotechnical Engineering*. Prentice-Hall, 733 pp.

Holtz R.D., Jamiolkowski M., Lancellotta R. (1986) Lessons from oedometer tests on high quality samples. *JGED, ASCE*, 768–776.

Holtz R.D., Jamiolkowski M., Lancellotta R. and Pedroni S. (1991) *Prefabricated Vertical Drains–Design and Performance*. CIRIA, Butterworth-Heinemann, London.

Houlsby G.T. (1985) The use of a variable shear modulus in elastic-plastic models for clays. *Computers and Geotechnics*, 1, 3–13.

Houlsby G.T. and Teh C.I. (1988) Analysis of the piezocone in clay. *Proc. Int. Symp. on Penetration Testing, ISOPT*-1, Orlando, 2, 777–782.

Houlsby G.T. and Withers N.J. (1988) Analysis of the cone pressuremeter test in clay. *Géotechnique*, 38, 575–587.

Houlsby G.T. and Wroth C.P. (1982) Determination of undrained strength by cone penetration tests. *Proc. 2nd European Symp. on Penetration Testing*, Amsterdam, 2, 585–590.

Hungr O., Morgan G.C. and Kellerhals R. (1984) Quantitative analysis of debris torrent hazards for design of remedial measures. *Canadian Geot. J.*, 21, 4, 663–677.

Hutchinson J.N. (1995) The significance of tectonically produced pre-existing shears. *Proc. 11th ECSMFE*, Copenhagen, 4, 59–67.

Hvorslev M.J. (1937) On the strength properties of remoulded cohesive soils. Ph.D. Thesis, Danmarks Naturvidenskabelige Samfund, Ingeniorvidenskabelige Skrifter, Copenhagen, 45, 159.

Hvorslev M.J. (1949) Subsurface exploration and sampling of soils for civil engineering purposes. *Waterways Experimental Station*. Vicksburg, Mississippi.

Hvorslev M.J. (1951) Time lag and soil permeability in ground-water measurements. *Waterways Experiment Station Bulletin N. 36*, Vicksburg, US Army Corps of Eng.

Ingold T.S. (1979) The effects of compaction on retaining walls. *Géotechnique*, 29, 265–283.

Jaky J. (1944) The coefficient of earth pressure at rest. *J. Soc. Hung. Arch. Eng., Budapest*, 355–358.

Jamiolkowski M., Ghionna V.N., Lancellotta R. and Pasqualini E. (1988) New correlations of penetration tests for design practice. *Proc. First Int. Symp. on Penetration Testing, ISOPT*-1, Balkema, 263–296.

Jamiolkowski M., Lancellotta R. and Wolski W. (1983) Precompression and speeding up consolidation, State of the Art Report, *VIII ECSMFE*, Helsinki, 3, 1201–1226.

Jamiolkowski M., Ladd C.C., Germaine J.T. and Lancellotta R. (1985) New developments in field and laboratory testing of soils. *Theme Lecture XI ICSMFE*, San Francisco, 1, 57–152.

Janbu N. (1969) The resistance concept applied to deformation of soils. *7th ICSMFE*, Mexico, 1, 191–196.

Janbu N. (1973) *Slope Stability Computations. Embankment Dam Engineering.* Casagrande volume. Wiley, 47–86.

Janbu N., Bjerrum L. and Kjaernsli L. (1956) Soil mechanics applied to some engineering problems. *Norwegian Geotechnical Institute*, Publ. 16, 93 pp.

Janbu N. and Senneset K. (1974) Effective stress interpretation of in situ static penetration tests. *ESOPT I*, Stockholm.

Jardine R.J., Potts D.M., Fourie A.B. and Burland J.B. (1986) Studies of the influence of non-linear stress-strain characteristics in soil structure interaction. *Géotechnique*, vol. 36, 377–396.

Jardine R.J., St. John H.D., Hight D.W. and Potts D.M. (1991) Some practical applications of a non-linear ground model. *Proc. 10th ECSMFE*, Florence.

Jefferies M.G. (1988) Determination of horizontal geostatic stress in clay with self-bored pressuremeter. *Canadian Geotechnical Journal*, 25, 559–573.

Jefferies M. and Been K. (2006) *Soil Liquefaction. A Critical State Approach.* Taylor & Francis.

Jezequel J.F., Lamy J.L. and Perrier M. (1982) The LPC-TLM pressiopenetrometer. *Proc. Symp. On Pressuremeter and Marine Applications.* Paris, Ed. Technip.

Jezequel J.F. and Mieussens C. (1975) In situ measurement of coefficient of permeability and consolidation in fine soils. *Proc. Spec. Conf. on In Situ Measurements of Soil Properties, ASCE*, Raleigh, N.C., 1, 208–224.

Jovicic V. and Coop M.R. (1998) The measurement of stiffness anisotropy in clays with bender element tests in the triaxial apparatus. *Geotechnical Testing Journal*, 21, 1, 3–10.

Jovicic V., Coop M.R. and Simic M. (1996) Objective criteria for determining G_{max} from bender elements tests. *Géotechnique*, Tech. Note, 46, 357–362.

Karlsrud K., Lunne T., Kort D.A. and Strandvik S. (2005) CPTU correlations for clays. *16th ICSMFE*, Osaka.

Kenney T.C. (1967) The influence of mineral composition on the residual strength of natural soils. *Proc. Geotechnical Conference on the shear strength properties of natural soils and rocks*, Oslo, 1, 123–129.

Kenney T.C. (1977) Residual strength of mineral mixtures. *10th ICSMFE*, 1, 155–160.

Khosla R.B.A.N., Bose N.K. and Taylor E. McK. (1954) Design of weirs on permeable foundations. *Central Board of Irrigation*, New Delhi, India.

Kjekstad O., Lunne T. and Clausen C.J.F. (1978) Comparison between in situ cone resistance and laboratory strength for overconsolidated North Sea clays. *Marine Geotechnology*, 3, 4, 23–26.

Kjellman W. (1948) Accelerating consolidation of fine grained soils by means of cardboard wicks. *2nd ICSMFE*, Rotterdam, 2, 302–305.

Kjellman W., Kallstenius T. and Wager O. (1950) Soil sampler with metal foils. *Royal Swedish Geotechnical Institute*, 1, 1–76.

Kolbuszewski J.J. (1948) An experimental study of the maximum and minimum porosities of sands. *2nd ICSMFE*, Rotterdam, 1, 158–165.

Kondner R.L. and Zelasko J.S. (1963) Void ratio effects on the hyperbolic stress–strain response of a sand. *ASTM, STP* 361, Laboratory Testing of Soils, 250–257.

Koutsoftas D.C. and Ladd C.C. (1985) Design strength of an offshore clay. *JGED, ASCE*, 3, 337–355.

Kramer S. L. (1996) *Geotechnical Earthquake Engineering*. Prentice-Hall, 653 pp.

Kulhawy F.H. and Mayne P.W. (1990) Manual on estimating soil properties for foundation design. *Electric Power Research Institute, EL-6800*, Research Project 1493–6.

Kumar J. (2001) Seismic passive earth pressure coefficients for sands. *Can. Geotech. J.*, 38, 876–881.

Kumar J. and Chitikela S. (2002) Seismic passive earth pressure coefficients using the method of characteristics. *Can. Geotech. J.*, 39, 463–471.

Kurup K.P. (2006) Innovations in Cone Penetration Testing. *ASCE, Geotechnical Special Publication 149, Site and Geomaterial Characterization*, 48–55.

Ladd C.C. (1971) Settlement analysis for cohesive soils. *MIT Research Report R71-2, Soil Publ. 272*.

Ladd C.C. (1975) Foundation design of embankments constructed on Connecticut Valley Varved Clays. *Research Report R75-7, Geotechnical Publication 343, MIT*.

Ladd C.C. (1976) Use of precompression and vertical sand drains for stabilization of foundation soils. *Short Course*, Un. California, Berkeley.

Ladd C.C. (1991) Stability evaluation during staged construction. *JGED, ASCE*, 4, 540–615.

Ladd C.C. and Edgers L. (1972) Consolidated-Undrained direct-simple shear tests on saturated clays. *Research Report R72-92, Soils Publication 284*, MIT.

Ladd C.C., Foott R., Ishihara K., Schlosser F. and Poulos H.G. (1977) Stress-deformation and strength characteristics. S.O.A. Report, *IX ICSMFE*, Tokyo, 2, 421–494.

Lai C.G. (1998) Simultaneous inversion of Rayleigh phase velocity and attenuation for near-surface site characterization. Ph.D. Thesis, Georgia Inst. of Technology, Atlanta.

Lambe T.W. (1951) *Soil Testing for Engineers*. Wiley.

Lambe T.W. (1967) Stress path method. *Journal of Soil Mechanics and Foundation Division, ASCE*, SM6, 309–331.

Lambe T.W. and Whitman R.V. (1969) *Soil Mechanics*. Wiley, 553 pp.

Lambrechts J.R. and Leonards G.A. (1978) Effects of stress history on deformation of sands. *JGED, ASCE*, 11, 1371–1387.

Lancellotta R. (1983) Analisi di affidabilità in Ingegneria Geotecnica. *Atti Istituto di Scienza delle Costruzioni, n. 625*, Politecnico di Torino.

Lancellotta R. (1993) Stability of a rigid column with non-linear restraint. *Géotechnique*, Technical Note, 43, 2, 331–332.

Lancellotta R. (1993) *Geotecnica*. (2nd edn), Zanichelli, 555 pp.

Lancellotta R. (2001) Coupling between the evolution of a deformable porous medium and the motion of the fluids in the connected porosity. In Springer Volume on *Porous Media* to honour R. de Boer (Ehlers, W. and Bluhm, J., eds). Springer.

Lancellotta R. (2002) Analytical solution of passive earth pressure. *Géotechnique*, Technical Note, 52, 8, 617–619.

Lancellotta R. (2007) Lower-bound approach for seismic passive earth resistance. *Géotechnique*, Technical Note, 57, 3, 319–321.

Lancellotta R. and Calavera J. (1999) *Fondazioni*. McGraw-Hill, 611 pp.

Lancellotta R., Costanzo D. and Pepe M. (1994) *The Leaning Tower of Pisa*, Geotechnical characterization of the Tower subsoil: a summary report. Technical University of Turin, Consorzio Torre di Pisa.

Lancellotta R. and Preziosi L. (1997) A general nonlinear mathematical model for soil consolidation problems. *Journal of Engineering Sciences*, 35, 1045–1063.

Leonards G.A. (1976) Estimating consolidation settlements of shallow foundations on overconsolidated clay. *TRB, Special Report* 163.

Leonards G.A. and Altshaeffl A.G. (1964) Compressibility of clay. *JSMFD, ASCE*, 90, SM5, 133–155.

Leonards G.A. and Frost J.D. (1988) Settlement of shallow foundations on granular soils. *JGE, ASCE*, 114, 7, 791–809.

Leonards G.A. and Girault P. (1961) A study of the one-dimensional consolidation test. *5th ICSMFE*, Paris, 1, 213–218.

Leonards G.A. and Ramiah B.K. (1959) Time effects in the consolidation of clay. *ASTM, STP* 254, 116–130.

Leroueil S. (2001) Natural slopes and cuts: movement and failure mechanisms. 39th Rankine Lecture. *Géotechnique*, 51, 3, 197–243.

Leroueil S., Guerriero G., Picarelli L. and Saihi F. (1997) Large deformation shear strength of two types of structured soils. *Proc. Symp. On Deformation and Progressive Failure in Geomechanics*. IS-Nagoya, 217–222.

Leroueil S. and Hight D.W. (2003) Behaviour and properties of natural soils and soft rocks. In *Characterization and Engineering Properties of Natural Soils*, vol. 1, 29–254, Tan *et al.* eds, Swets & Zeitlinger, Lisse.

Leroueil S., Magnan J.P. and Tavenas F. (1990) *Embankments on Soft Clays*. Ellis Horwood, 360 pp.

Leroueil S. and Vaughan P.R. (1990) The general and congruent effects of structure in natural soils and weak rocks, *Géotechnique*, 3, 467–488.

Levadoux Y.N. and Baligh M.M. (1980) Pore pressure during cone penetration in clays. *M.I.T. Report R80-15*, Cambridge, Mass.

Li X.S. (2003) Effective stress in unsaturated soil: a microstructural analysis. *Géotechnique*, 53, 2, 273–277.

Lings M.L., Pennington D.S. and Nash D.F.T. (2000) Anisotropic elastic parameters and their measurement in a stiff natural clay. *Géotechnique*, 50, 2, 109–125.

Lollino G. (1992) Automated Inclinometer System. *Sixth Int. Symposium on Landslides*. Christchurch, New Zealand.

Lo Presti D.C.F. (1987) Comportamento meccanico della sabbia del Ticino in colonna risonante. Ph.D. Thesis, Politecnico di Torino.

Lo Presti D.C.F., Shibuya S. and Rix G. (2001) Innovation in Soil Testing. *2nd Int. Conf. on Pre-failure Deformation Characteristics of Geomaterials*. Balkema, 2, 1027–1076.

Love A.E.H. (1927) *A Treatise on the Mathematical Theory of Elasticity*. Cambridge University Press (Dover ed., 1944).

Lowe J., Jonas E. and Obrician V. (1969) Controlled gradient consolidation test. *ASCE*, SM1, 77–97.

Lunne T., Eide O. and De Ruiter J. (1976) Correlations between cone resistance and vane shear strength in some Scandinavian soft to medium stiff clays. *Canadian Geotechnical Journal*, 13, 430–441.

Lunne T., Robertson P.K. and Powell J.J.M. (1997) *Cone Penetration Testing in Geotechnical Practice*. Blackie, London.

Lupini J.E., Shinner A.E. and Vaughan P.R. (1981) The drained residual strength of cohesive soils. *Géotechnique*, 31, 181–213.

Malvern L.E. (1969) *Introduction to the Mechanics of a Continuum Medium*. Prentice-Hall.

Manassero M. (1989) Stress–strain relationship from drained self-boring pressuremeter tests in sands. *Géotechnique*, 39, 2, 293–307.

Mandel J. (1957) Consolidation des couches d'argiles. *4th ICSMFE*, London, 1, 360–367.

Mandel J. (1961) Tassements produits par la consolidation d'une couche d'argile de grande épaisseur. *5th ICSMFE*, Paris, 1, 733–736.

Marchetti S. (1975) A new in situ test for the measurement of horizontal soil deformability. *ASCE Spec. Conf. on In Situ Measurement of Soil Properties*, Raleigh, N.C., 2.

Marchetti S. (1980) In situ tests by flat dilatometer. *Journal of Geotechnical Division, ASCE*, GT3, 299–324.

Marchetti S., Monaco P., Totani G. and Calabrese M. (2001) The flat dilatometer test (DMT) in soil investigations. A Report by ISSMGE Committee TC16. *Proc. Int. Conf. On In Situ Measurements of Soil Properties and Case Histories*, Bali, 95–131.

Marsland A. (1971) Laboratory and in situ measurements of deformation moduli of London clay. *Symp. on Interaction of Structure and Foundation*, Birmingham.

Marsland A. (1974) Comparison of results from static penetration tests and large in situ plate tests in London clay. *ESOPT I*, Stockholm.

Marsland A. and Powell J.J.M. (1979) Evaluating the large scale properties of glacial clays for foundation design. *Proc. BOOS 79*, 1, 193–214.

Martin G.K. and Mayne P.W. (1997) Seismic flat dilatometer tests in Connecticut Valley varved clay. *ASTM Geotechnical Testing Journal*, 20, 3, 357–361.

Mathias S.A. and Butler A.P. (2006) An improvement on Hvorslev's shape factors. Tech. Note, *Géotechnique*, 56, 10, 705–706.

Matsuoka H. and Nakai T. (1974) Stress deformation and strength characteristics under three different principal stresses. *Proc. JSCE*, 232, 59–70.

Matsuzawa H., Ishibashi I. and Kawamura M. (1985) Dynamic soil and water pressures of submerged soils. *JGE, ASCE*, 111, 10, 1161–1176.

Maugeri M., Castelli F., Massimino M.R. and Verona G. (1998) Observed and computed settlements of two shallow foundations on sands. *Journal of Geotechnical and Geonvironmental Engineering, ASCE*, v. 124, n.7.

Mayne P.W. (2006) The Second J.K. Mitchell Lecture. Undisturbed sand strength from seismic cone tests. *Geomechanics and Geoengineering: An International Journal*, 1, 4, 239–257.

Mayne P.W., Schneider J.A. and Martin G.K. (1999) Small and large-strain soil properties from seismic flat dilatometer tests. *Proc. 2nd Int. Symp. on Pre-Failure Deformation Characteristics of Geomaterials*, Torino, 1, 419–427.

McNamee J. and Gibson R.E. (1960) Plane strain and axially symmetric problems of the consolidation of a semi-infinite clay stratum. *Q. J. Mech. Appl. Math.*, 13, part II, 210–227.

Menard L.F. (1957) Mesure in situ des caracteristiques physiques des sols. *Ann. Ponts Chausses*, 14, 357–377.

Menard L.F. (1971) *Le tassements des fondations et les methods pressiometriques*. Ann. Inst. Technique Batiment Trav. Publ., Paris, 288, 105–121.

Mesri G. (1973) Coefficient of secondary compression. *JSMFD, ASCE*, 99, SM1, 123–137.

Mesri G. and Cepeda-Diaz A.F. (1986) Residual shear strength of clays and shales. *Géotechnique*, 36, 269–274.

Mesri G. and Choi Y.K. (1985) The uniqueness of the end of primary void ratio-effective stress relationships, *Proc. 11th ICSMFE*, San Francisco, Balkema, 2, 587–590.

Mesri G. and Godlewski P.M. (1977) Time and stress-compressibility interrelationship. *Journal of Geotechnical Eng. Division, ASCE*, GT5, 417–430.

Mesri G. and Shahien M. (2003) Residual shear strength mobilized in first-time slope failures. *Journal of Geotechnical and Geoenvironmental Engineering, ASCE*, 129, 1, 12–31.

Meyerhof G.G. (1951) The ultimate bearing capacity of foundations. *Géotechnique*, 2, 301–332.

Meyerhof G.G. (1953) Bearing capacity of foundations under eccentric and inclined load. *Proc. 3rd ICSMFE*, Rotterdam, 1, 440–445.

Meyerhof G.G. (1956) Discussion paper by Skempton *et al.*: Settlement analysis of six structures in Chicago and London. *Proc. ICE*, 5, 1, 170.

Mieussens C. and Ducasse P. (1977) Mesure en place des coefficients de perméabilité et des coefficients de consolidation horizontaux et verticaux. *Canadian Geotechnical Journal*, 1, 76–90.

Mindlin R.D. (1936) Force at a point in the interior of a semi-infinite solid. *Physics*, vol. 7, 195–202.

Mises R., von (1913) Mechanik der Festen Korper im Plastisch-deformablen Zustand – Nachrichten der Koniglicher Gesellschaft der Wissenschaft, Gottingen. *Mathematik-Physik Klasse*, 582–592.

Mitchell J.K. (1976) *Fundamentals of Soil Behaviour*. Wiley, 422 pp.

Mitchell J.K. (1988) New developments in penetration testing and equipment. *Penetration Testing* 1988, ISOPT-1, 1, 245–261.

Mitchell J.K. (1993) *Fundamentals of Soil Behaviour* (2nd edn). Wiley, 437 pp.

Mitchell J.K. and Brandon T.L. (1998) Analysis and use of CPT in earthquake and environmental engineering. *Geotechnical Site Characterization*, Mayne P.W. and Robertson P.K. eds, 1, 69–95, Balkema.

Mitchell J.K. and Soga K. (2005) *Fundamentals of Soil Behaviour* (3rd edn). Wiley, 577 pp.

Miura K., Yoshida N. and Kim Y.S. (2001) Frequency dependent property of waves in saturated soil. *Soils and Foundations*, 41, 2, 1–19.

Mohr O. (1882) Uber die Darstellung des Spannungszustandes und des Deformation-zustandes eines Korper-elements. *Zivilingenieur*, 113.

Mononobe N. and Matsuo H. (1929) On the determination of earth pressure during earthquake. *Proc. 2nd World Conf. Earthquake Engng.*

Mooney M.A., Finno R.J. and Viggiani M.G. (1998) A unique critical state for sands?, *Journal of Geotechnical and Geoenveronmental Engineering*, ASCE, 124, 11, 1100–1108.

Moore R. (1991) The chemical and mineralogical controls upon the residual strength of pure natural clays. *Géotechnique*, 41, 1, 35–47 (Discussion, *Géotechnique*, 42, 1, 151–153, 1992).

Morgenstern N. and Price V.E. (1965) The analysis of stability of generalised slip surfaces. *Géotechnique*, 15, 79–93.

Morgenstern N.R. and Tchalenko J.S. (1967) Microstructural observations on shear zones from slips in natural clays. *Proc. Geotech. Conf. on Shear Strength Properties of Natural Soils and Rocks*, Oslo, 1, 147–152.

Morton K. and Au E. (1975) Settlement observations on eight structures in London. Proc. British Geotechnical Society Conference Settlement of Structures, Cambridge, Pentech Press, 183–203.

Moum J. and Rosenqvist I. Th. (1957) On the weathering of young marine clay. *ICSMFE*, Londra, 1, 77–79.

Muir Wood D. (2004) *Geotechnical Modelling*. Spon Press, 488 pp.

Musso G. (2000) Electrokinetic phenomena in soils. Ph.D. Thesis, Politecnico di Torino.

Musso G., Romero E., Gens A. and Castellanos E. (2003) The role of structure in the chemically induced deformations of Febex bentonite. *Applied Clay Science*, 23, 229–237.

Nazarian S. (1984) In situ determination of elastic moduli of soil deposits and pavement system by spectral analysis of surface wave method. Ph.D. Thesis, The University of Texas at Austin.

Neuman S.P. (1972) Theory of flow in unconfined aquifers considering delayed response of the water table. *Water Resources Research*, 8, 1031–1045.

Neuman S.P. (1974) Effect of partial penetration on flow in unconfined aquifer considering delayed gravity response. *Water Resources Research*, 10–2, 303–312.

Neuman S.P. (1975) Analysis of pumping test data from anisotropic unconfined aquifers considering delayed gravity response. *Water Resources Research*, 11, 3.

Neuman S.P. (1977) Theoretical derivation of Darcy's law. *Acta Mechanica*, 25, 153–170.

Neuman S.P. (1988) Analysis of pumping tests. *Seminar at Technical University of Torino* (Italy).

Newmark N.M. (1942) Influence charts for computation of stresses in elastic foundations. University of Illinois, Bull. 338.

Newmark N.M. (1965) Effects of earthquakes on dams and embankments. *Géotechnique*, 15, 2, 139–160.

Nova R. (2002) *Fondamenti di Meccanica delle Terre*. McGraw-Hill, 373 pp.

Nova R. and Montrasio L. (1991) Settlements of shallow foundations on sand. *Géotechnique*, 41, 2, 243–256.

Okabe S. (1924) General theory of earth pressure and seismic stability of retaining wall and dam. *J. Japan Civil Engng Soc.*, 10, 6.

Olivares L. and Picarelli L. (2003) Shallow flowslides triggered by intense rainfalls on natural slopes covered by loose unsaturated pyroclastic soils. *Géotechnique*, Technical Note, 53, 2, 283–287.

Padfield C.J. and Mair R.J. (1984) Design of retaining walls embedded in stiff clays. *Report 104, Construction Industry Research and Information Association*, London.

Paik K., Salgado R. (2003) Arching effects on active earth pressure on retaining walls. *Géotechnique*, 53, 7, 643–653.

Pane V. (1985) Sedimentation and consolidation of clays, Ph.D. Thesis, University of Colorado, Boulder.

Pane V. and Burghignoli A. (1988) Determinazione in laboratorio delle caratteristiche dinamiche dell'argilla del Fucino. Convegno CNR su: Deformazioni dei terreni ed interazione terreno-struttura in condizioni di esercizio, Monselice, 115–139.

Pane V. and Schiffman R.L. (1985) A note on sedimentation and consolidation, *Géotechnique*, 35, 1, 69–72.

Papa V., Silvestri F. and Vinale F. (1988) Analisi delle proprietà di un tipico terreno piroclastico mediante prove dinamiche di taglio semplice. Convegno CNR su: Deformazioni dei terreni ed interazione terreno-struttura in condizioni di esercizio, Monselice, 265–286.

Parry R.H.G. (1958) Correspondence: on the yielding of soils. *Géotechnique*, 8, 4, 183–186.

Parry R.H.G. (1970) Overconsolidation in soft clay deposits. *Géotechnique*, 20, 442–446.

Parry R.H.G. (1978) Estimating foundation settlements in sand from plate bearing tests. *Géotechnique*, 28, 107–118.

Pasqualini E., Fratalocchi E. and Peroni N. (2001) Sonde penetrometriche per la caratterizzazione dei siti inquinati. *Atti delle Conferenze di Geotecnica di Torino*, XVIII ciclo.

Peck R.B. (1969) Advantages and limitations of the observational method in applied soil mechanics. 9th Rankine Lecture. *Géotechnique*, 2, 171–187.

Peck R.B. (1969) Deep excavations and tunneling in soft ground. State of the Art Report, *7th ICSMFE*, Mexico, 225–290.

Pellegrino A., Picarelli L. and Bilotta E. (1985) Geotechnical properties and slope stability in structurally complex clay soils. Geotechnical Engineering in Italy, Jubilee Volume XI *ICSMFE*, San Francisco, AGI, 195–214.

Pennington D.S., Nash D.F.T. and Lings M.L. (1997) Anisotropy of G_o shear stiffness in Gault clay. *Géotechnique*, 47, 3, 391–398.

Pepe M. (1995) La Torre pendente di Pisa. Analisi teorico-sperimentale della stabilità dell'equilibrio. Ph.D. Thesis, Politecnico di Torino.

Pickering D.J. (1970) Anisotropic elastic parameters for soil. *Géotechnique*, 20, 3, 271–276.

Pirulli M. (2005) Numerical modelling of landslide runout. Ph.D. Thesis, Politecnico di Torino.

Polshin D.E. and Tokar R.A. (1957) Maximum allowable non-uniform settlement of structures. *Proc. IV ICSMFE*, 1, 402.

Polubarinova-Kochina P. Ya. (1952) *Theory of the Motion of Ground Water*. Gostekhizdat, Moskow.

Porter O.J. (1936) Studies of fill construction over mud flats including a description of experimental construction using vertical drains to hasten stabilization. *1st ICSMFE*, Cambridge
Massachusetts, 1, 229–235.

Potts D.M. and Fourie A.B. (1986) A numerical study of the effects of wall deformation on
earth pressures. *Int. Journal for Numerical and Analytical Methods in Geomechanics*, 10,
383–405.

Potts D.M., Kovacevic N. and Vaughan P.R. (1997) Delayed collapse of cut slopes in stiff clay.
Géotechnique 47, 5, 953–982.

Poulos H.G. and Davis E.H. (1974) *Elastic solutions for Soil and Rock Mechanics*. Wiley.

Powrie W. (1996) Limit equilibrium analysis of embedded retaining walls. *Géotechnique*, 46,
4, 709–723.

Powrie W. (1997) *Soil Mechanics. Concepts and Application*. Spon Press, 420 pp.

Prandtl L. (1921) Uber die Eindringungs-festigkeit und festigkeit (Harte) plastischer Baustoffe
und die Festigkeit von Schneiden. Zeit. F. Angew. *Math. Mech.*, 1, 15–20.

Radenkovic D. (1961) Théorémes limites pour un matériau de Coulomb à dilatation non
standardiséè. *C.R. Ac. Sc. Paris* 252, 4103.

Radenkovic D. (1962) Théorie des charges limites. Séminaire de Plasticité, J. Mandel, ed. *P.S.T.*
116, 129–142.

Ramiah B.K., Dayalu N.K. and Purushothamaraj P. (1970) Influence of chemicals on the residual
strength of silty clay. *Soils and Foundations*, 10, 25–36.

Rampello S. (1989) Effetti del rigonfiamento sul comportamento meccanico di argille fortemente
sovraconsolidate. Ph.D. Thesis, Università di Roma La Sapienza.

Rampello S. and Pane V. (1988) Deformabilità non drenata statica e dinamica di un'argilla
fortemente sovraconsolidata, Convegno CNR su: Deformazioni dei terreni ed interazione
terreno-struttura in condizioni di esercizio, Monselice, 141–160.

Rampello S. and Callisto L. (1998) A study on the subsoil of the Tower of Pisa based on results
from standard and high quality samples. *Can. Geotechnical J.*, 35, 1074–1092.

Rampello S, Viggiani G.M.B. and Amorosi A. (1997) Small strain stiffness of reconstituted
clay compressed along constant triaxial effective stress ratio paths. *Géotechnique*, 47, 3,
475–489.

Randolph M.F. and Wroth C.P. (1979) An analytical solution for the consolidation around a
driven pile. *Int. Journal for Numerical and Analytical Methods in Geomechanics*, 3, 217–229.

Rankine W.J.M. (1857) On the stability of loose earth. *Phil. Trans. Royal Soc.*, London.

Ratnam S., Soga K. and Whittle R.W. (2001) Revisiting the Hvorslev's intake factors using the
finite element method. *Géotechnique*, 51, 7, 641–645.

Rendulic L. (1935) Ein Beitrag zur Bestimmung der Gleitsicherheit. *Der Bauingenieur*, 19/20.

Rendulic L. (1936) Porenziffer und porenwasserdruck in tonen. *Bauingenieur*, 17, 559.

Reynolds O. (1885) On the dilatancy of media composed of rigid particles in contact, with
experimental illustrations. *Phil. Mag.*, 20, 469–481.

Ricceri G., Soranzo M. (1985) An analysis on allowable settlement of structures. *Rivista Italiana
di Geotecnica*, 4, 177–188.

Richards R. and Elms D. (1979) Seismic behaviour of gravity retaining walls. *JGED, ASCE*,
105, 4, 449–464.

Richart F.E., Hall J.R. and Woods R.D. (1970) *Vibrations of Soils and Foundations*.
Prentice-Hall.

Riedel W. (1929) Zur Mechanik geologisher Brucherscheinungen. *Centralbl. F. Mineral*. Geol.
U. Pal., B, 354–368.

Rix G.J. (1988) Experimental study of factors affecting the Spectral Analysis of Surface Wave
method. Ph.D. Thesis, The University of Texas at Austin.

Robertson P.K. (1990) Soil classification using the cone penetration test. *Canadian Geotechnical
Journal*, 27, 151–158.

Robertson P.K., Campanella R.G., Gillespie D. and Rice A. (1985) Seismic CPT to measure in-situ shear wave velocity. *Proc. of Geotechnical Engineering Division, ASCE*, Measurement and use of shear wave velocity, Denver, 34–48.

Roesler S.K. (1979) Anisotropic shear modulus due to stress anisotropy. *JGED, ASCE*, 105, 5, 871–880.

Romano A., Lancellotta R. and Marasco A. (2006) *Continuum Mechanics using Mathematica: Fundamentals, Applications and Scientific Computing with Mathematica*. Birkhauser, 388 pp.

Roscoe P.W. and Burland J.B. (1968) *On the Generalised Stress-Strain Behaviour of an Idealised Wet Clay. Engineering Plasticity*. Heyman and Leckie eds, Cambridge University Press, 535–609.

Roscoe P.W., Schofield A.N. and Wroth C.P. (1958) On the yielding of soils. *Géotechnique*, 1, 22–52.

Rowe P.W. (1952) Anchored sheet-pile walls. *Proc. ICE*, Part 1, 1, 27–70.

Rowe P.W. (1955) A theoretical and experiment analysis of sheet-pile walls. *Proc. ICE*, Part 1, 4, 32–69.

Rowe P.W. (1962) The stress-dilatancy relation for static equilibrium of an assembly of particles in contact. *Proc. Royal Soc*. London, A269, 500–527.

Saada A.S., Townsend F.C. (1981) State of the Art: Laboratory Strength Testing of Soils. Laboratory Shear Strength of Soils, ASTM STP 740, ASTM, 7–77.

Salençon J. (1974) *Théorie de la plasticité pur les application à la Mécanique des Sols*. Eyrolles.

Salgado R. (2008) The Engineering of Foundations. McGraw-Hill, 882 pp.

Sampaolesi P. (1956) *Il Campanile di Pisa*. Opera della Primaziale di Pisa, 109 pp.

Sanchez-Salinero I. (1987) Analytical investigation of seismic methods used for engineering applications. Ph.D. Thesis, The University of Texas at Austin.

Santamarina J.C., Klein K.A. and Fam M.A. (2001) *Soils and Waves*. Wiley, New York, 488 pp.

Sarma S.K. (1973) Stability analysis of embankments and slopes. *Géotechnique*, 23, 423–433.

Scarpelli G. (1981) Shear bands in sands. MSc Thesis, Cambridge, UK.

Scarpelli G. and Wood D.M. (1982) Experimental observations of shear band patterns in direct shear test. *IUTAM Conf. Deformation and Failure of Granular Materials*, Delft, 473–484.

Scavia C. (1995) A method for the study of crack propagation in rock structures. *Géotechnique*, 45, 3, 447–463.

Schiffman R.L. (1958) Consolidation of soil under time dependent loading and varying permeability, *Proc. Highway Research Board*, 37, 584–617.

Schiffman R.L. and Fungaroli A. (1965) Consolidation due to tangential loads. *6th ICSMFE*, Montreal.

Schiffman R.L., Chen A.T.F. and Jordan J.C. (1969) An analysis of consolidation theories. *JSMFD, ASCE*, 95, SM1.

Schmertmann J.H. (1955) The undisturbed consolidation behaviour of clay, *Trans. ASCE*, 120, 1201–1233.

Schmertmann J.H. (1970) Static cone to compute static settlement over sand. *JSMFD, ASCE*, 96, SM3, 1011–1043.

Schmertmann J.H. (1986) Dilatometer to compute foundation settlement, *Proc. of 'In Situ '86'*, Geotechnical Special Publication n.6, *ASCE*, 303–321.

Schmertmann J.H., Hartman J.D. and Brown P.R. (1978) Improved strain influence factor diagrams. *JGED, ASCE*, Technical Note, 104, GT8, 1131–1135.

Schmidt B. (1966) Discussion on earth pressure at rest related to stress history, *Canadian Geotechnical Journal*, 3, 239–242.

Schofield A. and Wroth P. (1968) *Critical State Soil Mechanics*. McGraw-Hill, 310 pp.

Scotto di Santolo A. (2002) Le colate rapide. Hevelius, 118 pp.

Selvadurai A.P. (1979) *Elastic Analysis of Soil-Foundation Interaction. Development in Geotechnical Engineering*, Vol. 17, Elsevier.

Sevaldson R.A. (1956) The slide at Lodalen, October 6, 1954. *Géotechnique*, 6, 167–182.

Silvestri F. (1991) Il comportamento di terreni naturali in prove torsionali cicliche e dinamiche. Ph.D. Thesis, Universita' di Napoli.

Simeoni L. (1998) Fenomeni di scivolamento planare nelle Langhe. Ph.D. Thesis, Politecnico di Torino.

Simons N., Menzies B. and Matthews M. (2001) *Soil and Rock Slope Engineering*. Thomas Telford, 432 pp.

Simons N.E. and Som N. (1969) The influence of lateral stresses on the stress-deformation characteristics of London clay. *7th ICSMFE*, Mexico, 1, 369.

Simons N.E. and Som N.N. (1970) Settlement of structures on clay, with particular emphasis on London clay, *CIRIA Report 22*, London.

Simpson B. (1992) Retaining structures: displacement and design. 32nd Rankine Lecture. *Géotechnique*, 42, 4, 541–576.

Simpson B., Blower T., Craig R.N. and Wilkinson W.B. (1989) The engineering implications of rising groundwater levels in the deep aquifer below London, *CIRIA Special Publication 69*, London.

Simpson B., Calabresi G., Sommer H. and Wallays M. (1979) Design parameters for stiff clays. State of the Art Report. *7th ECSMFE*, Brighton, 5, 91–125.

Skempton A.W. (1948) The $\Phi = 0$ analysis of stability. Its theoretical basis. *2nd ICSMFE*, Rotterdam.

Skempton A.W. (1951) The bearing capacity of clays. *Proc. Building Research Congress*, London, 1, 180–189.

Skempton A.W. (1953) The colloidal activity of clays. *III ICSMFE*, Zurich, 1, 57–61.

Skempton A.W. (1954) The pore pressure coefficients A and B. *Géotechnique*, 4, 143–147.

Skempton A.W. (1961) Horizontal stresses in an overconsolidated Eocene clay. *5th Int. Conf. Soil Mech. Found. Eng.*, Paris, 1, 351–357.

Skempton A.W. (1964) Long term stability of clay slopes. *Géotechnique*, 14, 77–102.

Skempton A.W. (1966) Some observations on tectonic shear zones. *Proc. 1st Congr. Int. Soc. Rock Mech.*, Lisbon, 1, 329–335.

Skempton A.W. (1970) *The Consolidation of Clays by Gravitational Compaction*. Q.J. Geo. Soc., London, 125, 373–412.

Skempton A.W. (1970) First-time slides in overconsolidated clays. *Géotechnique*, 20, 320–324.

Skempton A.W. (1977) Slope stability of cuttings in brown London clay. *Proc. 9th ICSMFE*, Tokyo, 3, 261–270.

Skempton A.W. (1985) Residual strength of clays in landslides, folded strata and the laboratory. *Géotechnique*, 35, 3–18.

Skempton A.W. (1986) Standard penetration test procedures and the effects in sands of overburden pressure, relative density, particle size, aging and overconsolidation. *Géotechnique*, 36, 3, 425–447.

Skempton A.W. and Bjerrum L. (1957) A contribution to the settlement analysis of foundations on clay. *Géotechnique*, 4, 168–178.

Skempton A.W. and Brown J.D. (1961) Landslide in boulder clay at Selset. *Géotechnique*, 11, 280–293.

Skempton A.W. and Hutchinson J.N. (1969) Stability of natural slopes and embankment foundations. State of the Art report, *7th ICSMFE*, Mexico City, 2, 291–340.

Skempton A.W. and La Rochelle P. (1965) The Bradwell slip: a short-term failure in London clay. *Géotechnique*, 15, 221–242.

Skempton A.W. and MacDonald D.H. (1956) Allowable settlement of buildings. *Proc. ICE*, London, 5, 727–768.

Skempton A.W. and Northey R.D. (1953) The sensitivity of clays. *Géotechnique*, 1, 30–53.

Skempton A.W. and Petley D. (1967) The strength along structural discontinuities of stiff clays. *Proc. Geotechnical Conference*, Oslo, 2, 29–46.

Sloan S.W. and Yu H.S. (1996) Rigorous plasticity solutions for the bearing capacity factor N_γ. *Proc. 7th Australian-New Zealand Conference on Geomechanics*, Adelaide, 544–550.

Sokolovskii V.V. (1965) *Statics of Granular Media*. Pergamon Press, Oxford.

Somerville S.H. (1986) Control of groundwater for temporary works. *CIRIA*, Report 113.

Somerville S.H. and Shelton J.C. (1972) Observed settlements of multi-storey buildings on laminated clays and silts in Glasgow. *Géotechnique*, 22, 3, 513–520.

Spencer E. (1967) A method of analysis of the stability of embankments assuming parallel inter-slice forces. *Géotechnique*, 17, 11–26.

Stallebrass S.E. (1990) Modelling of the effect of recent stress history on the deformations of overconsolidated soils. Ph.D. thesis, City University, London.

Steedman R.S. and Zeng X. (1990a) The influence of phase on the calculation of pseudo-static earth pressure on a retaining wall. *Géotechnique*, 40, 1, 103–112.

Steedman R.S. and Zeng X. (1990b) Hydrodynamic pressures on a flexible quay wall. *Proc. European Conf. on Structural Dynamics*. Bochum, 843–850.

Steedman R.S. and Zeng X. (1990c) The seismic response of waterfront retaining walls. Design and performance of earth retaining structures. *Geotechnical Special Publication 25, ASCE*.

Steinbrenner W. (1934) Tafeln zur Setzungsberechnung, Die Strasse, v.1, 121–124. also see *Proc. 1st ICSMFE*, Cambridge, v.2, 142–143, 1936.

Stokoe K.H., Wright S.G., Bay J.A. and Roesset J.M. (1994) Characterization of geotechnical sites by SASW method. Special volume on geophysical characterization of sites, R.D. Woods ed., *XIII ICSMFE*, New Delhi, Balkema, 15–34.

Tajimi H. (1973) Dynamic earth pressures on basement wall. *Proc. 5th World Conf. Earthquake engng*, 2, 1560–1569.

Tarantino A. and Mongiovì L. (2000) Experimental investigations on the stress variables governing unsaturated soil behaviour at medium and high degrees of saturation, in Experimental evidence and theoretical approaches in unsaturated soils. Tarantino and Mancuso eds, Balkema, 3–19.

Taylor D.W. (1948) *Fundamentals of Soil Mechanics*. Wiley, 700 pp.

Taylor G.I. and Quinney H. (1931) The plastic distortion of metals. *Phil. Trans. Roy. Soc. A.*, 230, 323–362.

Teh C.I. and Houlsby G.T. (1991) An analytical study of the cone penetration test in clay. *Géotechnique*, 41, 1, 17–34.

Terracina F. (1962) Foundation of the leaning Tower of Pisa. *Géotechnique*, 12, 4, 336–339.

Terzaghi K. (1922) Der Grundbush an Stauwerken und seine Verhutung. *Die Wasserkraft*, 17, 445–449 (see *From Theory to Practice*. Wiley, 114–118, 1960).

Terzaghi K. (1923) Die Berechnung der Durchlassigkeitsziffer des Tones aus dem Verlauf der hydrodynamischen Spannungserscheinungen. *Sitz. Akad. Wissen*, Wien Math-naturw Kl. Abt. IIa, 132, 125–138.

Terzaghi K. (1925) Erdbaumechanik auf bodenphysikalischer Grundlage. Leipzig Deuticke (see from *Theory to Practice*. Wiley, 146–148, 1960).

Terzaghi K. (1934) Large retaining wall tests. *Eng. News Record*, Feb. 1, Feb. 22, Mar. 8, Mar. 20, Apr. 19.

Terzaghi K. (1936) The shearing resistance of saturated soils and the angle between the planes of shear. *1st ICSMFE*, Cambridge Mass., 1, 54–56.

Terzaghi K. (1943) *Theoretical Soil Mechanics*. Wiley.

Terzaghi K. (1950) Mechanism of landslides, in application of geology to engineering practice. *Geological Society of America*, 83–123.

Terzaghi K. and Peck R.B. (1948) *Soil Mechanics in Engineering Practice*. Wiley.

Terzaghi K. and Peck R.B. (1967) *Soil Mechanics in Engineering Practice* (2nd edn). Wiley.

Terzaghi K., Peck R.B. and Mesri G. (1996) *Soil Mechanics in Engineering Practice* (3rd edn). Wiley.

Theis C.V. (1935) The relation between the lowering of the piezometric surface and the rate and duration of discharge of a well using groundwater storage. *Transac. American Geophysical Union*, 16, 519–524.

Thompson W. (Lord Kelvin) (1848) On the equation of equilibrium of an elastic solid. Cambr. Dubl. Math. J., 3, 87–89.

Timoshenko S.P. (1983) *History of the Strength of Materials*, Dover (originally published by McGraw-Hill in 1953).

Timoshenko S.P. and Goodier J.N. (1971) *Theory of Elasticity* (3rd edn). McGraw-Hill, 567 pp.

Todd D.K. (1980) *Groundwater and Hydrology* (2nd edn). Wiley, 535 pp.

Tomlinson M.J. (1963) *Foundation Design and Construction*. Pitman.

Tornaghi R. (1981) *Indagini geotecniche in sito*. Ordine degli Ingegneri di Catania.

Torstensson B.A. (1975) Pore pressure sounding instrument. *ASCE Spec. Conf. on In situ Measurement of Soil Properties*, Raleigh, N.C.

Tresca H. (1864) Sur l'ecoulement des corps solids soumis à de fortes pression. *Comptes Rendus de l'Academie des Sciences*, Paris, 59, 754.

Truesdell C. and Noll W. (1965) *The Non-linear Field Theory of Mechanics*. Handbuch der Physik, III/3, Springer.

Tsytovich N. (1976) *Soil Mechanics*. MIR, 293 pp.

Tucker M.E. (2001) *Sedimentary Petrology*. Blackwell Publishing (3rd edn). pp. 262.

U.S. Bureau of Reclamation (1947) Laboratory tests on protective filters for hydraulic and static structures. *Earth Materials Laboratory Rept.*, EM-132, Denver.

Varnes D.J. (1978) Slope movement types and processes, Chapter 2 of Landslides- Analysis and Control (Schouster and Krizek, eds.), *Transportation Research Board*, S.R.176, 11–33.

Verruijt A. (1984) *The Theory of Consolidation. Fundamentals of Transport Phenomena in Porous Media*, J. Bear and Y.M. Corapcioglu eds, Martinus Nijhoff, Dordrecht, 330–350.

Vesic A.S. (1972) Expansion of cavities in infinite soil masses. *JSMFD, ASCE*, 98, SM3, 265–290.

Vesic A.S. (1973) Analysis of ultimate loads of shallow foundations. *JSMFD, ASCE*, 1, 45–73.

Vesic A.S. (1975) *Bearing capacity of shallow foundations. Foundation Engineering Handbook*, Winterkorn and Fang eds, Van Nostrand Reinhold.

Viggiani C. (1970) An analysis of consolidation theories. Discussion, *JGED, ASCE*, SM1, 331–334.

Viggiani G.M.B. (1992) Small strain stiffness of fine grained soils. Ph.D. Thesis, City University, London.

Viggiani G.M.B. and Atkinson J.H. (1995) The interpretation of the bender element tests. *Géotechnique*, 45, 1, 149–154.

Weltman A.J. and Head J.M. (1983) Site investigation manual. *CIRIA*, London.

Wesley L.D. (2002) Interpretation of calibration chamber tests involving cone penetrometer tests in sands. *Géotechnique*, 52, 4, 289–293.

Westergaard H.M. (1931) Water pressure on dams during earthquakes. *Transactions, ASCE*, 418–433.

Whitman R.V. and Liao S. (1985) Seismic design of retaining walls. *Miscellaneous Paper GL-85*, U.S. Army Engineers Waterways Experiment Station, Vicksburg, Mississippi.

Wilkinson W.B. (1968) Constant head in situ permeability tests in clay strata. *Géotechnique*, 2, 172–194.

Windle D. and Wroth C.P. (1977) In situ measurement of the properties of stiff clays. *IX ICSMFE*, Tokyo, 1, 347–352.

Wissa A., Christian J.T., Davis E.H. and Heiberg S. (1971) Consolidation at constant rate of strain. *Journal SMFD, ASCE*, 97, 10, 1393–1413.

Wissa A.E.Z., Martin R.T. and Garlanger J.E. (1975) The piezometer probe. *ASCE Spec. Conf. on In Situ Measurement of Soil Properties*, Raleigh, N.C.

Withers N.J., Schaap L.H.J. and Dalton C.P. (1986) The development of the full displacement pressuremeter. *Proc. 2nd Int. Symp. The Pressuremeter and Its Marine Applications*. Special Technical Publication 950, ASCE.

Wood D.M. (1984) On stress parameters. *Géotechnique*, 34, 2, 282–287.

Wood D.M. (1990) Soil Behaviour and Critical State Soil Mechanics. Cambridge University Press, 462 pp.

Wroth C.P. (1958) The behaviour of soils and other granular media when subjected to shear. Ph.D. Thesis, University of Cambridge.

Wroth C.P. (1984) Interpretation of in-situ soil tests. *Géotechnique*, 4, 449–489.

Wroth C.P. and Bassett R.H. (1965) A stress–strain relationship for the shearing behaviour of sand. *Géotechnique*, 15, 1, 32–56.

Wroth C.P. and Houlsby G.T. (1985) Soil mechanics. Property characterisation and analysis procedures. *Theme lecture, 11th ICSMFE*, San Francisco, 1–55.

Wroth C.P. and Hughes J.M.O. (1972) An instrument for the in situ measurement of the properties of soft clays. *Rep. CUED/D-Soils TR 13*, Dep. Eng., Cambridge, U.K.

Yoshimi Y., Hatanaka M. and Oh-Oka H. (1978) Undisturbed sampling of saturated sands by freezing. *Soils and Foundations*, 18, 3, 59–73.

Yoshimi Y., Hatanaka M., Oh-Oka H. and Makihara Y. (1985) Liquefaction of sands sampled by in situ freezing. *Proc. 11th ICSMFE*, San Francisco, 4, 1927–1930.

Yu H.S. (2000) *Cavity Expansion Methods in Geomechanics*. Kluwer Academic Publishers.

Yu H.S. (2004) In situ soil testing: from mechanics to interpretation. *J.K. Mitchell Lecture, Proc. ISC-2 on Geotechnical and Geophysical Site Characterization*. Millpress, Porto, 3–38.

Yu H.S. (2006) *Plasticity and Geotechnics*. Springer, 522 pp.

Yu H.S., Carter J.P. and Booker J.R. (1993) Analysis of the dilatometer test in undrained clay. In *Predictive Soil Mechanics*, Thomas Telford, London, 783–795.

Yu H.S. and Collins I.F. (1998) Analysis of self-boring pressuremeter tests in overconsolidated clays. *Géotechnique*, 48, 5, 689–693.

Yu H.S. and Houlsby G.T. (1991) Finite cavity expansion in dilatant soil: loading analysis. *Géotechnique*, 41, 2, 173–183.

Zdravkovic L. (1996) The stress-strain-strength anisotropy of a granular medium under general stress conditions. Ph.D. thesis, Imperial College of Technology, Science and Medicine, University of London.

Zdravkovic L., Potts D.M. and Hight D.W. (2002) The effect of strength anisotropy on the behaviour of embankments on soft ground. *Géotechnique*, 52, 6, 447–457.

Zeevaert L. (1983) *Foundation Engineering for Difficult Subsoil Conditions* (2nd edn). Van Nostrand Reinhold Company.

Zeng X. and Steedman S. (1993) On the behaviour of quay walls in earthquakes. *Géotechnique*, 43, 3, 417–431.

Zeinkiewicz O.C. and Taylor R.L. (1994) *The Finite Element Method* (4th edn). McGraw-Hill.

Zuidberg H.M., Schapp L.H.J. and Beringen F.L. (1982) A penetrometer for simultaneously measuring of cone resistance, sleeve friction and dynamic pore pressure. *ESOPT II*, Amsterdam.

Zytynski M., Randolph M.F., Nova R. and Wroth C.P. (1978) On modelling the unloading-reloading behaviour of soils. *Int. Journal for Numerical and Analytical Methods in Geomechanics*, 2, 87–94.

Index

page references followed by e indicate a boxed example; f indicates an illustrative figure; t indicates a table